艾｜诗｜德｜移｜动｜技｜术｜丛｜书

# iPhone 应用程序开发指南

## （基础篇）

丛书主编　王绪兵　彭楚夫

本书主编　张英锋　刘　超

山东科学技术出版社

丛书主编　王绪兵　彭楚夫
出版策划　高瑞东　胡　明

本书主编　张英锋　刘　超
本书编者　安凌靖　王杨杨　张建文　韩元涛　张华青
本书校阅　张　林　刘洪峰　于凡舒　杨　丽　刘向丽
　　　　　王　远　张　黎　曹咏志　侯　曼　刘　彪

# 内容简介

《艾诗德移动技术丛书》是国内第一套原创的 iPhone 开发书籍，作者均是一线的 iPhone 软件工程师，他们拥有丰富的项目开发经验，完全按照国人的思路和语言来编写，避免了翻译书籍因译者水平、思维方式和文化差异产生的阅读障碍。

本书是《艾诗德移动技术丛书》的一个分册，从最基础的 Objective-C 语言开始，逐渐深入地介绍 iPhone 软件开发技术。书中各章均结合案例进行讲解，读者可以在实现案例的过程中逐渐掌握 iPhone 开发技能。随书赠送的光盘中不仅有包含中文注释的程序源代码，还有丰富的 PPT 课件和视频，便于教育、培训机构的教师开展 iPhone 开发教学。

本书作者团队同时开发了 iPhone 电子书，结合 iPhone 的视频、动画、多触摸、重力感应等特性，为读者提供了更好的移动阅读体验，欢迎通过苹果软件商店购买。本书及电子书的内容均会随着 iPhone 开发技术的演进，不断地更新和升级。

读者如果在学习中遇到问题，可以到艾诗德独立软件开发者联盟的 iPhone 开发论坛(http://58.64.130.106/forum/index.php)，与本书作者和全国各地的 iPhone 开发者交流。

# 前言

20世纪80年代中期开始普及的个人电脑和90年代中期开始普及的基于固定平台的互联网,使人类社会真正进入了信息时代,人与人的联系更加方便,国与国之间的壁垒愈来愈少,不同的国家在贸易、资金方面的合作越来越多,在人才、知识及信息资源方面的交流也日益频繁,世界各国的人们开始在一个广泛而透明的平台上互相合作和竞争。

手机作为通话工具的应用虽然已有20年的历史了,但近几年开始普及的3G手机和无线网络赋予了手机新的、巨大的生命力。作为一种新兴的信息终端,手机在我们今后的生活中将扮演越来越重要的角色。如果说,上个世纪80年代的个人电脑和90年代的固定互联网掀起过改变我们生活的两次浪潮的话,那么新时期的由手机+互联网构成的移动互联网必将掀起一次新的浪潮,它将会在更加广泛的范围内,以更加密切的方式影响和改变我们今后的生活。从网络音乐、网络新闻和搜索引擎,到商务交易方面的网上支付、旅游预订、网络炒股、网上银行和网络购物等等,所有这一切都可以通过手机来完成。中国互联网络信息中心2010年1月15日发布的《第25次中国互联网络发展状况统计报告》显示,我国手机网民规模在2009年1年内增加了1.2亿,到2009年底已达到2.33亿人,占整体网民的60.8%。

作为移动互联网设备的领先者,苹果公司在2007年推出了革命性的科技结晶——iPhone。作为第一台操作便捷的触摸屏手机,iPhone仅用了短短的两年时间,便在智能手机的市场上占据了领先地位,这体现为5700万部手机、10万应用软件和2亿iTunes的使用者。摩根士丹利(Morgan Stanley)的研究报告表明,iPhone和iPod touch的独特之处是,虽然它们的用户只占全球智能手机用户的17%,但却占据了全球移动网络浏览量的65%和移动应用程序用户的50%。2009年10月底,iPhone手机正式登陆中国大陆,随即在各地掀起iPhone热,有的地方已经出现了iPhone俱乐部,iPhone手机玩家通过"网上俱乐部"和"实体俱乐部"交流iPhone手机知识和使用体验。

iPhone内置的软件商店App Store是iPhone成功的重要因素,它在市面推出的18个月的时间里,就拥有了超过10万个应用程序供用户下载,下载次数超过1亿次,并且软件的数量每年以163%的速度在增长。App Store因它的透明、公开及收费的合理性,深受开发者及用户的欢迎。App Store中的软件大部分都属于轻小的软件,范围包括游戏、音乐、学习及参考、企业管理、工作软件等等,可以说包罗万象。因为门框比较低,故吸引了越来越多的开发者的关注和参与。这些软件大部分是由个人开发者或人数不多的小型开发队伍创造而成的,很多开发者在业余时间利用自己的技术,发挥创意,开发了这些小软件并发布在 App Store 上。许多成功的软件为开发者带来了相当可观的回报，甚至创造了多达百万美元的收入。

苹果公司建立的良好的技术平台和完备的商业模式，吸引着众多的追随者，很多软件及电信企业纷纷推出自己的“App Store”，运作模式与苹果公司如出一辙，如微软的 Market Place，黑莓的 Appworld，谷歌的 Andriod Market，中国移动的 Mobile Market，诺基亚的 Ovi Store。其他品牌的手机如三星、索尼、爱立信、LG 等，也都宣布将推出自己的“App Store”。在这样的大趋势下，全球范围的对基于移动互联网的应用软件的需求愈见迫切，这必将为广大的软件开发者提供一次难得的商机。

在这样的大趋势下，我们推出了《艾诗德移动技术丛书》，希望能够帮助大家快速掌握移动互联网的开发技术，并通过自己的创意，开发出各式各样的移动软件；同时，我们发起了艾诗德独立软件开发者联盟（www. aisidechina. com），目的是帮助个人开发者进行从产品策划、技术支持，到产品包装、市场推广等一系列商业活动，实现个人创业的梦想。

《艾诗德移动技术丛书》包括 5 个分册，书名和读者对象见本书封底。

本书是《艾诗德移动技术丛书》的一个分册，内容分为 4 个部分：准备篇、语言篇、核心篇和扩展篇，各篇内容详见本书目录。

本书代码中，新添加的代码行采用黑体表示（如 **IBOutlet UITextField * password;**），删除的代码行用删除线表示（如~~IBOutlet UITextField * password;~~），代码过长无法在一行内显示而转行时使用↳标识，❶ ❷ ❸…是代码讲解编号。

读者在阅读完本书后，会对 iPhone 开发平台有比较全面的了解，可以开发出完整的 iPhone 应用程序，但要成长为 iPhone 软件工程师，还需要更多项目的锻炼和对 iPhone 平台高级特性的深入研究。为此，请关注和阅读《艾诗德移动技术丛书》的其他分册。

在本书编写过程中，得到了张林、刘洪峰、于凡舒、刘向丽、侯曼、杨丽、张黎、刘彪、曹咏志、王远、史庆保等人的帮助，在此表示诚挚的感谢。

由于编者水平所限，加之时间仓促，书中难免会有纰漏，欢迎广大读者批评指正。

**编者**

# 目 录

## 第1篇 准备篇



## 第2篇 语言篇

# 第 3 篇 核心篇

# 第 4 篇 扩展篇

# 第 1 篇　准备篇

本篇介绍了进行 iPhone 开发需要做的准备工作、必须具备的条件,以及开发工具的使用方法。

准备篇包括 2 章:

第 1 章 iPhone 开发前的准备,说明了进行 iPhone 开发应该具备的条件,并介绍了 iPhone 开发的特点和流程。

第 2 章开发工具介绍,通过一个名为 About 的项目对 iPhone 开发所用的工具进行了介绍,包括 Xcode、Interface Builder 和 iPhone 模拟器,并列举了 iPhone 开发中常用的快捷键。

# 第1章　iPhone开发前的准备

**本章内容**

- 需要具备的知识
- iPhone SDK 的下载与安装
- iPhone 软件的开发流程

iPhone 可以说是目前最有特色最新颖的移动开发平台。它拥有多点触摸、重力感应等强大的技术支持以及开发成本低、易于发布等特点，这些特点使 iPhone 软件开发成为代表了未来开发方向的一个新领域，全球越来越多的开发者将投身其中。

在正式进入 iPhone 开发的学习之前，本章将详细介绍 iPhone 开发所要做的准备工作，以及从注册成为一个 iPhone 开发人员到将自己的软件发布到 App Store 上的大体流程。

## 1.1　应该具备的条件

现在就来做实际开发前必要的准备工作。不用担心，相信这些准备工作对你而言将是非常轻松的。

### 1.1.1　需要掌握的知识

**1. 编程基础**

进行 iPhone 开发的主要语言是 Objective-C 语言。它是 C 语言的一个扩展集，在 C 语言的基础上添加了一些新的特性，其中最重要的一个方面是添加了面向对象的特性。在本书第二篇语言篇中将对 Objective-C 语言进行一定讲解。但在这之前，你应该已经具备了一定的其他语言的开发经验。

**2. 熟悉 Mac 系统**

iPhone 的开发需要在 Mac 系统下进行。虽然在下一章中会对 iPhone 开发工具进行一定讲解，但在这之前，关于 Mac 系统的一些基本操作，你最好有一定的了解。

### 1.1.2　开发环境

**1. 需要的设备**

由于 iPhone SDK 开发工具包必须运行在使用 Intel 处理器的 Mac 操作系统上，所以最好配备一台近几年出的 iMac/MacBook/MacBook Pro。当然也可以在自己的普通 PC 上安装 Mac 与 Windows 双系统，但由于其过程较为复杂，在此将不作讲解，其具体操作步骤可以参考艾思德开发论坛http://www.aisidechina.com/forum/上面的详细介绍。

此外，最好配有一个 iPhone 或者 iPod touch 来对开发的成果进行真实的测试与体验。因为在安装 SDK 后，虽然会有"iPhone 模拟器"这样一个方便的工具，但像加速计等效果并不能在模拟器上得到很好的体验。

**2. 下载最新的 iPhone SDK**

iPhone SDK 分为免费版、标准版与企业版。免费的 iPhone SDK 不支持将 iPhone 软件下载到实际的 iPhone 或 iPod touch 中使用或测试，也不支持将开发的软件发布到苹果公司的 App Store 上面，使用它只可以在自己的电脑上进行 iPhone 开发，但这对于我们进行开发学习已经足够了。如果想让软件能够使用真机测试和发布到 App Store 上，需要下载 iPhone SDK 的付费版本，iPhone SDK 的付费版本有标准版和企业版。三个版本的比较见表 1-1。

**表 1-1　iPhone SDK 三个版本的比较**

| iPhone SDK 版本 | 费用 | 说明 |
|---|---|---|
| 免费版 | 免费 | 任何注册 Apple ID 的开发人员均可免费下载，进行 iPhone 开发，但无法将软件下载到实际的 iPhone 上面，或在 App Store 上发布自己开发的软件 |
| 标准版 | 99 美元/年 | 可以将软件下载到实际的 iPhone 上面，以及在 App Store 上发布 |
| 企业版 | 299 美元/年 | 可供企业开发使用，用于内部的 iPhone 软件开发，一般面向 500 人以上的企业 |

免费 iPhone SDK 可以在苹果开发中心的中文网站上下载，下载地址：http://www.apple.com.cn/developer/iPhone/，如图 1-1 所示。

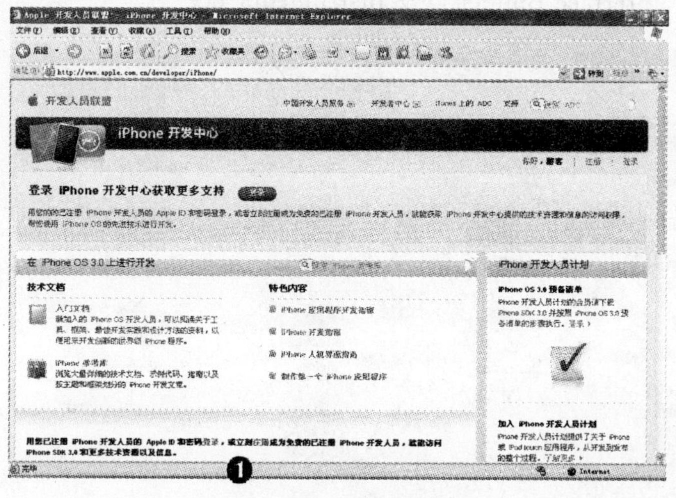


图 1-1　iPhone 开发中心首页

在这里会提供最新的 iPhone SDK 下载，目前的最新版本是 iPhone SDK 4.2。进行 iPhone SDK 下载前，需要先申请一个 Apple ID，见图 1-1 中❶处的提示。

注册完毕后登录，点击主页的下载图标进行下载，见图 1-2 中❷处的提示。

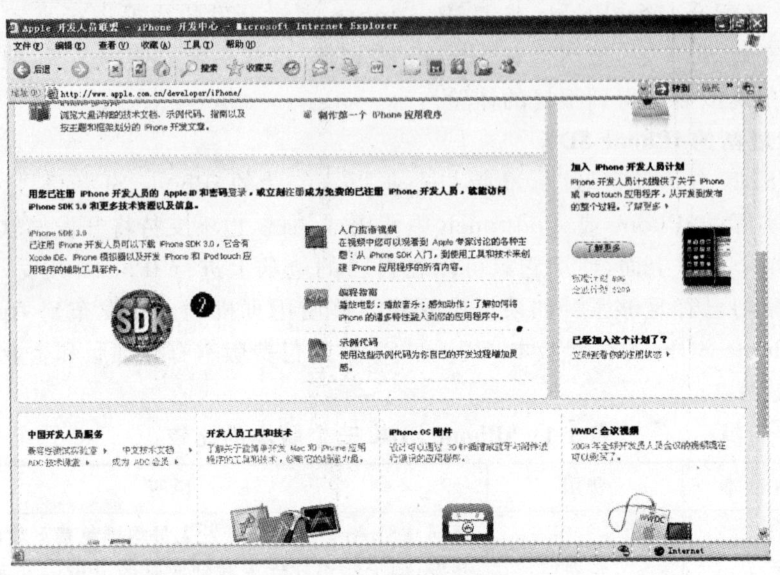


图 1-2　下载 iPhone SDK

在这里下载的是免费的 iPhone SDK。标准版或企业版的下载，可以通过连接到 http://developer.apple.com/iphone/program/apply.html进行注册与下载。

**3. SDK 的安装**

iPhone SDK 下载完毕后，安装的过程较为简单，只需要根据提示按步骤操作即可。

安装完成后，如图 1-3 所示，可以在/Developer/Applications 中找到 iPhone 开发工具的图标：Xcode、Interface Builder 以及 Instruments 等。

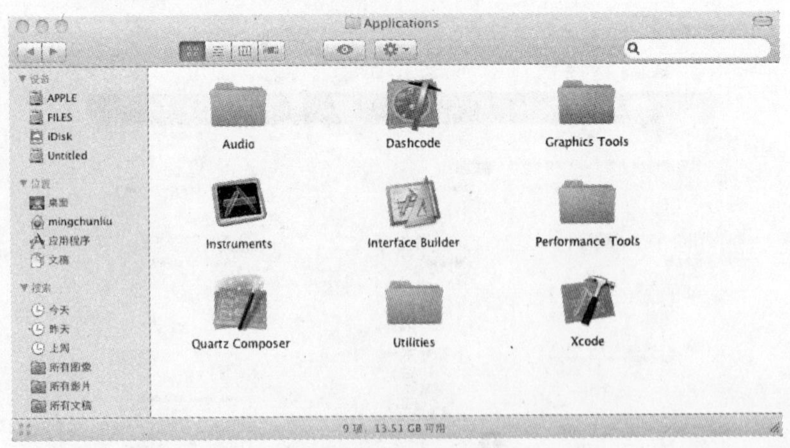


图 1-3　iPhone 开发工具图标

关于它们如何使用,将在下一章作详细介绍。为方便以后的开发,将这些图标拖放到桌面的Dock上,如图1-4所示。

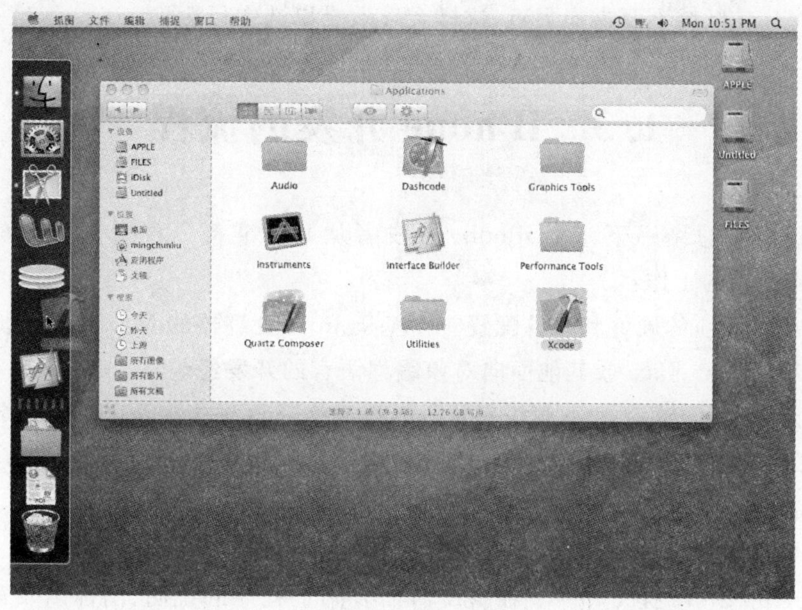

图1-4 拖放Xcode到Dock上

到此为止,进行开发前所需要的准备工作便全部完成了。是不是感觉很轻松?

## 1.2 iPhone开发的特点

iPhone开发与一般手机开发略有不同,比如苹果不允许有后台进程运行(但这只是限制非iPhone系统自带的程序,并不会限制系统进程)。

iPhone中的应用程序只可以访问自己文件系统中的资源,而不可以访问其他内容。

在开发应用于iPhone的软件时,不用太在意开发包的大小,目前iPhone最大可拥有32G的存储容量,以及128M的RAM。这允许应用程序的界面可以添加更多元素来进行丰富。

在iPhone上添加更多的元素来丰富界面时应该注意哪些问题呢?下面,我们结合iPhone的特点,将iPhone软件界面设计原则大体总结如下:

(1) 在界面上可以进行的操作,应尽量保持和iPhone系统自带程序中的操作一致。

(2) 要简化用户的操作,例如用户到达目标页面的过程中不要经过3个以上页面。

(3) 不需要在界面上添加退出程序按钮——整个iPhone上只有一个退出按钮,还不准备充分利用一下吗?

(4) 尽量让用户使用手势(如轻扫屏幕、晃动屏幕等)来进行操作,而不要在页面中增加按钮，也尽量不要使用菜单。

(5) 为方便用户点击，要使手指点击的目标区域足够大，建议使用大小为 44 * 44 像素的区域；而且各区域不要靠得太近，这样会导致难以选中目标。

# 1.3 iPhone 开发的流程

下面，我们来总结一下一个 iPhone 开发者从开始准备一直到将软件发布在 App Store 上所需要做的工作：

(1) 准备好相应的硬件和软件配置——基于 Intel 处理器的 Mac 操作系统。

(2) 具有 C++、Java 或其他面向对象编程语言的开发经验。

(3) 最新 iPhone SDK 的下载与安装。如果要将开发的软件发布到 App Store 上，还需要注册成为 iPhone 开发人员，并下载标准版 iPhone SDK，它的费用是 99 美元/年。

(4) 软件的设计。

(5) 软件的开发与测试修改。

(6) 向 App Store 提交程序。在提交时需要附上程序的说明、图标以及截图。苹果公司会对所提交的程序进行审核，顺利通过的话，很快就会在 App Store 上看到自己的软件，但如果在审核过程中发现明显的缺陷，则会被拒绝放到 App Store 上进行发布，这时便需要重新修改程序并再次提交。

(7) 发布后的修改与更新。在发布后，还需要不断完善和更新软件，这将有利于软件的质量与在 App Store 中的排名。

iPhone 软件开发流程如图 1-5 所示。

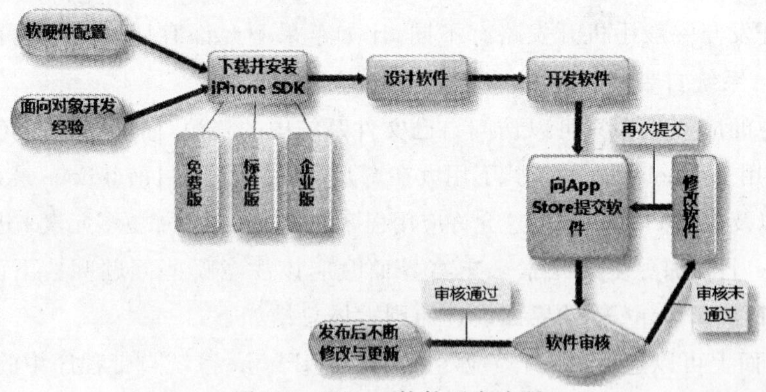


图 1-5 iPhone 软件开发流程

## 1.4 小结

本章中，介绍了进入 iPhone 开发学习前需要做的准备工作：

首先需要具备一定的面向对象编程经验，虽然本书在后面章节会对 Objective-C 语言进行一定讲解，但这需要建立在你以前编程知识的基础之上。

然后介绍了如何配置 iPhone 开发所需要的环境，比如 iPhone SDK 的下载与安装等。

本章还大体介绍了 iPhone 开发的特点、iPhone 界面设计的注意事项，以及从注册成为一个 iPhone 开发人员一直到将开发的软件发布到 App Store 上的大体流程。

至此，准备工作便做好了。下面就开始我们的开发旅程吧！

# 第2章 开发工具介绍

**本章内容**

- Xcode 窗口介绍
- 用 Interface Builder 构建界面
- Xcode 中常用的快捷键

"工欲善其事,必先利其器",为了更好地进行后面章节的学习,我们需要先对几个重要的开发工具有一定的了解。在这一章中,将着重介绍用于 iPhone 开发的最常用的几个工具,快速有效地掌握它们,为后面 iPhone 开发的学习打下基础。

## 2.1 开发工具简介

下面是在进行 iPhone 开发的过程中,我们最常用的两个工具 Xcode 和 Interface Builder 的图标。

Xcode

Interface Builder

Xcode 是苹果 Mac OS X 的集成开发环境的发动机,它负责绝大多数的开发工作,包括代码的编写以及项目中各种资源的管理。

第二个主要的开发工具是 Interface Builder,它是一个用于创建用户界面的图形工具。在开发过程中,Interface Builder 与 Xcode 是紧密结合的。关于它们的使用方法,我们会在本书以后章节逐步为你介绍。

此外,还有一个 iPhone 模拟器工具。当临时需要查看效果时,可以很方便地使用模拟器进行查看。

## 2.2 About项目的创建

下面就通过一个简单的项目来讲解一下这些开发工具应如何使用。如果要将软件放到App Store上面销售，我们需要注明软件的版权信息，所以本章将会制作一个用来说明软件版权的“关于”界面，如图2-1所示。在本章结束后，你就可以设计自己的版权声明，并将它放到以后要发布的软件当中了。

图2-1 About的运行效果

首先启动Xcode。如果是第一次启动Xcode，将会出现欢迎界面，如图2-2所示，但在实际开发过程中，并不需要它，所以取消选择左下角的Show at launch复选框，这样，在以后启动Xcode时，欢迎界面就不会再出现了。

关闭欢迎界面，Xcode就启动成功了。

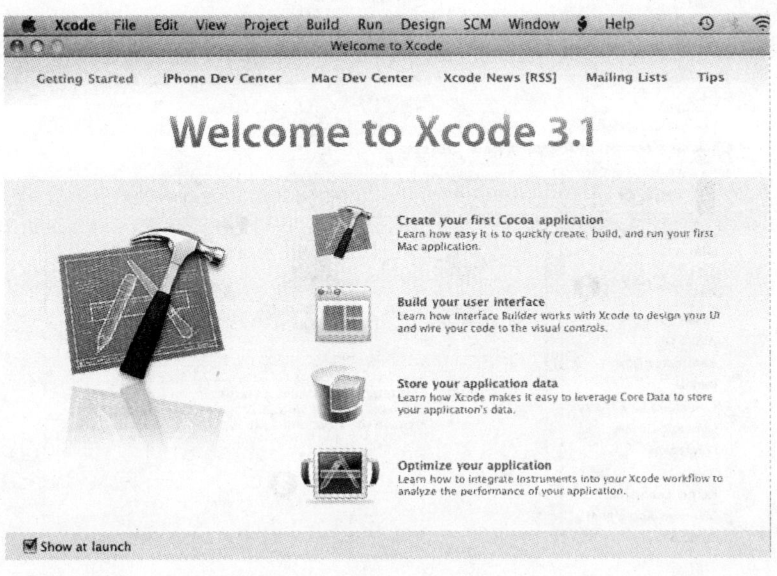


图2-2 Xcode欢迎界面

启动完成后，创建一个新项目About。首先选择File的New Project选项，如图2-3所示。

图 2-3　选择构建新项目

将会出现如图 2-4 所示的 New Project 窗口。

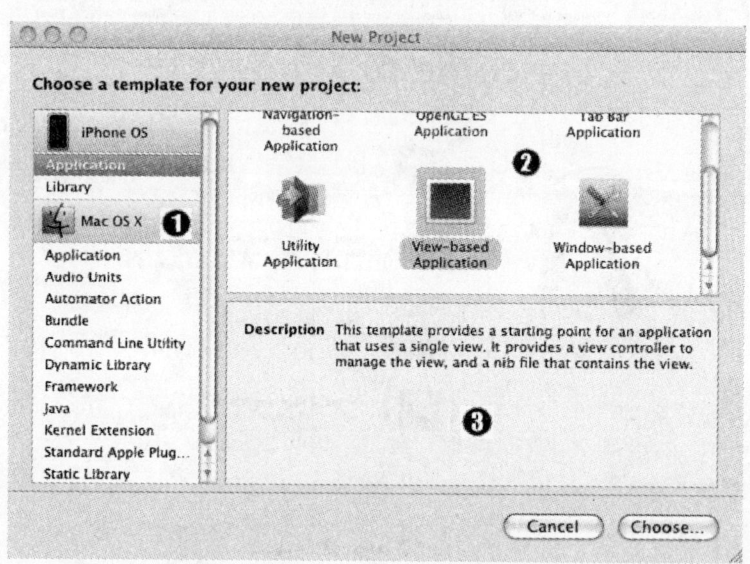


图 2-4　New Project 帮助窗口

如图 2-4 所示，整个窗口主要分为三个部分。项目模板类别部分❶又分为 iPhone OS 和 Mac OS X 两个层级。iPhone OS 层级下面的选项包含用于 iPhone 开发的模板，Mac OS X 层级下的选项用于一般的 Mac 系统开发。

在本项目中，选择 iPhone OS 层级下的 Application 选项。选择之后，可以看到在窗格❷中显示出了一系列对应的下级选项图标，其中的每个图标均代表一个 iPhone 应用程序的项目模板，并且每种项目模板均提供对应的 iPhone 应用程序所必需的各种基本功能和资源，为开发者提供了极大的方便。在这里，选择最基本的 View-based Application 作为这个项目的模板，因为它在一开始便准备好了一个视图，而且没有其他控件与功能，不会增加程序的大小。窗格❸负责显示对所选模板的描述，图 2-4 中显示的正是对 View-based Application 模板的描述信息。

选择好项目模板之后，点击右下角的 Choose 按钮，弹出为新建项目命名和指定存放路径的窗口，如图 2-5 所示。

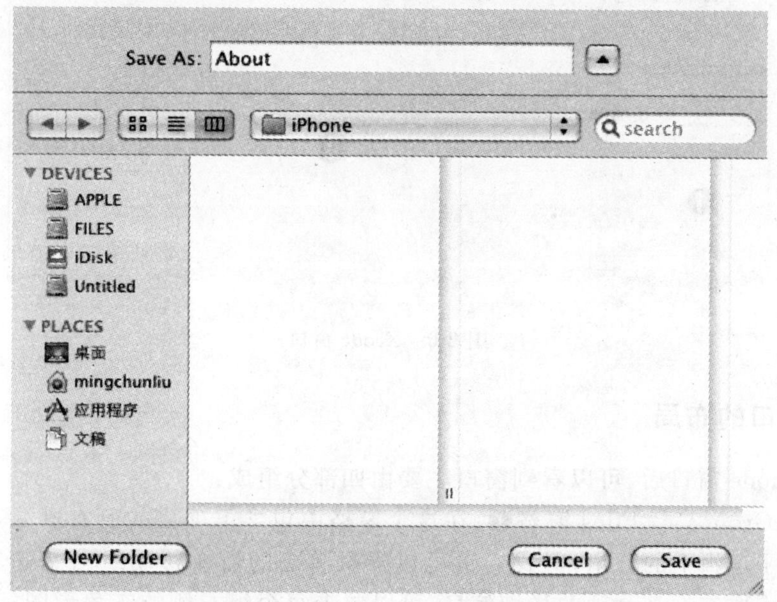


图 2-5 指定项目的名称和位置

为该项目命名为 About，并为它选择合适的存放路径。一切设置好之后，点击该窗口右下角的 Save 按钮进行确定，项目就新建成功了，之后将进入 Xcode 窗口进行项目的开发。

## 2.3 Xcode 窗口

在上面的 New Project 窗口中点击 Save 按钮确定保存之后，界面自动进入 Xcode 窗口，如图 2-6 所示。

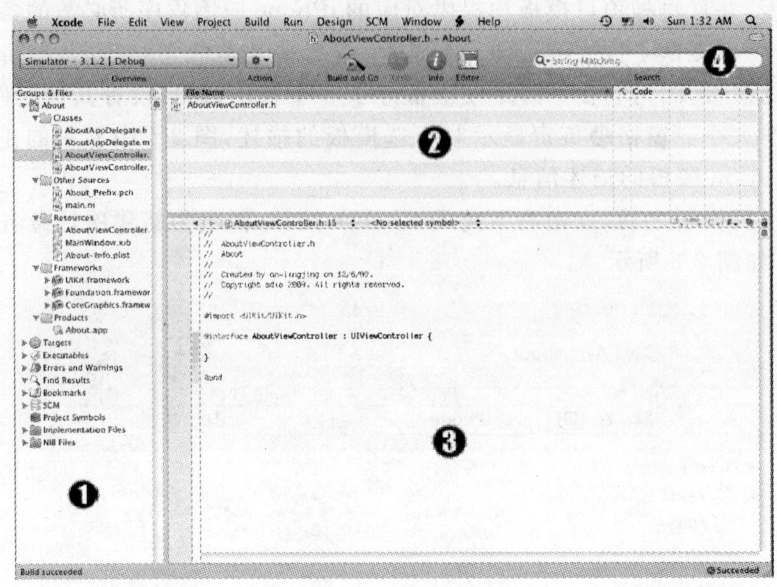

图 2-6　Xcode 窗口

### 2.3.1　窗口的布局

打开 Xcode 窗口后，可以看到窗口主要由四部分组成：

左侧的 Groups & Files 窗格❶：在这个窗格中包含了项目的所有资源，这些资源已经自动进行了逻辑上的分组，但我们也可以根据自己的需要移动组内文件、删除组或重命名组。若需要查看某一项的详细信息，可以单击每个组左侧的三角形图标来展示下一级子目录，查看完毕后可以再次单击三角形图标来隐藏子目录。这里显示的分组只是逻辑上的，存储的物理结构并不一定和在这里看到的相同。

右上方的细节显示窗格❷：该窗格中显示了 Groups & Files 窗格中被选择项的详细信息。如在图 2-6 中，在 Groups & Files 中选择了 AboutViewController. h 文件，该窗格中便只显示了 AboutViewController. h 文件。如果在 Groups & Files 中选择的是 Classes 文件夹，那么其中显示的内容将是：AboutAppDelegate. h、AboutAppDelegate. m、AboutViewController. h 以及 AboutViewController. m 四个文件。

右下方的文件编辑窗格❸：该窗格可根据在 Groups & Files 中所选文件，来显示文件的具体内容，并且可以在该区域进行文件内容的编辑工作。图 2-6 中，在 Groups & Files 中选中了 AboutViewController. h 文件，在右下方的文件编辑窗格中就对应显示了该文件的代码。

窗口最顶部一栏是工具栏❹，其中提供了许多常用的命令，如运行、搜索等。

以上，就是 Xcode 窗口的整体布局。

### 2.3.2 常用资源管理

下面我们来重点看一下 Xcode 是如何对开发中的一些常用资源进行管理的。在开发过程中常用资源都放在了 Groups & Files 窗格的第一项中，如本例的 About 项。

点击 About 左侧的小三角形图标，可以对它的子目录进行展开或折叠。现在展开 About 的子目录，可以看到 5 个子文件夹：Classes、Other Sources、Resources、Frameworks 和 Products。下面就来分别介绍一下这 5 个文件夹在资源管理方面的主要职责。

(1) Classes 文件夹：该文件夹主要用来保存代码文件，绝大多数代码文件要放在这里进行管理。如果项目较大、文件较多的话，也可以在这里继续向下设置子目录进行系统的管理。

(2) Other Sources 文件夹：该文件夹也是用来管理项目的代码文件，但只包含非 Objective-C 类的源代码文件。例如图 2-6 所示的一个预编译头文件 About_Prefix. pch 和整个程序的入口文件 main. m。

(3) Resources 文件夹：这里主要用来管理和保存项目的非代码文件，比如图像、声音等文件。因为 iPhone 程序可以访问的所有资源都放在自己的应用程序沙盒中，其他程序资源对本程序来说都是不可见的，所以需要将程序所需的一切资源都保存到这里来。

**Tips**

什么是应用程序沙盒呢？在 iPhone 中，每个程序只可以读写系统为本程序所创建的文件系统中的文件，而无法访问其他应用程序。这个文件系统就称为该应用程序的沙盒。

Resources 文件夹下面，有两个后缀名为. xib 的文件。. xib 是 iPhone 开发当中非常重要的一个文件类型，双击它会启动 Interface Builder。这类文件中包含了所有关于界面设计的信息。

About-Info. plist 文件提供了一个列表，利用它，可以对程序的属性进行一些修改。

(4) Frameworks 文件夹：这里主要用来包含程序要使用的框架和库，类似于 C++ 中的 lib 文件与 Java 中的 jar 文件。在我们的程序创建时，项目已默认添加了最常用的框架和库，即如图 2-6 所示的三个：UIKit. framework、Foundation. framework 以及 CoreGraphics. framework。如果要使用其他未默认添加进来的框架和库，比如数据库功能或 OpengGL 的相关功能，则需要手动将对应的框架链接进来。

(5) Products 文件夹：此文件夹包含本项目已经编译好的文件，如本例中的 About. app。

这五个文件夹，就是项目常用资源的管理者。通过后面的学习，你将会对它们有更深的体会。

## 2.4 用 Interface Builder 构建 About 的界面

为了轻松地掌握这些开发工具的使用，在这个项目中不会涉及太复杂的东西，只需要设计一个界面来向人们展示要声明的版权信息。

下面，我们就来看一下如何使用 Interface Builder 来设计这个界面。

**Tips**

Interface Builder 已经有较长的使用历史，以前的 Interface Builder 文件使用 .nib 扩展名，在不久前才改变为 .xib 扩展名。因此，可以将 Interface Builder 文件称为 Nib 文件或 Xib 文件。

首先熟悉一下 Xib 主窗口的构成。双击打开 Groups & Files 窗格中的 AboutViewController.xib 文件，我们将使用该文件来对程序界面进行可视化设计，如图 2-7 所示。

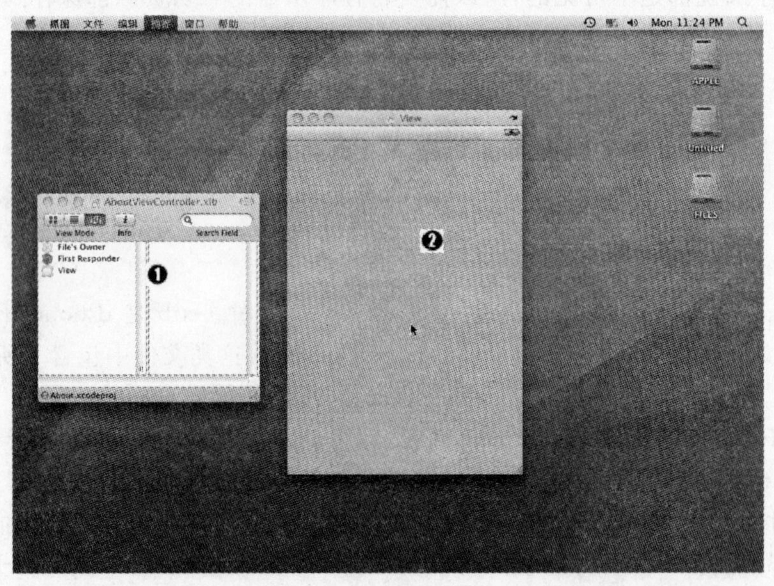


图 2-7 Interface Builder 的基本构成

在图中可以看到两个窗口：AboutViewController.xib 窗口❶与 View 窗口❷。

View 窗口就是进行界面设计的舞台，界面的一切设计都可以直接在上面操作，并且它可以有一个直观的反馈效果。如果 View 窗口没有显示的话，可以双击 Xib 窗口中的 View 项来打开 View 窗口。

AboutViewController.xib 窗口是进行界面设计的主窗口，它系统地管理着一切界面资源。每个扩展名为 .xib 的文件的窗口中都有 File's Owner 和 First Responder 两个元素，它们是在应用程序创建后自动创建的。但在这里，还没有必要对它们进行深究，因为随着今后学习的深入，会在后面的章节对它们进行具体的解释，在那里来理解它们，会感觉轻松很多。

除了这两个元素，Xib 窗口中的其他任何元素都是界面中的一个控件实例，并将在程序启动加载 Xib 文件时被创建。比如本例的 Xib 主窗口中的 View 图标，它就表示该项目界面中现在唯一的一个控件，也就是视图 View。

下面就来看一下，应该如何在这个视图中显示需要声明的版权信息。

### 2.4.1 添加需要的控件

在 View 窗口中，将通过添加一个图片来显示版权信息，正如图 2-1 显示的效果。那么，应该怎样添加这个图片呢？

**1. 在 Library 中选取控件**

首先选择 Tools→Library 选项（或者按下快捷键⌘＋Shift＋L），这样就会打开 Library 窗口，在这里面将会提供可以使用的各种控件，如图 2-8 所示。

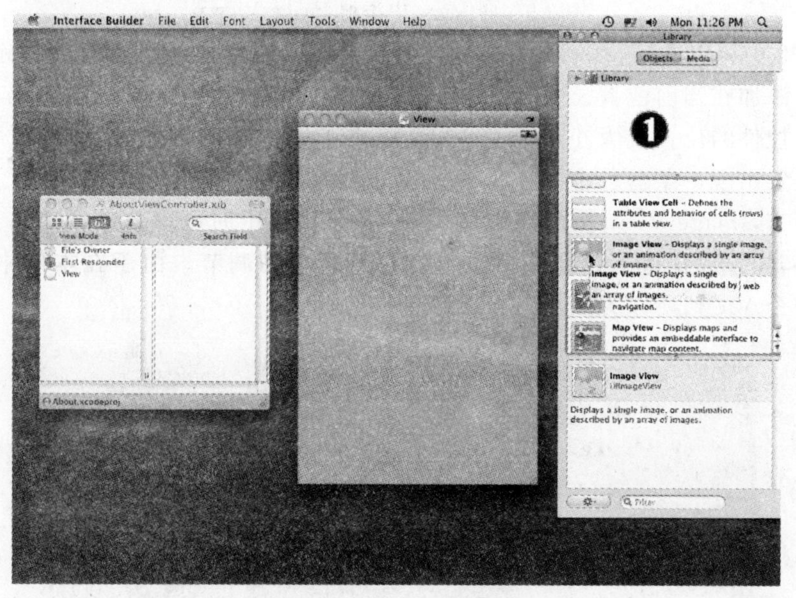


图 2-8 Library 窗口

从图中可以看到，在 Library 窗口❶中，摆放着开发中需要的各种控件。在本程序中，需要的是一个 Image View 控件。

**2. 向 View 窗口中添加选取的控件**

在 Library 窗口中选择 Image View 控件，并将它拖拽到 View 窗口中，如图 2-9 所示。

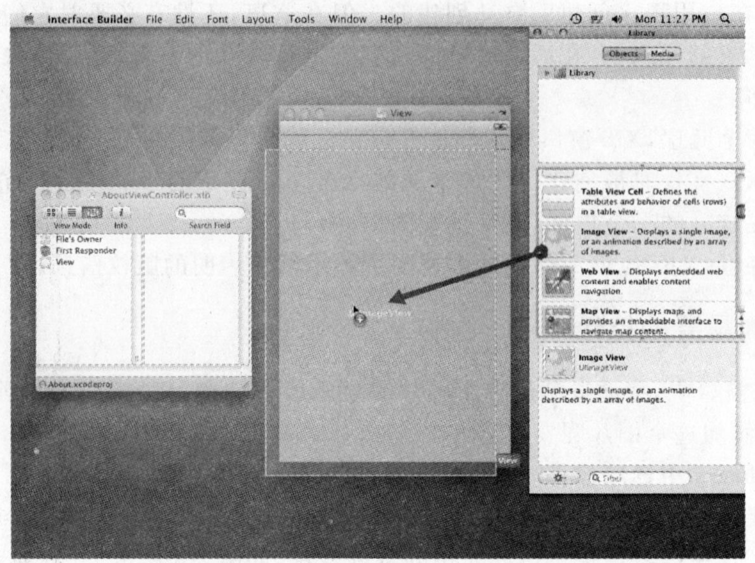

图 2-9　向 View 中添加 Image View

从图 2-9 中可以看到，拖动控件的过程中，当鼠标上出现一个绿色"＋"标志时，表示该控件可以添加在当前位置。拖放到合适的位置，松开鼠标，控件便添加成功了。如果摆放后感觉控件的位置或大小不合适，也可以再进一步进行调节，在调节时，View 窗口会在适当位置出现一条蓝色虚线作为辅助线来帮助定位控件位置，如图 2-10 所示。

一般而言，Image View 控件在刚加入到 View 窗口中时可以自动调节自身尺寸来适应窗口大小，如果你觉得不合适的话也可以再进行手动调节。由于在本例中，正是需要图像占用整个 View，所以无需再进行大小的调节。

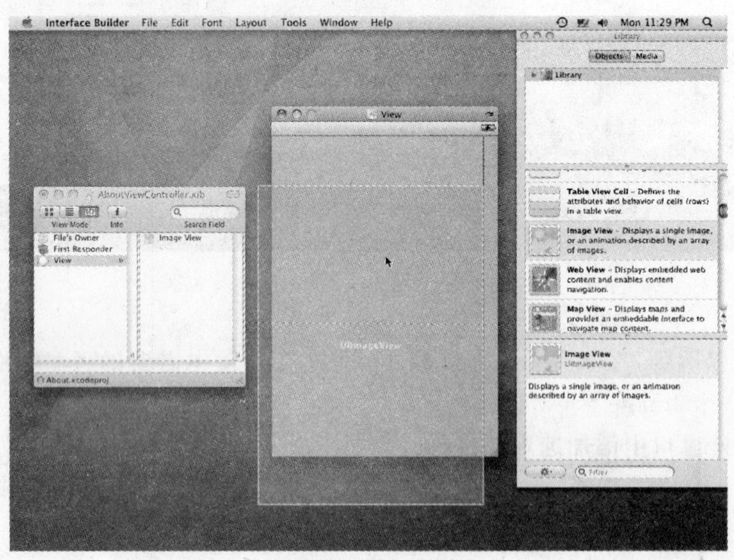

图 2-10　利用辅助线调节控件位置

现在就将控件添加成功了，下面就来看一下如何利用 Image View 控件来向界面中添加需要的图片。

Tips

在 iPhone 开发中，虽然大多数图片格式都是可用的，但建议使用 png 格式的图片，因为 Xcode 在构建应用程序时会自动优化. png 图像，使它们成为 iPone 应用程序中最快速最有效的图像类型。除此之外，相同质量的图片 png 格式容量最小，而且压缩失真度也是非常小的。

### 2.4.2 在 Inspector 中设置控件属性

**1. 导入需要的图片**

在利用 Image View 添加视图之前，要先将需要的图片添加进来。前面讲过，为方便资源管理，图片这类资源我们都放在 Groups & Files 窗格中的 Resources 文件夹下。所以，返回 Xcode 窗口，然后使 Resources 文件夹保持选中状态，通过在菜单栏中选择 Project→Add to Project，就会弹出一个选择图片文件路径的窗口。

图片 About. png 位于随书光盘 Chapter 2 About 文件夹下，你也可以使用其他图片。选择需要的图片之后，则会弹出一个如图 2-11 所示的窗口，用于设置加载图片的具体信息。

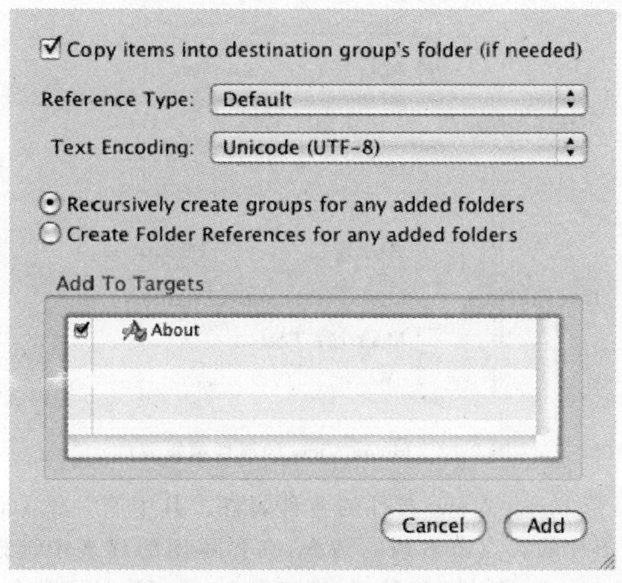


图 2-11 加载图片的具体信息

在这个窗口中，建议选中最上方的 Copy items into destination group's floder 复选框，这样可以保证将图片复制到项目目录中，而不只是引用该图片。

**2. 添加导入的图片到 Image View 中**

将需要的图片复制到了项目当中之后，选中 Image View 控件（可通过在 Xib 窗口中 View 选项的下级列表中选中 Image View 选项来实现），然后选择 Tools→Inspector 选项（或按下⌘+1），便可打开 Image View 的 Inspector 的 Attributes 窗口，如图 2-12 所示。

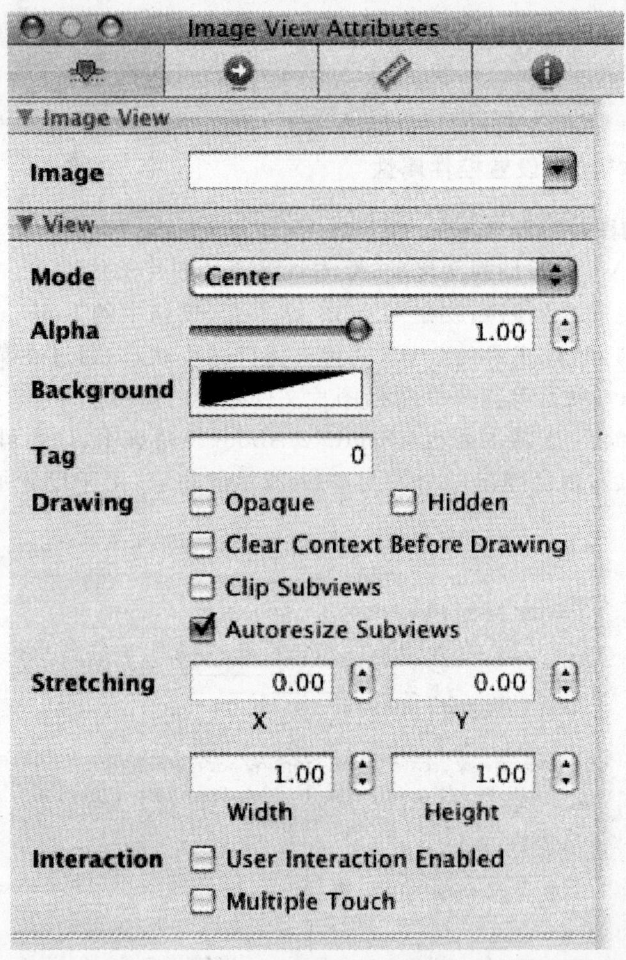


图 2-12 Image View 的属性编辑窗口

在图中，可以看到 Image View 控件的各种属性。其中第一项 Image 属性就是用来选择所需要显示的图片的。点击右边的箭头，在显示出的列表中选择想要添加的图片（如本例中的 About. png）。这样，在 View 窗口的 Image View 中就会相应显示出需要的图片了，如图 2-13 所示。

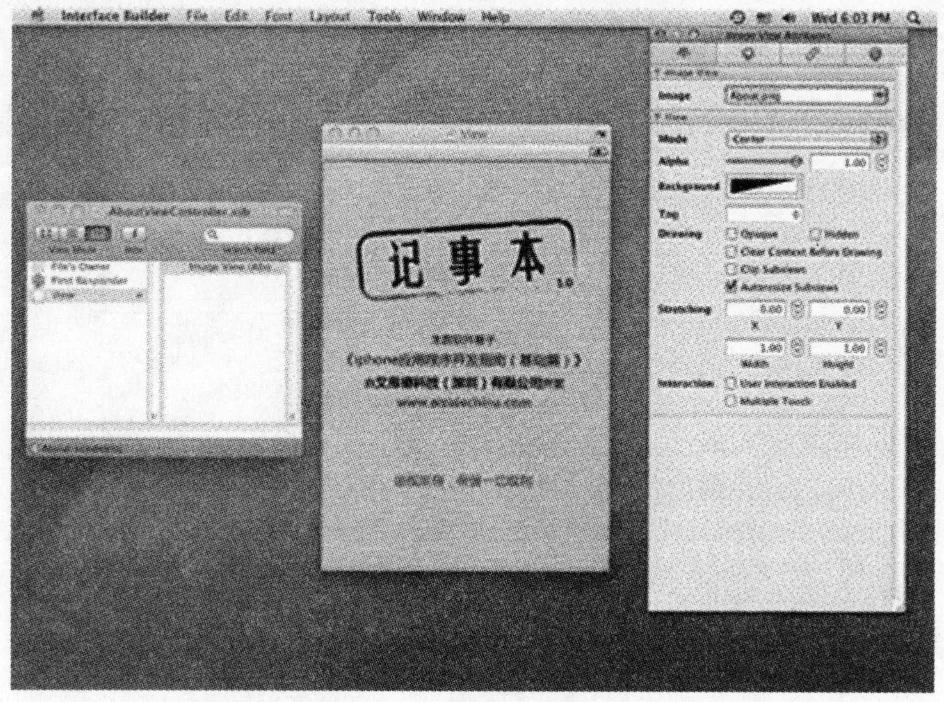


图 2-13 About 界面设置结果

在这里只设置了 Image View 的 Image 属性，关于 Image View 控件所特有的属性只有这一个，至于其他属性如 Mode、Alpha 等，是任何控件都具备的，我们可以在今后的学习过程中不断地加深了解。

同其他很多开发环境一样，Interface Builder 以图形化的方式构建用户界面，但值得一提的是：Interface Builder 并没有生成一行需要我们来维护的代码！这就避免了与代码生成相关的很多问题，为开发提供了极大的方便。

### 2.4.3 为程序添加图标

下面，再来给 About 项目添加一个图标，这可以让它看起来更加正式、美观一些。

首先，同样需要先将图片添加到项目中（图片 About.png 在随书光盘 Chapter 2 About文件夹下），在这里需要一个 57 * 57 像素的图片，这正是在 iPhone 中一个应用程序图标的大小。

图片导入成功之后，打开 Resources 文件夹下的 About-Info.list 文件来修改项目的图标信息。双击 Icon file 右列的空单元格，键入刚刚导入的图像文件的名称，如图 2-14 所示。

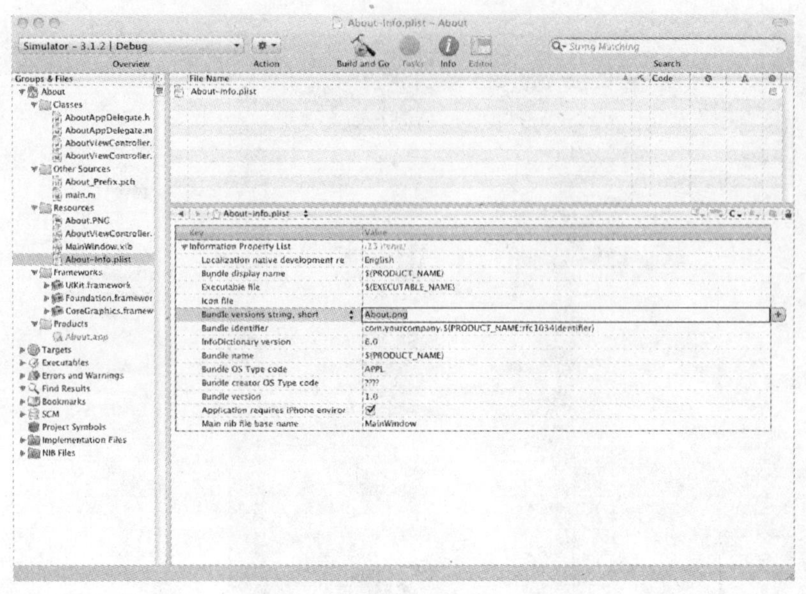


图 2-14　修改项目的图标信息

这样，项目图标便添加成功了。

## 2.5　在 iPhone 模拟器中运行程序

现在就可以编译并运行程序来查看一下结果了。返回到 Xcode 窗口，点击工具栏中的运行程序按钮或者在 Build 菜单中选择 Build and Run，程序便会在 iPhone 模拟器中运行，最终效果正如本章一开始在图 2-1中所展示的。之后点击模拟器上面的 Home 退出按钮退出程序，是不是看到了程序的图标？如图 2-15 所示。

怎么样？整个开发过程中一行代码都没有写，是不是感觉很轻松很方便？运用好这些开发工具，就会给你的开发旅程带来这样轻松的感觉！

## 2.6　常用的快捷键

在开发过程中我们往往会用到一些快捷键，像前面我们使用快捷键⌘＋Shift＋L 打开 Library 窗口、用⌘＋1 打开 Attributes 窗口，熟练地使用这些快捷键将会提高我们的开发效率。下面我们就总结一下比较常用的几个快捷键，供你在今后进行开发时查阅与参考，见表 2-1。

图 2-15　About 的图标

表 2-1 Interface Builder 快捷键及功能

| 快捷键 | 功能 |
|---|---|
| ⌘+1 | 打开 Attributes 窗口 |
| ⌘+2 | 打开 Connections 窗口 |
| ⌘+3 | 打开 Size 窗口 |
| ⌘+4 | 打开 Identity 窗口 |
| ⌘+Shift+L | 打开 Library 窗口 |
| ⌘+双击 | 跳到当前符号的定义 |
| ⌘+B | 构建程序 |
| ⌘+R | 编译并运行 |

注:⌘表示苹果键盘上的 Command 键。

## 2.7 小结

在这一章中,我们结合 About 这个简单的实例,形象地向你讲述了 iPhone 开发工具的具体使用方法与步骤。

首先,我们讲解了如何使用 View-based Application 模板创建一个 iPhone 应用程序。

然后,重点介绍了 Xcode 窗口的布局,其中 Groups & Files 窗格负责管理项目的所有资源,右下角的窗格即图 2-6 中的窗格❸是我们对项目文件进行编辑的区域。Interface Builder 用来进行程序界面的可视化设计,在本章我们讲解了如何使用 Interface Builder 来设计程序的界面。一切设计完成后,运行程序,我们的 About 程序成功地运行在了 iPhone 模拟器上。

最后,我们总结了在 Xcode 中一些常用的快捷键,熟练地使用它们将会提高我们今后的开发效率。

以上就是这一章所介绍的主要内容,下一章我们将开始语言篇的学习。

# 第 2 篇　语言篇

本篇对 iPhone 开发所用的标准语言Objective-C进行了讲解，读者可以通过本篇来学习 iPhone 的编程语言，为进一步学习 iPhone 开发知识打下基础。

语言篇包括 3 章：

第 3 章 Objective-C 基础，通过创建项目 Note，讲解了 Objective-C 的基本语法、如何创建一个类，以及在 iPhone 开发中比较重要的内存管理问题。

第 4 章几个重要的 Cocoa 类，讲解了在使用 Objective-C 进行 iPhone 开发时常用的几个类，包括 NSObject、NSString、NSArray 和 NSDictionary 等。

第 5 章类别和协议，讲解了 Objective-C 中重要而又有特色的两方面内容：类别和协议。

# 第 3 章　Objective-C 基础

**本章内容**

- Objective-C 的基本语法
- Objective-C 中类的创建
- 初始化方法
- 内存管理

Objective-C 是进行 iPhone 软件开发的标准语言。这一章将主要介绍一些 Objective-C 语言的基本知识，为后面 iPhone 开发的学习打下良好的基础。

在本章开始，首先对 Objective-C 这门语言进行了简要的介绍，然后通过一个例子对 Objective-C 的基本语法进行了讲解。

了解基本的知识之后，本章又对在 Objective-C 中如何定义类进行了讲解，其中包括类的声明与实现、对象的初始化以及属性等重要内容。

最后介绍了在 iPhone 开发中占有重要地位的内存管理方面的知识。

由于这本书并不是学习程序开发的入门书籍，所以在学习 Objective-C 之前，你应该已经具备一定的程序开发经验。

## 3.1　Objective-C 简介

Objective-C 语言是 C 语言的一个扩展集，它在标准 C 语言的基础上添加了面向对象的特性，是一个拥有面向对象层的 C。Objective-C 的运行环境库完全是用 C 编写的，任何一个 Objective-C 方法的调用，在运行环境中都会被替换成某些对应的 C 函数。

**1. Objective-C 与 C＋＋的比较**

表 3-1 将 Objective-C 与 C＋＋做了一下比较。

**表 3-1 Objective-C 与 C++的比较表**

| 比较方面 | Objective-C | C++ |
|---|---|---|
| 类的声明形式 | @interface 类名{<br>//变量声明<br>……<br>}<br>//方法声明<br>……<br>@end | Class 类名{<br>//类的声明(包括变量与方法)<br>……<br>}; |
| 类的实现形式 | @implement 类名<br>类的实现<br>……<br>@end | 类的实现<br>…… |
| 初始化方法 | 实际上是以"init"字样开头的普通成员方法 | 构造函数 |
| 方法声明形式 | +/-(返回值类型)函数名:(参数 A 类型)参数 A 形参名 函数名:(参数 B 类型)参数 B 形参名 …; | (static) 返回值类型 函数名(参数列表); |
| 方法调用形式 | [对象 方法名:参数 A 方法名:参数 B …]; | 对象. 方法名(参数列表); |
| 对象的创建 | 类名 对象指针=[[类名 alloc] 初始化方法]; | 类名 对象指针=new 构造方法; |
| 对象的销毁 | dealloc 方法 | 析构函数 |
| 继承 | 不支持多继承 | 支持多继承 |
| 成员变量存取权限控制 | @public、@protected、@private | public、protected、private |
| 嵌套调用方法 | [[[对象 方法 1] 方法 2] 方法 3]; | 对象. 方法 1(). 方法 2(). 方法 3(); |
| 头文件引入 | #import "FileName. h" | #include "FileName. h" |
| 空指针 | nil | NULL |
| 简单变量定义 | 与 C++相同 | int i=0; |
| 循环控制语句 | 与 C++相同 | for(int i=0; i<n; i++) |
| 条件控制语句 | 与 C++相同 | if(判断条件)<br>……<br>else<br>…… |
| 避免命名冲突方法 | 使用名称前缀,如 NS | 使用命名空间机制 |

表 3-1 通过 Objective-C 与 C++的比较，对 Objective-C 的基本语法进行了讲解。如果其中的某些知识你暂时还无法深刻理解也不用担心，本章下面的内容将对这些知识进行深入的讲解。

**2. 关于 Cocoa**

在今后的学习中还会经常看到一个概念 Cocoa。Cocoa 是苹果公司的面向对象的开发环境，该环境下的任何类都要继承自 NSObject，只有这样，该类的对象才可以获得运行时的基本能力。Cocoa 的主要开发语言是 Objective-C。Cocoa 包含了两个核心框架：Foundation 框架和 AppKit 框架。其中 Foundation 框架包含了 Cocoa 中最基本的一些类，如 NSString、NSArray 等，它们在一个应用程序中通常负责对象管理、内存管理、容器等相关数据结构的操作。AppKit 框架负责界面的设计，但在 iPhone 应用程序开发中，将使用 UIKit 框架来完成界面设计这一功能。如果你对.NET 和 C# 的概念比较清楚，那么在这里，你可以通过类比它们之间的关系来对 Cocoa 与 Objective-C 的关系进行理解。

下面我们就来编写第一个程序，体会一下 Objective-C 这门语言。

## 3.2 创建项目

首先创建一个名为 Note 的项目，它将用来模拟我们对自己一天所做事情的记录，并将这些记录在控制台中输出。

选择 File→New Project，进入 New Project 窗口。这一次，使用 Command Line Utility中的 Foundation Tool 模板来建立该项目，如图 3-1 所示。

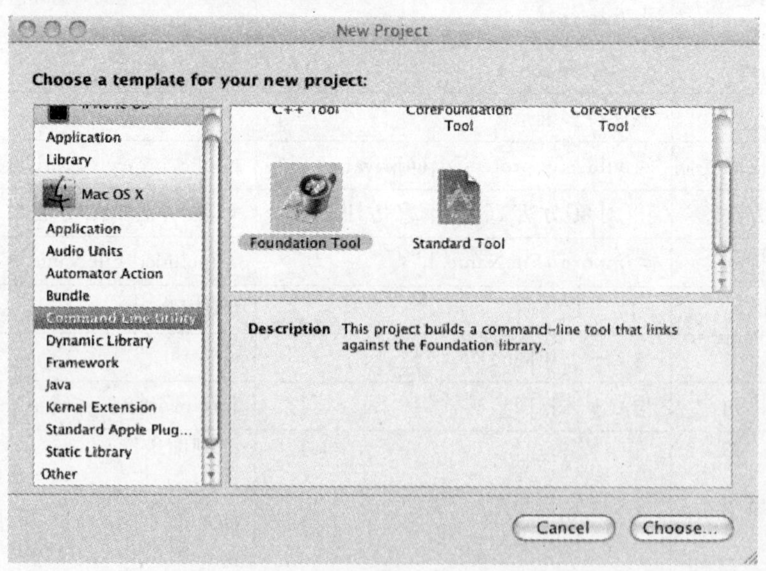


图 3-1　New Project 帮助窗口

给项目命名为Note，然后点击右下角的Choose按钮，就打开了该项目的Xcode窗口，如图3-2所示。

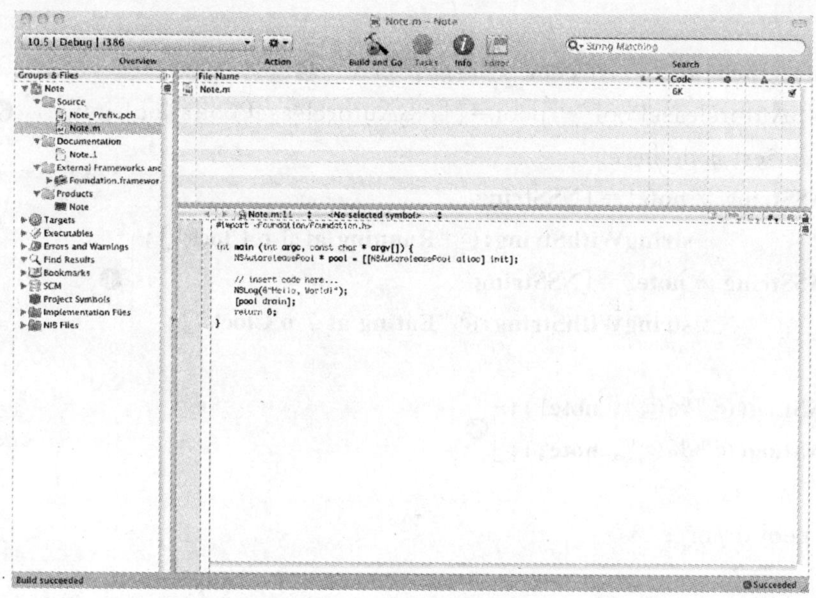

图3-2 Note的Xcode窗口

打开Xcode后，查看Groups & Files窗格中的Note.m文件。可以发现，Xcode已经为该项目准备好了一些代码：

### 代码3.1 Note.m文件

```
#import <Foundation/Foundation.h>
int main (int argc, const char * argv[]) {
    NSAutoreleasePool * pool=[[NSAutoreleasePool alloc] init];
    //insert code here...
    NSLog(@"Hello, World!");
    [pool drain];
    return 0;
}
```

这是Note程序的main函数，程序将从这里进入并开始运行。不用添加任何代码，现在本程序已经可以运行了。在菜单栏中选择Run→Console打开控制台，然后点击Build and Go编译并运行程序，将看到在控制台中输出了一个熟悉的字符串"Hello, World!"。

下面修改代码，记录自己的事情。

Note程序在这里要实现一个输出记录的功能，所以，添加如下黑体字所示的代码：

**代码 3.2 修改后的 Note.m 文件**

```
#import <Foundation/Foundation.h>   ❶

int main (int argc, const char * argv[]) {   ❷
    NSAutoreleasePool * pool=[[NSAutoreleasePool alloc] init];   ❸
    //insert code here...
    NSString * note1=[NSString
            stringWithString:@"Running at 6 o'Clock"];   ┐
    NSString * note2=[NSString                            ❹
            stringWithString:@"Eating at 7 o'Clock"];    ┘

    NSLog(@"%@", note1);  ┐
    NSLog(@"%@", note2);  ┘❺

    [pool drain];
    return 0;
}
@end
```

再次运行,程序在控制台中的输出就改变了,如图 3-3 所示。

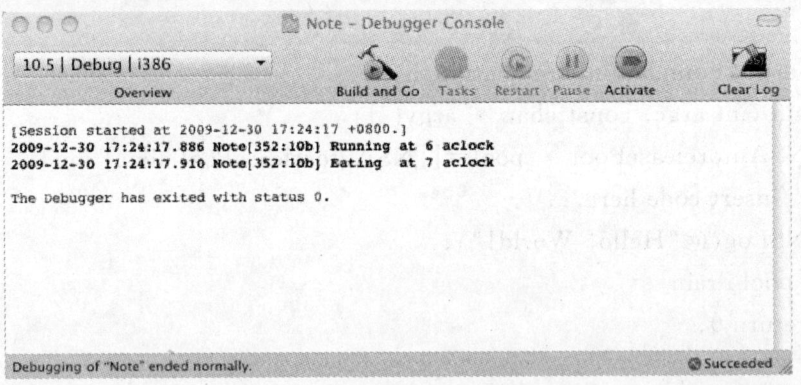


图 3-3 Notepad 运行结果

下面我们通过分析这个程序,来讲解一下 Objective-C 的基本语法知识,以期对 Objective-C 建立一个大体的印象。

## 3.3 解析 Note

**1. 文件类型**

首先来看一下打开文件的名称：Note. m。在 Objective-C 中，扩展名为. m 的文件与 C＋＋语言中扩展名为. cpp 的文件相同，均是程序的源文件。而程序的头文件与 C＋＋语言中的头文件一样，扩展名同样是. h。

**2. 引入头文件**

代码 3.2 中❶处的＃import ＜Foundation/Foundation. h＞表示程序在此处要引用 Foundation 框架中的 Foundation. h 头文件。

Objective-C 中的“＃import”语句同 C＋＋中的“＃include”语句类似，表示头文件的引入。但是还记得在 C/C＋＋中是如何避免同一个头文件被引入两次的吗？是在文件的一开始使用＃ifdef 命令来判断一个头文件是否已被加载过了。然而使用“＃import”的时候就无需如此，使用“＃import”引入一个头文件时，无论在同一个地方导入同一个头文件几次，都可保证该文件只被包含一次。

Tips

所谓框架，就是一个集合，其中包含着头文件、库以及图像和声音等资源文件。每个框架中均有一个主头文件，该主头文件包含了框架中的各个头文件，一旦引入了主头文件，就可以在自己的程序里使用任何在该框架里声明的类。

＃import ＜Foundation/Foundation. h＞表示要引入的文件是 Foundation 框架中的主头文件 Foundation. h。这样，我们就可以使用所有 Foundation 框架中的类了，比如 NSAutoreleasePool 就是 Foundation 框架中的一个类。

**3. 程序的入口——main 方法**

在❷处看到 int main(int argc, const char ＊ argv[ ])和下面的 return 0，跟 C 语言中的 main 函数一样，这里表示该应用程序的入口。

**4. 创建自动释放池对象**

在❸处的 NSAutoreleasePool ＊ pool＝[[NSAutoreleasePool alloc] init]，声明并初始化了一个可以用于自动释放对象的缓冲池，下面以 C＋＋的语法形式将它拆分来表现一下，理解起来可能就容易一些了：

(1) NSAutoreleasePool ＊ pool：同在 C＋＋中一样，这里表示声明了一个 NSAutoreleasePool 类型的指针，并命名为 pool。NSAutoreleasePool 类型用于构建一个缓冲池来存放各种对象和变量，缓冲池是用于内存管理的。内存管理方面，在iPhone开发中可以说是非常重要且较为复杂的，为使你可以系统牢固地掌握，内存管理的知识将放在后面进行详细讲解，在这里只需知道，为了程序的正常运行，pool 申请了一块公用的内存。

(2) [[NSAutoreleasePool alloc] init]：其中[NSAutoreleasePool alloc]可以理解为C++中的 NSAutoreleasePool：alloc()形式。

alloc 在这里，就是一个用于建立指定对象（也就是这里的 pool）的方法，它负责为该对象分配内存，并返回该对象的指针，但为指定对象开辟的内存中并无任何数据。下面紧跟的 init 相信你应该很容易理解了——为开辟了内存的对象调用初始化函数来进行初始化。这两句结合起来就相当于 C++中的 new 运算符。

对上面代码中的中括号“[ ]”是不是有些奇怪？在表 3-1 中所见的[对象　方法名：参数 A　方法名：参数 B…]，用一对中括号“[ ]”将对象与要调用的方法括起来，并在中间用空格隔开，便是 Objective-C 中调用方法的形式。而且在 Objective-C 的方法中，方法所需要的参数可以放在函数名的任何位置，使用“：”来表示要插入参数的位置，紧跟在“：”之后添加需要的参数，正如上面所示：“方法名：参数 A　方法名：参数 B”。

(3) [pool drain]：销毁缓冲池及其中的内容，并释放内存。

**5. 创建 NSString 类型对象——我们的记录**

❹处的 NSString * note1=[NSString stringWithString：@"Running at 6 o'Clock"]与 NSString * note2=[NSString stringWithString：@"Eating at 7 o'Clock"]，同前面的 NSAutoreleasePool * pool=[[NSAutoreleasePool alloc] init]一样用于创建对象，只是在这里，使用了 NSString 的 stringWithString：方法创建了两个 NSString 类型的指针 note1 与 note2。这样程序便不再是单纯地输出“Hello，World!”了，它将可以使用这两个 NSString 类型的对象来记录我们自己的事情。

NSString 类型相当于 C++中的 String 类型，它是 Foundation 框架中的一个用于处理字符串的类型，其中提供了很多用于字符串处理的方法，其相关内容将在后面进行详细讲解。

**6. NSLog 输出**

下面就来讲解一下在本程序中❺处的用于向控制台输出记录的两句代码：NSLog(@"%@"，note1)与 NSLog(@"%@"，note2)。NSLog 相当于 C++中的 printf 方法，在这里通过 NSLog 方法向控制台输出了 note1 与 note2 对象中所包含的信息。下面就来仔细分析一下这两条语句。

Tips

NSLog 中的 NS 是 NextSTEP 的缩写，它表示这个函数是来自 Cocoa 工具包的。Cocoa 对其所有函数、常量和类型的名称均添加 NS 前缀。

(1) NSLog 第一个参数是@"%@"，在 Objective-C 中，@"sometext"的形式表示要把引号内的字符串（如 sometext）当作一个 NSString 类的对象来进行处理。该字符串同 printf 中的一样，可以接受格式说明符，如%d 等，用来表示在此处要插入一个整数类型

(或者其他格式说明符对应的类型)的数值,并需要在该函数的后续参数中提供具体的相应数值。最后,NSLog将会在控制台中输出该字符串。

(2) 在程序要输出的字符串中,使用了类型说明符"%@",表示在此处输出一个对象的描述,但是如何输出一个对象的描述呢? 在Objective-C中,每个对象都有-(NSString *)description方法,在使用"%@"类型说明符来输出某种类型的对象时,就会调用这个方法,并在"%@"的对应位置输出-(NSString *)description方法的返回值。如果需要某类型的对象按照自己的意图进行输出,就需要重写-(NSString *)description方法。但事实上,如果今后使用"%@"类型说明符的话,大多数情况应该是需要输出一个NSString类型的对象,即一个字符串,NSString的-(NSString *)description方法将会输出这个字符串对象本身所包含的内容。

(3) 在今后开发过程中,会经常用到需要格式化的字符串,所以现在总结一下几种常用的格式说明符,见表3-2。

**表3-2 类型说明符**

| 类型说明符 | 代表类型 |
| --- | --- |
| %@ | 对象 |
| %s | 字符型 |
| %d | 整型 |
| %u | 无符号整型 |
| %f | 浮点类型 |

上面讲解了一些Objective-C与C/C++的差异。但因为Objective-C是C的一个超集,所以在C中声明变量的关键字、控制程序的分支循环结构、运算符等,在Objective-C中也同样适用,比如用int声明整数类型,用float声明浮点类型,用for语句来进行固定次数的循环,用if/else语句来进行条件选择等。

下面是一段与C/C++无异的小程序,可放在Objective-C程序中顺利运行,并将在控制台输出数字0到9。

**代码3.3 Objective-C基本语法示例**

```
int i=0;
for(i=0 ;i<10; i++){
    NSLog(@"%d", i);
}
```

# 3.4 Objective-C 中的面向对象

面向对象编程(Object-Oriented Programming,简称 OOP)如今已不算是什么新鲜的术语了。Objective-C 就是一种面向对象的编程语言。

在 Objective-C 中创建一个新类时,通常先要在一个与类名相同的.h 文件中,使用关键字@interface 来声明该类;然后在与类名相同的一个.m 文件中,使用关键字@implement 来实现该类。

本节将会对如何在 Objective-C 中声明和实现一个类进行讲解。

## 3.4.1 类的声明

在 Objective-C 中声明一个类时,需要使用@interface 关键字,它类似于 C++中的 class 关键字。声明一个类的格式如下:

**代码 3.4 Objective-C 中类的声明**

```
@interface ClassName : SuperClass {   ❶
@private
    vType variable1;   ❷
    ……
}
+ (rType)classMethodWithArg1: (arg1Type)arg1,      ┐
                  andArg2: (arg2Type)arg2;          │
- (rType)instanceMethodWithArg1:(arg1Type)arg1      │ ❸
                 andArg2:(arg2Type)arg2;            ┘

    @end   ❹
```

**1. 类的声明❶**

(1) @interface ClassName 表示要在此声明一个类,类名是 ClassName。

(2) :SuperClass 表示 ClassName 要继承于 SuperClass,即 SuperClass 是 ClassName 的父类。在 Objective-C 中只支持单继承,所以在只有单继承的情况下,Objective-C 与 C++基本无异:在类声明语句中的子类名称后用":"表示要继承的父类,子类继承父类的变量与方法。但在 Objective-C 中是不支持多继承的,所以在紧跟着 ClassName 的":"之后,只可以有一个父类。虽然不支持多继承,但 Objective-C 有类似方法来实现这一功能,相关的知识我们将在后面章节进行讲解。

**2. 成员变量的声明❷**

大括号{}内是类的成员变量的声明。@public、@private、@protected 是成员变量的访问修饰符，此处用@private 表示下面被修饰的成员变量是私有的，如果某成员变量不加任何修饰符的话，则 Objective-C 默认该成员变量为私有成员变量。修饰符下面的 vType 是成员变量的类型，variable1 是变量的名称。

**3. 成员方法的声明❸**

"}"与最后的@end 之间是类的成员方法的声明。可以看到在这里成员方法声明的形式有些不同寻常。

(1) ＋(rType) classMethodWithArg1:(arg1Type) arg1, andArg2:(arg2Type) arg2 方法声明中，在最前面有一个"＋"，这个"＋"用来表示这是一个类方法。并且类方法不可以对成员变量进行存取，因为各成员变量均存放在具体对象所对应的内存当中。-(rType) instanceMethodWithArg1: (arg1Type)arg1, andArg2: (arg2Type)arg2 方法声明前面的"-"，表示该方法是一个实例方法，应该由实现该类的某个实例对象来进行调用。

(2) ＋(rType)classMethodWithArg1:(arg1Type)arg1, andArg2: (arg2Type)arg2 方法声明与前面所讲的方法调用类似。在 Objective-C 的方法声明中，方法所需要的参数可以放在任何位置——使用":"来表示要插入参数的位置，紧跟在":"之后添加需要的参数。但一般会将参数放置在有助于理解函数功能和使用方法的位置，如在本方法中看到的。其中 arg1Type 表示参数 arg1 的类型，arg2Type 表示参数 arg2 的类型。

在这里声明的两个方法均具有两个参数，但在实际应用中，可以根据需要在方法声明的任何适当的位置添加参数。利用这种方法声明形式，我们将可以直接通过函数名，便清晰地知道每个参数的作用，而不用像在 Java 或 C＋＋当中一样，需要对每个参数的作用进行猜测。

**4. 声明结束❹**

使用@end 关键字来表示上面的 ClassName 类声明结束。

这样，便完成了对一个自定义类的声明。

### 3.4.2 类的实现

对一个类进行了声明之后，还必须要实现该类的方法才可以创建对象。

**代码 3.5 Objective-C 中类的实现**

```
@implement ClassName
    Methods Implementation
    ……
@end
```

在@implement 与@end 之间实现类的各个方法，方法可以是从父类继承下来的，也可以是自己声明的。

### 3.4.3　用 NoteClass 类封装记录

了解了如何在 Objective-C 中声明并实现一个类后，我们返回修改上面的 Note 程序，这次将用类来对记录进行封装。

下面，就来看看如何逐步实现对记录的封装。

**1. 添加一个新类"NoteClass"**

选中 Xcode 窗口中的 Classes 文件夹，右击选择 Add→New File 或者使用快捷方式⌘＋N 来新建一个文件，如图 3-4 所示。

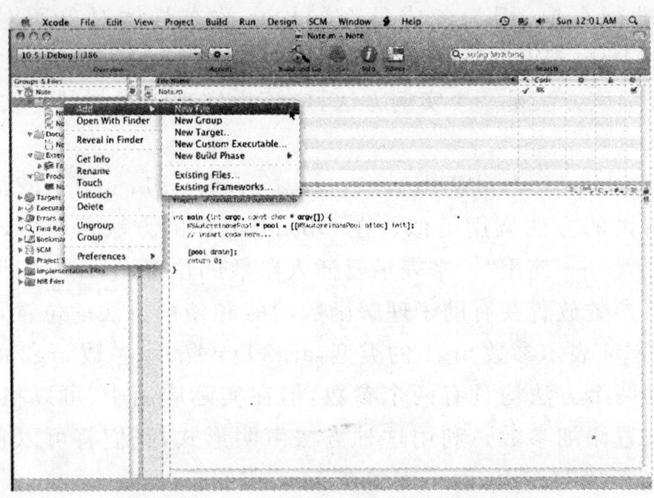


图 3-4　新建文件

选择 Cocoa 项下面的 Objective-C class 模板，如图 3-5 所示。

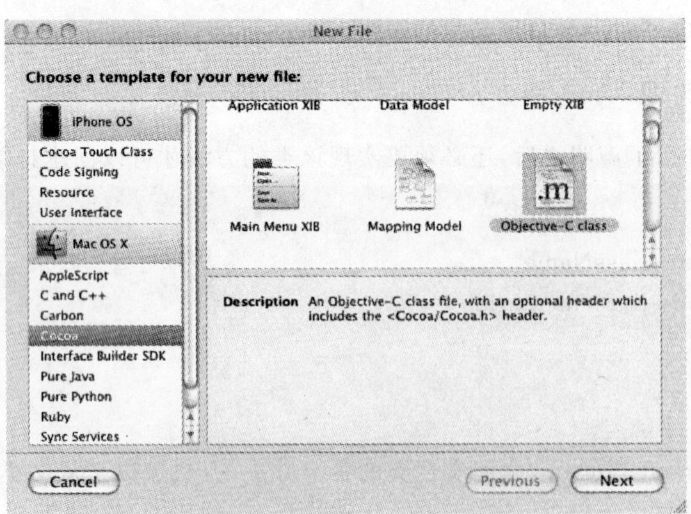


图 3-5　为新文件选择模板

为新文件命名为 NoteClass. m，同时注意选中 File Name 下面的 Also create "NoteClass. h"复选框，这样，Xcode 会为新文件自动创建对应的头文件"NoteClass. h"，如图 3-6 所示。

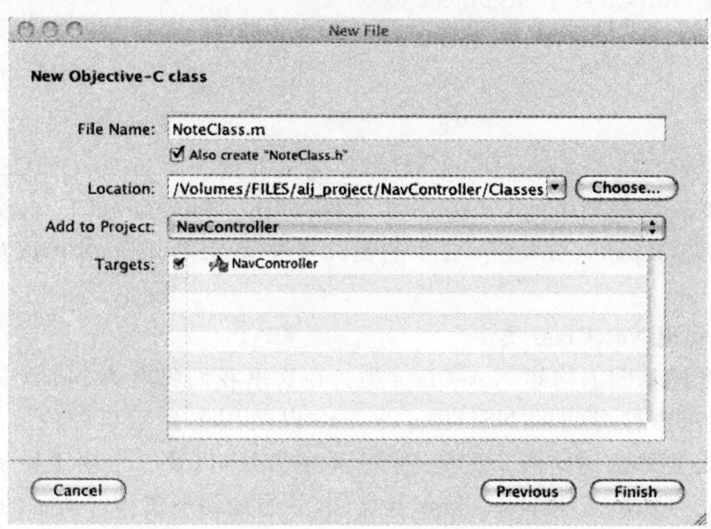


图 3-6 新文件命名窗口

点击右下角的 Finish 按钮确定，新文件便建立好了，返回 Xcode 窗口，在 Classes 文件夹下可以看到新增加的两个文件 NoteClass. h 和 NoteClass. m，如图 3-7 所示。

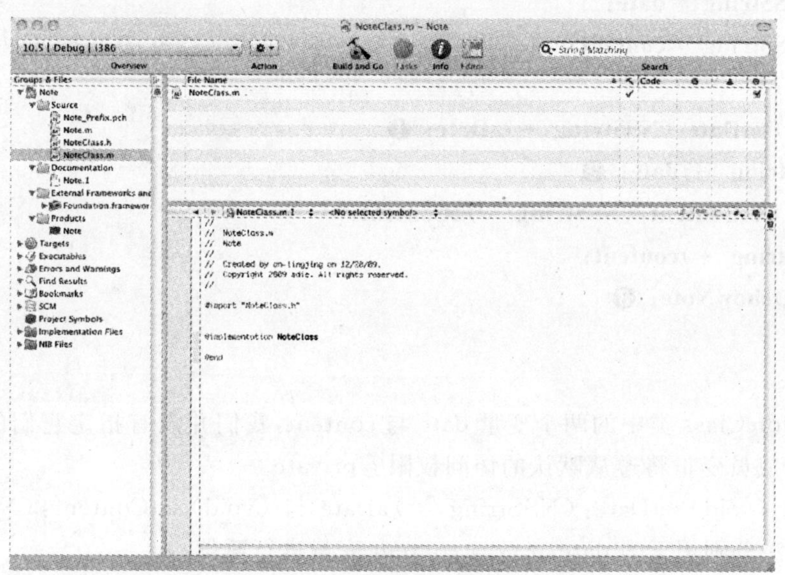

图 3-7 新增加的文件

选中新增加的 NoteClass. h 文件，可以看到，该文件已经有以下代码了：

**代码 3.6　NoteClass.h 文件**

```
#import <Cocoa/Cocoa.h>
@interface NoteClass : NSObject {
}
@end
```

这些代码是任何一个 Objective-C 类都必须具有的。#import <Cocoa/Cocoa.h> 语句表示引入 Cocoa 框架中的 Cocoa.h 头文件；@interface NoteClass : NSObject 表示该类要继承自 NSObject，Objective-C 中的类只有继承 NSObject，才可以具备运行时的各种基本能力。

**2. 声明并实现 NoteClass 类**

下面就在上述基础上构建 NoteClass 类。在这里我们将会为 NoteClass 添加两个成员变量：用于显示事件时间的 NSString * date 以及用于显示事件内容的 NSString * content，然后，分别添加对这两个变量进行存取的方法，以及一个用于显示这两个变量的 showNote 方法。所以，选中 NoteClass.h 文件，并添加如下黑体字所示的代码：

**代码 3.7　修改后的 NoteClass.h 文件**

```
#import <Cocoa/Cocoa.h>
@interface NoteClass : NSObject {
    NSString * date;
    NSString * content;
}
- (void)setDate:(NSString * )aDate; ❶
- (NSString * )date; ❷
- (void)setContent:(NSString * )aContent;
- (NSString * )content;
- (void)showNote; ❸
@end
```

对于 NoteClass 类中的两个变量 date 与 content，我们并没有指定它们的访问权限，所以这两个成员变量将遵从默认的访问权限@private。

❶处的-(void)setDate:(NSString * )aDate 与-(void)setContent:(NSString * )aContent方法分别用于为 date 变量与 content 变量进行赋值。

❷处的-(NSString * )date 方法与-(NSString * )content 方法分别用于取出 date 变量与 content 变量的值。

❸处的-(void)showNote 方法用于显示我们的记录。

可以看到，上面文件中的 5 个方法前面全部是“-”，这表示它们全部是实例方法，需要具体的 NoteClass 类的对象来进行调用。

这样，NoteClass 的声明工作就完成了，下面来看一下应该如何实现 NoteClass 的各个方法。选中 NoteClass. m 文件并添加如下黑体字所示的代码：

**代码 3.8 NoteClass. m 文件**

```
#import "NoteClass.h" ❶
@implementation NoteClass//开始声明 NoteClass 类

-(void)setDate:(NSString *)aDate {
    if (! [date isEqualToString:aDate ]){
        [date release ];
        date=[aDate retain ];                    ❷
    }
}
- (NSString *)date {
    return date;            ❸
}
-(void)setContent:(NSString *)aContent {
    if (! [content isEqualToString:aContent ]) {
        [content release ];
        content=[aContent retain ];
    }
}
- (NSString *)content {
    return content;
}
- (vo id)showNote {
    NSLog(@"%@ at %@.", content, date);   ❹
}
- (void)dealloc {
    [content release];
    [date release];         ❺
    [super dealloc];
}
@end
```

❶处的 #import "NoteClass. h"：同其他语言中一样，在类的实现文件里需要先将该

类的头文件包含进来。

❷处的一段代码用来为变量 date 赋值。首先检查 date 与新值 aDate 是否相同，如果不相同则使用 aDate 为 date 赋值。在赋值过程中，首先释放 date 对原来数据的所有权，然后再申请对新数据 aDate 的所有权(这些关系到 Objective-C 中内存管理的相关知识，这方面的知识将在本章最后做详细讲解)。这样，就实现了 setDate:方法。对于变量 content 的 setContent:方法，采用相同的方式实现。

❸处实现变量 date 的 getter 方法，在此定义了一个返回值为 NSString 类型的方法，用于返回 date 变量。

❹处便是用于显示记录的 showNote 方法，它通过 NSLog 方法定义了一个格式化的字符串，并用 date 与 content 两个变量来组合该格式化字符串，最终将字符串在控制台输出。

❺处相当于该类的析构方法，当某对象要被销毁时将自动调用该方法。该方法有两个必须完成的任务：释放本类变量的所有权从而释放内存；调用父类的 dealloc 方法来完成父类的释放工作。这样，就可以保证对象的销毁工作切实完成。本例使用[content release]与[date release]来分别释放对变量 date 与 content 的所有权，之后使用[super dealloc]来调用父类的 dealloc 方法。

这样，就完成了对 NoteClass 类的声明与实现。下一节将介绍关于对象初始化方面的知识。

### 3.4.4 初始化方法

前面我们讲到过要创建一个可用的对象，在 Objective-C 中需要两个步骤来完成：创建对象与初始化对象。创建对象即我们前面讲到过的 alloc 方法，alloc 方法负责为该类对象分配一个足够大的内存，然后将整块内存区域全部初始化为 0，也就是使被创建的对象不包含任何数据，比如 int 类型变量全被初始化为 0，float 类型变量全被初始化为 0.0，指针类型则全被初始化为 nil 等。

创建之后便可以调用初始化函数来对对象进行初始化。由于任何 Objective-C 类都是 NSObject 类的子类，所以它们都从 NSObject 类中继承了初始化方法 init，但该方法对于各子类基本没有任何作用与意义，子类需要根据自己类的具体情况来重写初始化方法，从而达到想要的初始化效果。

但要通过初始化方法来对对象进行初始化，也并不是一定要使用继承自 NSObject 的 init 方法，通常情况下对对象进行初始化时往往需要一些参数。如 NSString 类的一个初始化方法：initWithString: (NSString *)aString，它表示要接收一个字符串 aString，并用该字符串来对调用该方法的 NSString 对象进行初始化。

**Tips**

事实上，初始化方法与一般方法没有本质的区别，它只是在名称上添加“init”字样来表示这是一个初始化方法。它将负责为类中某些成员变量赋值。我们可以将初始化方法理解为该类的一个特殊的成员方法。

对于 NoteClass，我们可以在 NoteClass.h 文件中添加下面的一个初始化方法来对 NoteClass 类的对象进行初始化：

**-(id)initWithDate:(NSString *)aDate andContent:(NSString *)aContent;**

在 NoteClass.m 文件中，该方法的实现代码如下：

**代码 3.9 -(id)initWithDate: andContent:的实现**

```
- (id)initWithDate:(NSString *)aDate
      andContent:(NSString *)aContent { ❶
      if(self=[super init]){ ❷
            [self setDate:aDate];            ]
            [self setContent:aContent];      ]❸
      }
      return self;❹
}
```

在❶处可以看到，该方法的返回值为 id 类型，Objective-C 中，可以使用 id 类型来表示任意类型的指针，在本方法中，可以看到最后返回的是"self"，即调用该方法的对象本身。因为 NoteClass 可能会被继承，那么当子类使用该方法时，返回的类型就是子类的类型，所以，在实现该方法时，我们并不能确定它确切的返回值，因而使用 id 类型作为方法返回值，从而可以接受任何类型作为返回值。

在❷处的 if(self=[super init])，表示在初始化对象前，要先调用父类的初始化方法来完成父类自己的初始化工作。在这里也就是继承自 NSObject 类的 init 方法，从而让对象可以响应消息并处理引用计数器。因为实例变量所在的内存位置与隐藏的 self 参数是相对固定的，所以我们需要更新 self，以便任何变量的引用都可以被映射到正确的内存位置。这样，一旦父类初始化成功，便可以开始进行对子类的初始化工作了。

在❸处，分别调用对 date 变量与 content 变量进行设置的方法来为它们赋值，从而完成对这两个变量的初始化。

最后，使用❹处的返回语句 return self 来将初始化完成的对象返回自身。这样，便完成了对一个对象的初始化工作。

一般情况下，在对一个对象进行初始化时所需要的参数并不固定。比如，由于时间匆忙，在要记录某事件时，我们只记录下了当时的时间，准备以后再补录上具体事件，那么就需要在 NoteClass.h 文件中添加如下的一个初始化方法：

**-(id)initWithDate:(NSString *)aDate;**

该初始化方法只提供了一个用于初始化 date 变量的 aDate 参数，那么 content 变量将如何处理呢？对这种情况，我们往往是通过使用参数较少的初始化方法调用参数较多的初始化方法来完成初始化工作。

这正是 Cocoa 的设计人员所提出的 Disignated Initializer 概念。所谓 Disignated Initializer 就是将类中某个方法(一般是参数最多的一个方法)指定为 Disignated Initializer,其他初始化方法均通过调用该方法来执行初始化操作。子类调用父类的初始化方法时,也是调用该方法。

所以,可以在 NoteClass.m 文件中使用下面的代码实现 initWithDate: 方法:

**代码 3.10　-(id)initWithDate:(NSString *)aDate 的实现**

```
- (id)initWithDate:(NSString *)aDate {
        self=[self initWithDate:aDate andContent:@"I did something"]
        return self;
}
```

通过调用[self initWithDate:aDate andContent:@"I did something"],实现了对 date 变量与 content 变量的赋值,从而完成了对象的初始化。

加入 NoteClass 用于初始化对象的方法-(id)initWithDate:(NSString *)aDate andContent:(NSString *)aContent 与-(id)initWithDate:(NSString *)aDate 之后,我们对 NoteClass.h 文件添加如下黑体字所示的代码:

**代码 3.11　NoteClass.h 文件**

```
#import <Cocoa/Cocoa.h>
@interface NoteClass : NSObject {
        NSString *date;
        NSString *content;
}
- (id)initWithDate:(NSString *)aDate
        andContent:(NSString *)aContent;
- (id)initWithDate:(NSString *)aDate;
- (void)setDate:(NSString *)aDate;
- (NSString *)date;
- (void)setContent:(NSString *)aContent;
- (NSString *)content;
- (void)showNote;

@end
```

同时,还要修改 NoteClass.m 文件,添加如下黑体字所示的代码:

**代码 3.12　NoteClass.m 文件**

```
#import "NoteClass.h"
```

```
@implementation NoteClass//开始声明 NoteClass 类

- (id)initWithDate:(NSString * )aDate
        andContent:(NSString * )aContent {
    if(self=[super init]){
        [self setDate:aDate];
        [self setContent:aContent];
    }
    return self;
}
- (id)initWithDate:(NSString * )aDate {
    self=[self initWithDate:aDate andContent:@"I did something"];
    return self;
}
-(void)setDate:(NSString * )aDate {
    if(! [date isEqualToString:aDate]) {
        [date release];
        date=[aDate retain];
    }
}
- (NSString * )date {
    return date;
}
-(void)setContent:(NSString * )aContent {
    if(! [content isEqualToString:aContent]) {
        [content release];
        content=[aContent retain];
}
}
- (NSString * )content {
    return content;
}
- (void)showNote {
    NSLog(@"%@ at%@.", content, date);
}
- (void)dealloc {
    [content release];
```

```
        [date release];
        [super dealloc];
}
@end
```

下面，再来修改一下 Note.m 文件，添加如下黑体字所示的代码：

**代码 3.13 Note.m 文件**

```
#import <Foundation/Foundation.h>
#import "NoteClass.h"

int main (int argc, const char * argv[]) {
    NSAutoreleasePool * pool=[[NSAutoreleasePool alloc] init];
    //insert code here...

    NoteClass *note1=[[NoteClass alloc]
                        initWithDate:@"6:00"
                          andContent:@"Running"];   ❶

    NoteClass *note2=[[NoteClass alloc] initWithDate:@"7:00"]; ❷
    [note1 showNote];
    [note2 showNote];   ❸
    [note1 release];
    [note2 release];    ❹

    [pool drain];
    return 0;
}
```

首先，通过在❶处调用 NoteClass 的-(id)initWithDate:andContent:方法对 note1 对象进行了初始化，使 note1 记录“Running at 6:00”这一事件。

通过 NoteClass *note2=[[NoteClass alloc] initWithDate:@"7:00"]，note2 调用-(id)initWithDate:方法记录了一条时间信息“7:00”，如❷处所示。

接着在❸处通过调用 NoteClass 中用来输出记录的方法 showNote 来显示刚刚记录的信息，输出结果如图 3-8 所示。最后在❹处对 note1 与 note2 的所有权进行了释放。

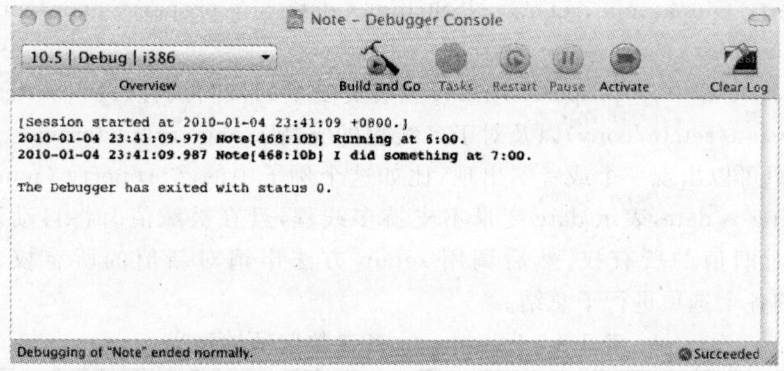


图 3-8 封装记录后的 Note 输出结果

### 3.4.5 属性

实现一个类时，还要为每一个变量实现其访问器，这是一个相当繁琐的工作。事实上，在 Objective-C 2.0 发布后，我们便可以很方便地使每一个变量自动生成其访问器，并使用点操作符"."来实现对变量的存取操作。

要实现这个功能，可以将 NoteClass.h 文件修改如下：

**代码 3.14 NoteClass.h 文件**

```
#import <Cocoa/Cocoa.h>
@interface NoteClass : NSObject {
    NSString *date;
    NSString *content;
}
@property (nonatomic, retain) NSString *date;       ❶
@property (nonatomic, retain) NSString *content;    ❶
-(id)initWithDate:(NSString *)aDate andContent:(NSString *)aContent;
-(id)initWithDate:(NSString *)aDate;
~~-(void)setDate:(NSString *)aDate;~~
~~-(NSString *)date;~~
~~-(void)setContent:(NSString *)aContent;~~
~~-(NSString *)content;~~

-(void)showNote;

@end
```

在上面的代码中注释掉了 date 变量与 content 变量的访问器方法，取而代之的是❶

处的@property (nonatomic,retain) NSString * date 与@property (nonatomic, retain) NSString * content。

在 Objective-C 中,@property 用来设置成员变量的读写属性(readwrite/readonly)、赋值语义(assign/retain/copy)以及对于多线程的支持 atomicity(nonatomic)。并且每组的各个选项只可以出现一个或者不出现,比如这个例子中的@property (nonatomic, retain) NSString * date,表示 date 变量不支持多线程,且在被赋值时将自动调用 release 方法来释放对旧值的所有权,然后调用 retain 方法申请对新值的所有权。表 3-3 对@property的各个选项进行了总结。

**表 3-3　@property 各参数选项的意义**

| 参数 | 意义 |
|---|---|
| readwrite | 设置该变量可读可写 |
| readonly | 设置该变量为只读 |
| assign | 为变量直接赋值,一般用于 int 等基本类型变量 |
| retain | 先释放变量对旧值的所有权,即先对变量进行 release 操作;然后使变量占有对新赋给的值的所有权,即 variable=[newData retain] |
| copy | 对变量进行 copy 操作,与 retain 处理流程相似,先使用 release 方法释放对旧值的所有权,再复制作为参数的新值,并将复制完成的副本赋值给该变量。它与 retain 的区别,将会在后面章节进行深入讲解 |
| nonatomic/atomicity | 关于线程的设置,在 iPhone 中只用 nonatomic,表示程序不需要多线程,从而可以避免生成额外代码,从而提高效率 |

下面来看一下对类的实现文件 NoteClass.m 的修改:

**代码 3.15　NoteClass.m 文件**

```
#import "NoteClass.h"
@implementation NoteClass
@synthesize date;   ❶
@synthesize content;

-(id)initWithDate:(NSString *)aDate andContent:(NSString *)aContent{
    if(self=[super init]){
        self.date=aDate;          ┐ ❷
        self.content=aContent;    ┘
    }
        return self;
}
-(id)initWithDate:(NSString *)aDate {
```

```
    self=[self initWithDate:aDate andContent:@"I did something"]
    return self;
}
-(void)setDate:(NSString * )aDate {
    if(date ! =aDate) {
    [date release];
    date=[aDate retain];
    }
}
-(NSString * )date {
    return date;
}
-(void)setContent:(NSString * )aContent {
    if(content ! =aContent) {
        [content release];
        content=[aContent retain];
    }
}
-(NSString * )content {
    return content;
}
- (void)showNote {
    NSLog(@"%@ at %@.", content, date);
}
- (void)dealloc {
    [content release];
    [date release];
    [super dealloc];
}

@end
```

在 NoteClass.m 中，同样删除了 date 变量与 content 变量访问器方法的实现，取而代之的是❶处的两行代码。在 Objective-C 中，@synthesize 负责根据前面@property 的设置，自动生成成员变量相应的存取方法，从而可以用点操作符来方便地存取该成员变量。

❷处-(id)initWithDate:(NSString * )aDate andContent:(NSString *) aContent 初始化方法的实现中，不再调用 date 与 content 的访问器，而是使用点操作符来为 date 变

量与 content 变量进行赋值：self. date＝aDate 与 self. content＝aContent。

self. date 并不是直接存取 NoteClass 的 date 变量。由于所有变量均默认为私有变量，在这里，date 的存取属性也依然是私有的。点操作符"."的作用，只是调用了由@synthesize自动生成的存取方法，从而实现对对应变量的存取。同时注意，self. date＝aDate 才是对 date 变量存取方法的调用，如果使用 date＝aDate 则只是单纯地对指针进行赋值，它不会执行 date 的存取方法，它既不会释放对旧值的所有权，也不会申请对新值 aDate 的所有权。

对于 Note. m 文件，无需做任何修改，运行程序，输出结果与图 3-8 一样。

## 3.5　内存管理

在 iPhone 开发中，对内存进行正确管理是非常重要的一个方面。iPhone 有 128M RAM，但其中约有一半的容量要用于屏幕缓冲和其他系统进程，同时 iPhone 不支持将内存写到交换文件，所以 iPhone 只有大约 64M 的内存用来运行应用程序，且严格受到物理内存量的限制。这样，基本上不容许我们开发的软件存在任何的内存泄露。

### 3.5.1　iPhone 中的内存管理

由于 iPhone 对内存严格的要求，所以当一个对象不再需要时，要及时释放它所占用的内存空间。

Objective-C 的内存管理采用了基于引用计数（Reference Count）这种非常常用的技术。简单讲，每个对象都有一个与之关联的整数，可以将它称为引用计数器或保留计数器，如果要使用一个对象，并确保在使用期间对象不被释放，需要通过函数调用来取得"所有权"，即引用计数器加 1，使用结束后再调用函数释放"所有权"，使引用计数器减 1。"所有权"的获得和释放，对应引用计数的增加和减少。引用计数为正数时代表对象还有引用，为 0 时代表可以释放。

回想前面的例子，[NoteClass alloc]就是为对象分配内存并使它的引用计数器的值为 1，而[note1 release]则是负责将 note1 对象的引用计数值减 1，如果检测到 note1 的引用计数值为 0，则 note1 对象会被释放掉，同时调用 dealloc 方法。但在这里要牢记的是，在 Objective-C 中，必须严格使用引用计数机制来控制内存的分配与释放，当对象的引用计数为 0 时，对象会自动进行释放，并调用 dealloc 方法，我们绝不可以自行调用 dealloc 方法。

下面就来详细说明一下内存是如何管理的。

### 3.5.2　用于内存管理的方法

在开发过程中，主要使用以下方法来对内存进行管理。

**1. 获得对象当前引用计数的方法**

-(NSUInteger)retainCount:方法返回当前对象的引用计数值。

**2. 获得对象所有权的方法**

(1) +(id)alloc:这是一个类方法,相信现在你已对它非常熟悉了,它负责为新建的对象分配内存并将新对象返回,同时将该对象的引用计数值设置为1。如果使用了该方法,则被创建的对象要负责调用 release 方法或 autorelease 方法来释放对象所有权,从而避免内存泄露。

(2) -(id)copy:拷贝引用该方法的对象,返回新复制的对象,被赋值的对象引用计数值加1。像 obj1=[obj2 copy],该语句表示复制对象 obj2,并将复制结果赋给 obj1,同时将 obj1 的引用计数值加1。

(3) -(id)retain:将调用该方法的对象的引用计数值加1。

另外,名字中带有 alloc、copy、retain 字样的函数,一般都会为引用计数值加1,这是一个在 Cocoa 中对方法的命名习惯。

**3. 释放对象所有权的方法**

(1) -(oneway void)release:负责将对象的引用计数值减1。如果对象的引用计数值减为0,则会调用该对象的 dealloc 方法,对该对象进行释放。

(2) -(id)autorelease:某些情况下,我们并不想取得对象所有权,又不希望对象被释放,例如在一个函数中生成了一个新对象并返回:

```
-(NSString *)description {
    NSString *desc=[[NSString alloc] initWithString:@"iPhone!"];
    return (desc);
}
```

像其中的 desc 对象就是这种情况,description 方法需要返回 desc 变量,所以不能释放它,但在方法之外又没有获取 desc 的所有权,无法释放。那么应该怎么办呢?

还记得前面的 NSAutoreleasePool * pool 吗?该对象被称为自动释放池,如果某个对象 obj 在自动释放池创建 NSAutoreleasePool * pool=[[NSAutoreleasePool alloc] init]与销毁[pool drain]之间使用[obj autorelease],则该对象就会被放在自动释放池中,表示会被自动释放,当自动释放池被销毁时,将该对象引用计数值减1。所以,我们可以对上面的 desc 对象做如下修改,将它放到自动释放池中:添加[desc autorelease]语句或将用于初始化的一句改为[[[NSString alloc] initWithString:@"iPhone!"] autorelease]。

同时,对于没有使用上面讲的三个获得对象所有权方法的对象,将自动放在自动释放池中,如本章一开始的例子中的 NSString * note1 =[NSString stringWithString:@"Running at 6 o'Clock"]中的 note1 对象。

自动释放池的效率并不高,所以在 iPhone 开发时,并不建议使用。

**4. 修改 Note.m 文件**

下面,来修改一下 **Note.m** 文件,这次所做的修改对于实际记录事情也许并没有多大意义,但它可以很好地帮助我们来理解内存的管理。

**代码3.16 Note.m 文件**

```
#import <Foundation/Foundation.h>
#import "NoteClass.h"

int main (int argc, const char * argv[]) {
    NSAutoreleasePool * pool=[[NSAutoreleasePool alloc] init];

    //insert code here...
    NSLog(@"alloc...");
    NoteClass *note1=[[NoteClass alloc] initWithDate:@"6:00"
                        andContent:@"Running"];
    NoteClass *note2=[[NoteClass alloc] initWithDate:@"7:00"];
    NSLog(@"note1's retaincount: %d", [note1 retainCount]); ❶
    [note1 showNote];
    [note2 showNote];
    //note1 retain
    NSLog(@"note1 retain...");
    [note1 retain]; ❷
    NSLog(@"note1's retaincount: %d", [note1 retainCount]);
    //note1 release
    NSLog (@"note1 first release...");
    [note1 release ];❸
    NSLog(@"note1's retaincount: %d", [note1 retainCount]);
    [note1 release];                                         ]❹
    NSLog(@"note1's retaincount: %d", [note1 retainCount]);  ]
    [note2 release];
    [pool drain];
    return 0;
}
```

下面来分析一下修改后的 Note.m 文件：

(1) 在❶处使用 alloc 创建了 note1 对象，然后输出 note1 对象的 retaincount 值，按照前面所讲的，使用 alloc 创建后 note1 的 retaincount 值应该为 1。

(2) 在❷处使用[note1 retain]，则调用 retain 方法的 note1 对象的引用计数值应该再加 1，所以此时的 note1 对象的 retaincount 值应该为 2。

(3) 在❸处使用[note1 release]对 note1 的所有权进行了释放，所以此时 note1 对象的 retaincount 应该为 1。

(4) 在❹处再次调用[note1 release],然后输出 note1 对象的 retaincount 值,此时应该为 0,对吗?

下面就运行程序来验证一下我们前面的猜测是否正确,如图 3-9 所示。

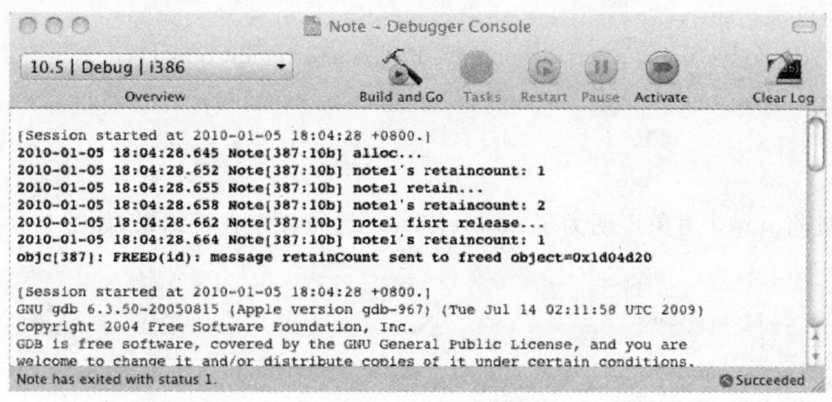


图 3-9 note1 的 retaincount 值

在图中可以看到,我们前三条猜测都被证实非常正确,但在第四条猜测的地方却出现了一条警告:向一个已经被释放的对象发送了 retainCount 消息。

原来 note1 已经被释放了,所以我们无法再向 note1 对象发送 retainCount 消息来获取它的引用计数值了。我们认为这条警告是一个好消息:当对象的引用计数值为 0 时,它竟然被释放得这么及时。

**5. 使用自动释放池**

那么当一个对象自动释放时,情况是怎样的呢?下面就让 note2 对象自动释放来观察一下。为了方便观察,删除刚才用于对 note1 对象内存进行操作的代码,修改 Note. m 文件如下:

**代码 3.17 Note. m 文件**

```
#import <Foundation/Foundation.h>
#import "NoteClass.h"
int main (int argc, const char * argv[]) {
    NSAutoreleasePool * pool=[[NSAutoreleasePool alloc] init];
    //insert code here...
    NSLog(@"alloc...");
    NoteClass * note1=[[NoteClass alloc] initWithDate:@"6:00"
                    andContent:@"Running"];
    NoteClass * note2=[[[NoteClass alloc] initWithDate:@"7:00"] ❶
                   autorelease];
    NSLog(@"note2's retaincount: %d", [note2 retainCount]); ❷
```

```
    [note1 release];
    [note2 release];
    NSLog(@"after pool drain...");

    [pool drain];
    NSLog(@"note2's retaincount: %d", [note2 retainCount]);
    return 0;
}
```

在❶处将 note2 对象声明为了 autorelease。运行程序查看结果，如图 3-10 所示。

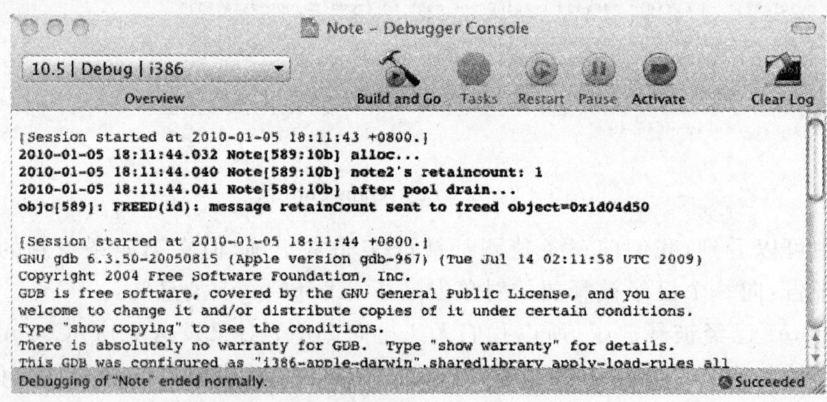


图 3-10 使用 autorelease 时的

在运行结果中可以看到，开始时输出 note2 对象的 retaincount 值为 1。当自动释放池被释放，[pool drain]被调用，再次输出 note2 对象的 retaincount 值时，可以发现"向一个已经被释放的对象发送了 retainCount 消息"这条提示又出现了——note2 对象在自动释放池销毁后，也随着被销毁了。

图 3-11 简单形象地描述了 Objective-C 中内存管理的流程。

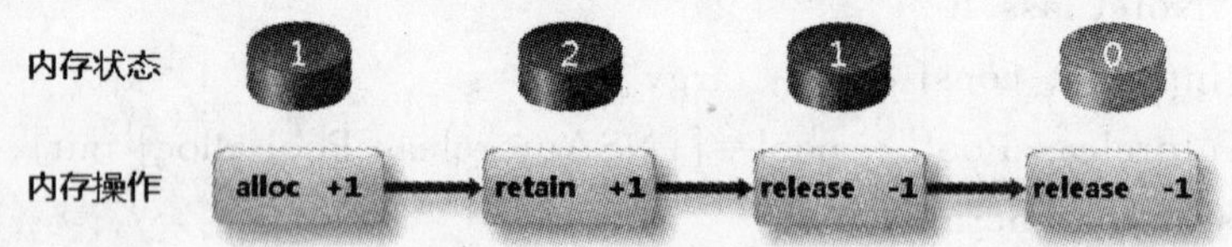


图 3-11 内存管理流程

**6. 其他注意事项**

以上便是关于内存管理大体的方法，下面来看一些关于内存管理的细微之处与实际使用时的小技巧：

(1) 各种容器类如 NSArray、NSDictionary、NSSet 等，会将引入的元素的引用计数值加 1 来获得所有权，而在元素被移除或者整个容器对象被释放时，释放容器内元素的所有权。

(2) iPhone 开发中不支持垃圾回收机制。

(3) 对象最好在需要时再创建，从而节省内存开销。

(4) 绝不可以发送 release 消息给 autorelease 对象。

(5) 在需要频繁分配与释放内存的地方(如 for 循环)，可以创建自己的 NSAutoReleasePool。

### 3.5.3 内存管理规则

下面总结一下内存管理的主要规则，在今后的开发中遵循这些规则，会使我们少出很多错误。

(1) 获得所有权的函数要和释放所有权的函数一一对应。

(2) 在对象的 dealloc 函数中释放对象所拥有的变量并调用父类的 dealloc 方法。

(3) 永远不要直接调用 dealloc 来释放对象，只使用引用计数来完成对象的释放。

大道至简，我们应首先牢固掌握这三条，这样，在今后的开发过程中，程序在内存方面应该就会平安无事了。

## 3.6 小结

本章介绍了关于 Objective-C 的一些基本知识，包括基本语法、类的声明与实现以及关于内存的管理。

关于基本语法，为了可以轻松快速地理解，我们演示了一个简单的例子，并结合 C++语言进行了对比讲解。

之后，着重介绍了在 Objective-C 中如何声明并实现一个类。在 Objective-C 中，基本上任何一个类都要继承 NSObject，才可以具有作为一个 Objective-C 类的基本行为能力。然后具体讲解了如何声明、实现一个类，如何初始化对象，以及如何对属性进行操作。

本章的最后对 Objective-C 中的内存管理进行了讲解。其中详细介绍了常用于内存管理的几个函数：alloc、retain、release 等；之后通过实例讲解了如何手动管理内存与使用自动释放池管理内存；最后，在 3.5.3 小节列出了内存管理中最为重要的三条规则，需要在今后的开发中牢记！

# 第 4 章　几个重要的 Cocoa 类

**本章内容**

- NSObject
- NSString 与 NSMutableString
- NSArray 与 NSMutableArray
- NSDictionary

Foundation 框架包含 100 多个类，本章将会为你介绍其中一些重要而且常用的类，比如 NSString、NSArray 等。完成本章学习后，你就完全可以在今后的开发过程中通过 iPhone SDK 来继续进行学习并解决开发中的问题了。

在 Foundation 框架中所包含的类，大体可按以下类别进行划分：

(1) 用于基本编程类型和操作的类，包括字符串、数组、字典、数字、异常处理等。

(2) 用于内核环境实体和服务的类，包括任务、运行循环、计时器、线程等。

(3) 对象管理类，包括内存管理、远程调用、存档等。

(4) 文件系统和 I/O 功能的类，包括 URL 处理、文件查找等。

(5) 其他功能类别，包括格式化数据、使用系统日期和时间等。

**Tips**

> Foundation 框架不会涉及到界面的操作，关于界面操作的部分都在 Application 框架中，但在 iPhone 开发时，将由 UIKit 替代 Application 框架来进行界面设计，UIKit 正是本书将要在后面的 iPhone 开发章节详细介绍的内容。

这一章，我们就来学习一下 Foundation 框架中比较重要而且常用的几个类。

## 4.1　NSObject

NSObject 是绝大多数 Cocoa 类的基类。通过 NSObject，各 Cocoa 类继承了运行时系统的基本接口，以及作为一个 Objective-C 类的基本行为能力。

当需要自己创建一个类时，也要直接或间接地继承 NSObject 类，只有这样，所创建的类才能具有作为一个 Objective-C 类所需要的基本行为能力。NSObject 类的子类从

NSObject 所继承的方法中最为常用的是用来初始化对象的，而且这些方法大多需要被子类重写，下面就来总结几个最为常用的方法见表 4-1。

**表 4-1 NSObject 的常用方法**

| 方法 | 介绍 | 示例 |
|---|---|---|
| + (id)alloc | 该方法是一个类方法，它负责创建并返回调用该方法的类的一个实例 | NSString * strDem=<br>[NSString alloc]; |
| - (id)init | 该方法负责对对象进行初始化，一般会以[[类名 alloc] init] 的形式调用，在对象被创建后马上进行初始化。<br>在 NSObject 类中，该方法只是返回“self”，并未进行任何初始化工作，子类如果要使用这个方法进行对象的初始化工作的话，需要进行重写 | NSString * strDem=<br>[[NSString alloc] init]; |
| - (void)dealloc | 该方法负责释放对象的内存，并做一些在释放对象内存前需要处理的工作，如释放某些成员变量的所有权。这就需要继承自 NSObject 的子类重写适合于自己的 dealloc 方法。<br>在上一章内存管理章节中提到过，对象绝不可以直接调用该方法，当对象的保留计数器的值为 0 时，该方法自动被调用 | [obj retainCount]为1<br>[obj release]<br>[obj retainCount]为0<br>自动调用<br>[obj dealloc] |

表 4-1 中的方法是在自定义一个类时最为重要与常用的 3 个方法，在今后的学习与开发中，你将会发现它们非常常见。下一节，将进行 Cocoa 中另一个非常重要与常用的类——NSString 类的学习。

# 4.2 NSString

在 Cocoa 中，NSString 是用来存储和处理字符串的类，它提供了多种方法让开发者创建字符串和获取其中的数据。但是 NSString 类的对象一旦被创建，便无法再对其中的内容进行修改。如果要使用可变的字符串，需要创建 NSMutableString 类的对象，我们将在 4.2.4 小节对 NSMutableString 的相关内容进行讲解。

下面，再来回顾一下前面的 Note 程序，同时增加一个新功能：查看是否有重复的记录，如果有重复记录，则程序不会重复输出相同记录；没有重复的话，程序便会将所有记录正常显示。

### 4.2.1　修改 Note

首先打开 Xcode，在菜单栏下选择 File→Open，如图 4-1 所示。

图 4-1　打开“Note”

选择 Open 菜单项后，则弹出如图 4-2 所示的 Open 窗口，选择 Note 的正确路径，然后打开 Note. xcodeproj 文件，就会出现 Note 的 Xcode 窗口了。

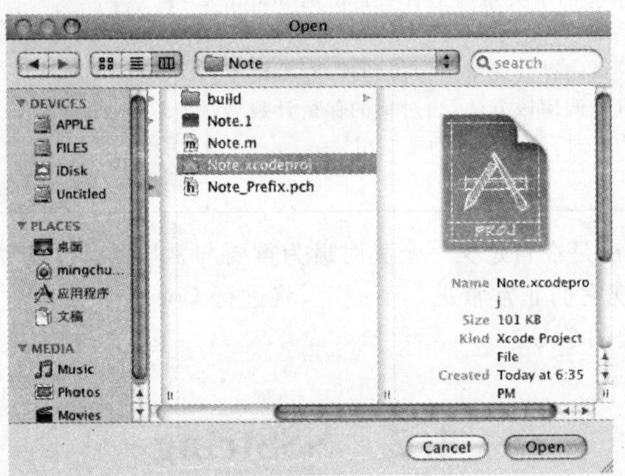


图 4-2　用来打开项目的 Open 窗口

在 Xcode 中，将 Note. m 文件修改如下：

**代码 4.1　Note. m 文件**

```
#import <Foundation/Foundation.h>
#import "NoteClass.h"

int main (int argc, const char * argv[]) {
```

```
NSAutoreleasePool * pool=[[NSAutoreleasePool alloc] init];

//insert code here...
//创建字符串
NSString * note1Date=[[NSString alloc]
                        initWithString:@"6:00"];
NSString * note1Content=[[NSString alloc]
                        initWithString:@"Running"];      ❶
NSString * note2Date=[[NSString alloc]
                        initWithString:@"7:00"];
//创建记录
NoteClass * note1=[[NoteClass alloc]
                    initWithDate:note1Date
                    andContent:note1Content];
NoteClass * note2=[[NoteClass alloc]
                    initWithDate:note2Date];
NoteClass * note3=[[NoteClass alloc]
                    initWithDate:note1Date
                    andContent:note1Content];
NSLog(@"note1:");
[note1 showNote];

//检查 note1 与 note2 是否是相同记录
if([note1.date isEqualToString:note2.date]
    &&[note1.content isEqualToString:note2.content]){
    NSLog(@"note2 is same with note1!");
}else{
    NSLog(@"note2 and note1 is different,note2:");
    [note2 showNote];
}
//检查 note1 与 note3 是否是相同记录                  ❷
if([note1.date isEqualToString:note3.date]
    &&[note1.content isEqualToString:note3.content]){
    NSLog(@"note3 is same with note1!");
}else{
    NSLog(@"note3 and note1 is different,note3:");
    [note3 showNote];
}
```

```
    [note1Date release];
    [note1Content release];
    [note2Date release];

    [note1 release];
    [note2 release];
    [note3 release];
    [pool drain];
    return 0;
}
```

运行程序，在控制台中将出现如图 4-3 所示的运行结果。

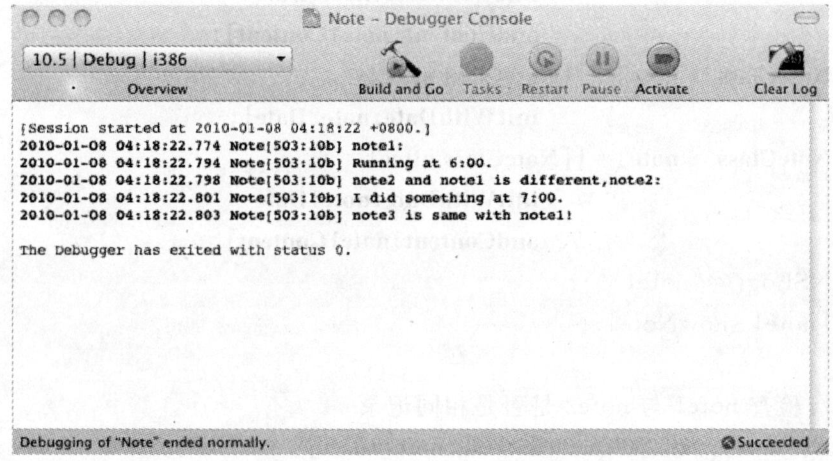


图 4-3　运行结果

### 4.2.2　字符串的创建

下面将通过分析在 Note 程序中所做的修改，来对关于创建 NSString 对象的知识内容进行讲解。

在代码 4.1 的❶处，首先创建并初始化了 3 个 NSString 的对象，并在后面使用这 3 个对象来初始化 NoteClass 类的对象：

**代码 4.2　在 Note.m 文件中创建 NSString 的对象**

```
NSString *note1Date=[[NSString alloc]
                        initWithString:@"6:00"];
NSString *note1Content=[[NSString alloc]
                        initWithString:@"Running"];
```

NSString ∗note2Date=[[NSString alloc]
　　　　　　　　initWithString:@"7:00"];

这 3 条初始化语句均通过调用 NSString 的实例方法 initWithString:来初始化 NSString 对象。方法的声明如下：

**-(id) initWithString:(NSString ∗)aString**

该方法通过复制另一个该类的对象 aString,来对调用该方法的对象进行初始化。

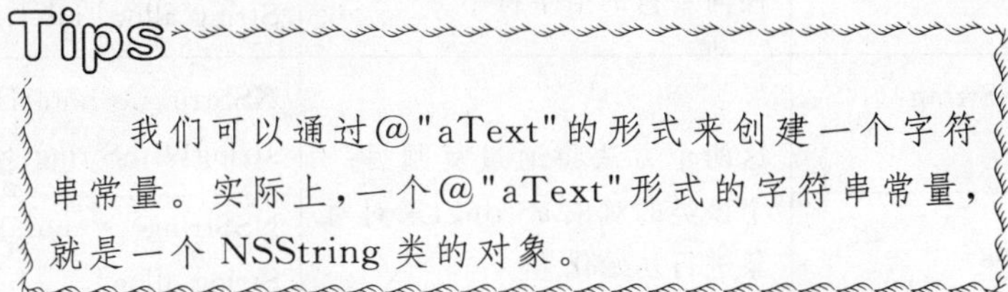
Tips

我们可以通过@"aText"的形式来创建一个字符串常量。实际上,一个@"aText"形式的字符串常量,就是一个 NSString 类的对象。

NSString 除了上面那个初始化方法外,还提供了多种初始化方法,以方便根据具体情况用不同形式初始化对象。下面就用一个 NSString 的类方法来实现相同功能,更改代码如下：

**代码 4.3 使用 stringWithString:方法初始化 NSString 对象**

~~NSString ∗note1Date=[[NSString alloc]~~
~~　　　　　　　　initWithString:@"6:00"];~~
~~NSString ∗note1Content=[[NSString alloc]~~
~~　　　　　　　　initWithString:@"Running"];~~
~~NSString ∗note2Date=[[NSString alloc]~~
~~　　　　　　　　initWithString:@"7:00"];~~
**NSString ∗note1Date=[NSString stringWithString:@"6:00"];**
**NSString ∗note1Content=[NSString stringWithString:@"Running"];**
**NSString ∗note2Date=[NSString stringWithString:@"7:00"];**
~~[note1Date release];~~
~~[note1Content release];~~
~~[note2Date release];~~

运行程序,可以得到与前面的例子同样的结果。在这里,将 NSString 的初始化方法由实例方法 initWithString:换成了类方法 stringWithString:。

同时注意,这里取消了用于释放 note1Date、note1Content 与 note2Date 三个 NSString 对象所有权的语句。正如在上一章内存管理中所讲的,在这里没有使用 alloc 为这三个对象分配指定的内存,它们被放在了自动释放池当中,所以不可以手动对其所有权进行释放。

下面,我们就来总结一下几个常用的 NSString 的初始化方法,见表 4-2。

表 4-2　NSString 常用的初始化方法

| 方法 | 介绍 | 示例 |
| --- | --- | --- |
| + (id)string | 这两个方法一个是类方法，一个是实例方法，使用它们均会将对象初始化为一个不包含任何字符的空字符串 | NSString * strClass = [NSString string]; |
| - (id)init | | NSString * strInstance = [[NSString alloc] init]; |
| +(id)stringWithString:(NSString *)aString; | 这两个方法将通过复制另一个该类的对象 aString，来对对象进行初始化 | NSString * note1Date=[NSString stringWithString:@"6:00"]; |
| -(id)initWithString:(NSString *)aString; | | NSString * note1Date = [[NSString alloc] initWithString:@"6:00"]; |
| + (id) stringWithFormat:(NSString *)format, ...; | 这两个方法均提供了一个格式化的字符串模板 format，并通过用后面的参数替换模板 format 中的格式转换符，形成需要的格式化字符串来初始化对象 | NSString * strDate = [NSString stringWithFormat: @ "%d: 00", 6]; |
| -(id)initWithFormat:(NSString *)format ...; | | NSString * strDate =[[NSString alloc] initWithFormat: @ "% d: 00",6]; |

这三组方法，会在今后的开发过程中经常遇到。

### 4.2.3　字符串的使用

通过上一小节，我们掌握了如何创建一个 NSString 类的对象，这一小节就来学习一下如何对 NSString 对象进行操作。

NSString 类的对象是不可变的，也就是说，NSString 的对象一旦创建，就不可以再更改它的值。所以，创建之后可以对它进行的操作并不多。

为实现检测重复记录这一功能，在代码 4.1 的❷处，我们添加了如下语句：

**代码 4.4　检查相同字符串**

```
if([note1.date isEqualToString:note2.date]
    &&[note1.content isEqualToString:note2.content]){
    NSLog(@"note2 is same with note1!");
}else{
    NSLog(@"note2 and note1 is different,note2:");
    [note2 showNote];
}
```

其中[note1. date isEqualToString: note2. date]调用了 NSString 类的实例方法

-(BOOL)isEqualToString:(NSString *)aString,来判断 note1. date 与 note2. date 是否相同,如果两者相同,则该方法返回 YES,否则返回 NO。

**Tips**

> 如果要判断两个字符串是否相同,我们需要使用 isEqualToString:方法,而不可以简单地用"= =",因为用"= ="判断的是两个指针所指内存地址是否相等,而不是两个字符串的内容是否相等。

进行判断后,程序就会根据判断的结果进行相应输出,结果如上面的图 4-3 所示。

以上就是 NSString 的 isEqualToString:方法的使用,下面再来总结一下对一个创建好的 NSString 对象进行操作的几种常用方法,见表 4-3。

**表 4-3 NSString 常用方法**

| 方法 | 介绍 | 示例 |
| --- | --- | --- |
| -(NSUInteger)length | 返回字符串中 Unicode 字符的个数 | int numOfStr= [note1Date length]; |
| -(unichar)characterAtIndex:(NSUInteger)index | 获取在字符串给定位置 index 处的一个字符 | NSLog(@"%c",[note1Date characterAtIndex:2]); |
| -(const char *)UTF8String | 以一个 C 字符串形式返回调用该方法的对象,返回的字符串以 null 结尾 | printf("%s",[note1 Date UTF8String]); |
| -(NSString *) stringByAppendingString:(NSString *)aString | 该方法将字符串 aString 追加在调用该方法的 NSString 对象后面形成一个新的 NSString 对象,并返回该对象 | NSString * newNote1Date= [note1Date stringByAppendingString:@"o'Clock"]; |
| -(NSString *) substringToIndex:(NSUInteger)anIndex | 返回一个调用该方法的对象的子字符串,在该子字符串中包含调用方法的对象从第一个字符一直到 anIndex(包括 anIndex)位置中间的所有字符 | NSString * newNote1Date=[note1Date substringToIndex: 1]; |
| -(NSString *) uppercaseString | 通过将调用该方法的字符串对象中的小写字母全部替换为大写字母,形成新字符串,并返回该字符串 | NSString * upNote1Content=[note1Content uppercaseString]; |
| - (float)floatValue | 将调用该方法的对象转化为浮点类型的值 | Float myPIE=[@"3. 14" floatValue]; |

### 4.2.4 可变字符串

刚刚创建的字符串是不可变的，但是如果我写错了记录怎么办？难道我每写错记录中的一个字符就要重新将整条记录再写一遍吗？这当然不合适，Cocoa 中的 NSMutableString 类便是 NSString 的可变版本，同时它也是 NSString 的子类。除了继承 NSString 的所有属性与方法外，NSMutableString 还提供了许多可以用来对字符串进行修改的方法。

下面我们就来把 Note 程序修改一下：今天早上时间匆忙，不仅没有及时记录 7:00 时做的事情，连 6:00 时做的事情也记错了，现在我要修改一下之前的记录。

**代码 4.5 Note.m 文件**

```
#import <Foundation/Foundation.h>
#import "NoteClass.h"

int main (int argc, const char * argv[]) {
    NSAutoreleasePool * pool=[[NSAutoreleasePool alloc] init];

    //insert code here...
    //创建可变字符串
    NSMutableString *note1Date=[[NSMutableString alloc]
                                initWithString:@"6:00"]; ❶
    NSMutableString *note1Content =[[NSMutableString alloc]
                                initWithString:@"Running"];
    NSMutableString *note2Date=[[NSMutableString alloc]
                                initWithString:@"7:00"];

    //创建记录
    NoteClass *note1=[[NoteClass alloc]
                        initWithDate:note1Date
                        andContent:note1Content];
    NoteClass *note2=[[NoteClass alloc]
                        initWithDate:note2Date];
    NoteClass *note3=[[NoteClass alloc]
                        initWithDate:note1Date
                        andContent:note1Content];

    //输出 note1
    NSLog(@"note1:");
    [note1 showNote];
```

```
    //修改 note1 记录中的事件内容
    [note1Content setString:@"Reading"]; ❷
    //输出修改内容后的 note1
    NSLog(@"note1:");
    [note1 showNote];

    //检查 note1 与 note2 是否是相同记录
    if([note1.date isEqualToString:note2.date]
        &&[note1.content isEqualToString:note2.content]){
        NSLog(@"note2 is same with note1!");
}else {
    NSLog(@"note2 and note1 is different,note2:");
    [note2 showNote];
    }
    //检查 note1 与 note3 是否是相同记录
    if([note1.date isEqualToString:note3.date]
        &&[note1.content isEqualToString:note3.content]){
        NSLog(@"note3 is same with note1!");
    }else {
        NSLog(@"note3 and note1 is different,note3:");
        [note3 showNote];
    }

    [note1Date release];
    [note1Content release];
    [note2Date release];
    [note1 release];
    [note2 release];
    [note3 release];
    [pool drain];
    return 0;
}
```

运行程序，将会出现如图 4-4 所示的输出结果。

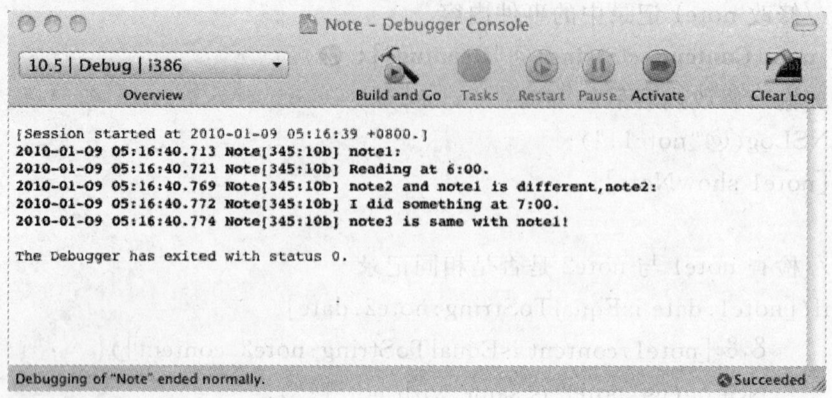


图 4-4　可修改记录的"Note"运行结果

下面来分析一下上面的程序。

NSMutableString 是 NSString 的子类，继承了 NSString 的一切方法，所以在❶处我们使用继承自 NSString 中的 initWithString: 方法对 note1Date、note1Content 和 note2Date 对象进行初始化。

在❷处对错误的记录 note1Content 进行了修改：

[note1Content setString:@"Reading"];

上面这行代码使用了 NSMutableString 的-(void)setString:(NSString *)aString 方法，使用这个方法，可以将调用该方法的 NSMutableString 对象的内容替换为 aString。在本例中，就将先前记录的事件信息"Running"替换成了"Reading"。

下面，我们再来看一下 NSMutableString 类中一些常用的方法，见表 4-4。

**表 4-4　NSMutableString 常用方法**

| 方法 | 介绍 | 示例 |
|---|---|---|
| +(id)stringWithCapacity:(NSUInteger)capacity | 该方法负责创建并返回一个可以容纳 capacity 个字符的空字符串来初始化 NSMutableString 对象。capacity 只是起优化存储的作用，它并不会真正限制 NSMutableString 对象所能容纳字符串的长度 | NSMutableString * note1Date = [NSMutableString stringWithCapacity:5]; |
| -(void)appendString:(NSString *)aString | 将一个新字符串 aString 添加到调用该方法对象的后面，从而改变该对象 | [note1Date appendString:@" o'Clock"]; |
| -(void)insertString:(NSString *) aString atIndex:(NSUInteger) anIndex | 在原 NSMutableString 对象的 anIndex 位置，添加字符串 aString来改变该对象 | [[note1Content insertString:@"I am "] atIndex:0]; |

表 4-4 中介绍的三个方法的作用分别为：使用类方法创建一个规定了大小的NSMutableString对象，向 NSMutableString 对象末尾增加一个字符串来改变原有对象，以及向NSMutableString对象的指定位置添加一个字符串来改变原有对象。除此之外，NSMutableString 的其他方法还可以完成删除指定位置字符、向对象末尾增加一个格式化字符串来改变原有对象等功能。关于这些方法的详细使用，你可以通过查阅 iPhone SDK 帮助文档来进行了解和学习。

## 4.3 NSArray

假设这一天我做了 100 件事，我当然可以用 100 个 NSLog 语句逐个输出它们，但通常我希望可以将这些语句组织得更有条理，而不是像 100 个 NSLog 语句这样冗余和杂乱无章。

Cocoa 提供了许多集合类来对数据进行组织，NSArray 就是进行开发时较为常用的一个。它实现了数组的功能，但使用 NSArray 时有两点需要注意：

(1) NSArray 只能存储 Objective-C 的对象，而不能存储像 int、float 这些基本的数据类型。但由于 Objective-C 对 C 兼容，所以在 Objective-C 程序中，仍然可以使用 C 的数组来存储普通类型的数据。

> **Tips**
>
> NSArray 不能存储基本类型的数据，可以用 NSNumber 类来将 int、float 等各种基本数据类型进行封装从而使它们“变为”标准的 Cocoa 对象，这样便可以将需要的数据存储进 NSArray 了。

(2) NSArray 一旦创建便不可以再对它进行更改了，如果要进行对数组的增、删、改等操作的话，需要使用 NSArray 的子类 NSMutableArray 来创建对象，这部分内容将会在后面进行讲解。

NSArray 提供了大量方便的方法来创建对象和组织其中的元素，下面就使用 NSArray 来组织一下 Note。

### 4.3.1 用数组组织多个记录

Note 程序可以检测出相同的记录并避免重复输出，但它们真的是相同的记录吗？如果不放心，就通过这一节做一下验证：先将各条记录全部输出来查看一下，之后再进行检测重复记录的操作。

修改 Note.m 文件如下：

代码 4.6　Note.m 文件

```
#import <Foundation/Foundation.h>
#import "NoteClass.h"

int main (int argc, const char * argv[]) {
    NSAutoreleasePool * pool=[[NSAutoreleasePool alloc] init];

    //insert code here...
    //创建可变字符串
    NSMutableString * note1Date=[[NSMutableString alloc]
                                 initWithString:@"6:00"];
    NSMutableString * note1Content=[[NSMutableString alloc]
                                    initWithString:@"Running"];
    NSMutableString * note2Date=[[NSMutableString alloc]
                                 initWithString:@"7:00"];
    //创建记录
    NoteClass * note1=[[NoteClass alloc]
                       initWithDate:note1Date
                       andContent:note1Content];
    NoteClass * note2=[[NoteClass alloc]
                       initWithDate:note2Date];
    NoteClass * note3=[[NoteClass alloc]
                       initWithDate:note1Date
                       andContent:note1Content];

    NSArray * noteArray=[[NSArray alloc]
                         initWithObjects:note1,
                         note2, note3, nil];
    for(int i=0; i<[noteArray count]; i++){
        [[noteArray objectAtIndex:i] showNote];
    }
    NSLog(@" ");
    NSLog(@"test same note...");
    //输出 note1
    NSLog(@"note1:");
```

```
        [note1 showNote];
        //修改 note1 记录中的事件内容
        [note1Content setString:@"Reading"];
        //输出修改内容后的 note1
        NSLog(@"note1:");
        [note1 showNote];
        //检查 note1 与 note2 是否是相同记录
        if([note1.date isEqualToString:note2.date]
            &&[note1.content isEqualToString:note2.content]){
            NSLog(@"note2 is same with note1!");
        }else {
            NSLog(@"note2 and note1 is different,note2:");
            [note2 showNote];
        }
        //检查 note1 与 note3 是否是相同记录
        if([note1.date isEqualToString:note3.date]
            &&[note1.content isEqualToString:note3.content]){
            NSLog(@"note3 is same with note1!");
        }else {
            NSLog(@"note3 and note1 is different,note3:");
            [note3 showNote];
        }

        [note1Date release];
        [note1Content release];
        [note2Date release];

        [note1 release];
        [note2 release];
        [note3 release];
        [pool drain];
        return 0;
}
@end
```

运行程序，将会得到与以前一样的效果，如图 4-5 所示。

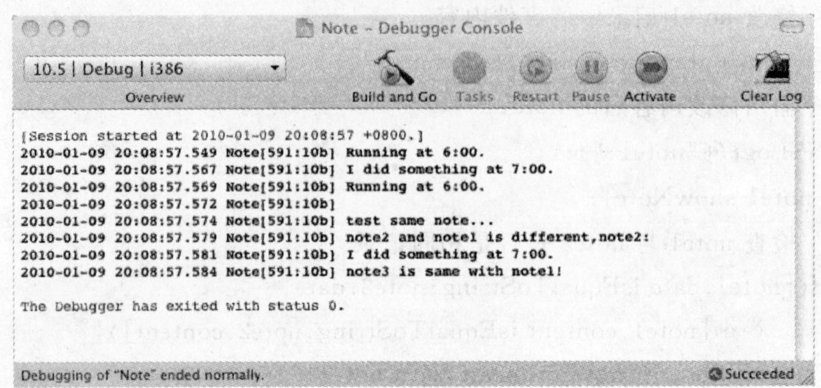


图 4-5 使用 NSArray 后的运行结果

下面就来具体分析一下修改后的程序，并学习 NSArray 的相关知识。

### 4.3.2 NSArray 对象的创建

代码 4.6 中首先创建了一个 NSArray 的对象 noteArray，并用前面的三条记录来初始化它：NSArray * noteArray =[[NSArray alloc] initWithObjects: note1, note2, note3, nil]。可以看到，在这一句代码中，我们使用了 NSArray 的 initWithObjects:方法来对 noteArray 进行了初始化，该方法声明如下：

```
-(id)initWithObjects:(id)firstObj,…
```

该方法的参数为不定参数，它使用一组对象来初始化 NSArray 对象，这组对象可以是任意多个，也可是不同类型的，比如你可以将 NSString 类型的对象与 NSNumber 类型的对象同时包含在一个 NSArray 对象中，这不会有任何问题。最后一个参数必须是 nil，以此表示初始化参数列表的结束。

这样就用一组对象初始化好了需要的 NSArray 对象。此外，另外一个较为常用的初始化方法是通过另一个 NSArray 对象作为参数来进行初始化工作：-(id)initWithArray:(NSArray *)anArray。该方法通过 anArray 对象中的元素对新对象进行初始化，比如在初始化好了 noteArray 的基础上，可以这样初始化另外一个 NSArray 对象：NSArray * myNoteArray=[[NSArray alloc] initWithArray: noteArray]。

### 4.3.3 获取 NSArray 指定索引处的元素

创建好 NSArray 的对象 noteArray 后，使用下面语句将其中所包含的记录进行一一输出：

```
for(int i=0 ; i<[noteArray count]; i++){
        [[noteArray objectAtIndex:i] showNote];
}
```

(1) 首先，通过[noteArray count]可以得到 noteArray 中元素的个数，程序中根据这个值来设置 for 循环的次数。

(2) 然后由[[noteArray objectAtIndex:i]可以获得 noteArray 中指定索引值为 i 的元素,在本例中便是 NoteClass 类的对象,也就是前面创建的 note1、note2 与 note3。

(3) 最后,通过调用 NoteClass 的显示消息的方法 showNote,将三条记录进行了一一输出。

在这个 for 循环中,我们用到了获取 NSArray 中元素个数的方法-(NSUInteger)count 和获取指定索引位置对象的方法-(id)objectAtIndex:(NSUInteger)index。下面,我们在表 4-5 中再列举几条 NSArray 的常用方法。

**表 4-5　NSArray 常用方法**

| 方法 | 介绍 | 示例 |
|---|---|---|
| -(NSArray *) arrayByAddingObject:(id)anObject | 将一个对象 anObject 添加到原 NSArray 对象的末尾形成新的 NSArray 对象,并返回该对象 | NSArray * newArray = [noteArray arrayByAddingObject:note4 ]; |
| -(BOOL) containsObject:(id) anObject | 检测调用该方法的 NSArray 对象的所有元素中是否包含 anObject 对象 | if([noteArray containsObject:note1]){<br>[note1 showNote];<br>} |
| -( BOOL ) isEqualToArray:(NSArray *)otherArray | 检测调用该方法的 NSArray 对象与 otherArray 是否相同 | if([noteArray isEqualToArray newArray]){<br>NSLog(@"%d",<br>[newArray count]);<br>} |
| -( NSString * ) componentsJoinedByString: ( NSString *)separator | 该方法使用字符串 separator 将 NSArray 对象中各个元素连接成为一个 NSString 类的对象,并将该对象返回 | [ noteArray componentsJoinedByString: @" and "]; |

NSArray 只是一个不可变的数组,如何既有条理地组织各条记录,又能随时添加记录呢?下一小节就向你介绍可变数组 NSMutableArray 的相关知识。

### 4.3.4　NSMutableArray

同 NSMutableString 与 NSString 的关系一样,NSMutableArray 是 NSArray 的子类,也是 NSArray 的可变版本,它继承了 NSArray 的所有属性及方法,同时又增加了新的方法,从而可以方便地对已创建的 NSMutableArray 对象进行修改。

这一小节将向你介绍可变数组 NSMutableArray 的使用,比如今天结束工作,刚回到家,我想记录下这件事,那么在 Note 程序中应该怎样做呢?现在,来对程序做以下修改:

代码 4.7 Note.m 文件

```
#import <Foundation/Foundation.h>
#import "NoteClass.h"

int main (int argc, const char * argv[]) {
    NSAutoreleasePool * pool=[[NSAutoreleasePool alloc] init];

    //insert code here...
    //创建可变字符串
    NSMutableString *note1Date=[[NSMutableString alloc]
                                initWithString:@"6:00"];
    NSMutableString *note1Content=[[NSMutableString alloc]
                                initWithString:@"Running"];
    NSMutableString *note2Date=[[NSMutableString alloc]
                                initWithString:@"7:00"];
    NSMutableString *note4Date=[[NSMutableString alloc]
                                initWithString:@"18:00"];
    NSMutableString *note4Content=[[NSMutableString alloc]
                                initWithString:@"Getting home"];
    //创建记录
    NoteClass *note1=[[NoteClass alloc]
                      initWithDate:note1Date
                          andContent:note1Content];
    NoteClass *note2=[[NoteClass alloc]
                      initWithDate:note2Date];
    NoteClass *note3=[[NoteClass alloc]
                      initWithDate:note1Date
                      andContent:note1Content];
    NoteClass *note4=[[NoteClass alloc]
                      initWithDate:note4Date
                      andContent:note4Content];

    NSMutableArray *noteArray=[[NSMutableArray alloc]
                               initWithObjects:note1,
                               note2, note3, nil]; ❶
    for(int i=0; i<[noteArray count]; i++){
```

```
        [[noteArray objectAtIndex:i] showNote];
    }
    NSLog(@"after add a new note:");
    [noteArray addObject:note4]; ❷
    for(int i=0; i<[noteArray count]; i++){
        [[noteArray objectAtIndex:i] showNote];
    }
    NSLog(@" ");
    NSLog(@"test same note...");
    //检查 note1 与 note2 是否是相同记录
    if([note1.date isEqualToString:note2.date]
        &&[note1.content isEqualToString:note2.content]){
        NSLog(@"note2 is same with note1!");
    }else {
        NSLog(@"note2 and note1 is different,note2:");
        [note2 showNote];
    }
    //检查 note1 与 note3 是否是相同记录
    if([note1.date isEqualToString:note3.date]
        &&[note1.content isEqualToString:note3.content]){
        NSLog(@"note3 is same with note1!");
    }else {
        NSLog(@"note3 and note1 is different,note3:");
        [note3 showNote];
    }
    [note1Date release];
    [note1Content release];
    [note2Date release];
    [note4Date release];
    [note4Content release];
    [note1 release];
    [note2 release];
    [note3 release];
    [note4 release];
    [pool drain];
    return 0;
}
```

@end

运行修改后的程序，可看到如图 4-6 所示的运行结果。

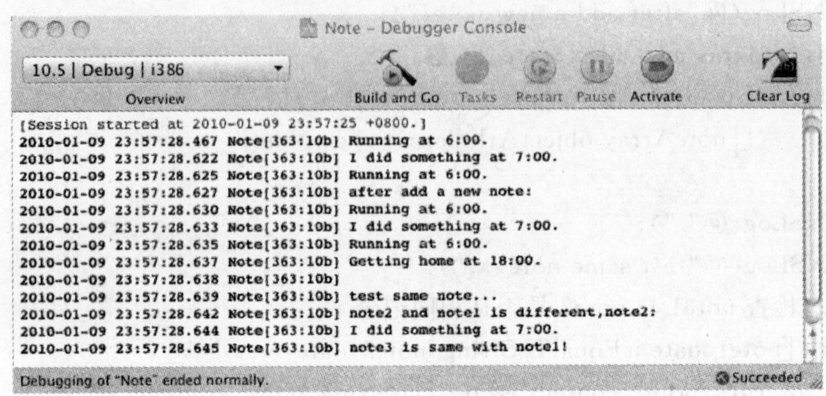


图 4-6 可修改内容的“Note”的运行结果

下面，我们就来分析一下这个程序。

首先，在❶处可以看到，我们换用 NSMutableArray 类声明了用来组织记录的数组对象：

```
NSMutableArray * noteArray =[[NSMutableArray alloc]
                        initWithObjects:note1, note2, note3, nil];
```

NSMutableArray 是 NSArray 的子类，因此依然可用 NSArray 中的初始化方法 initWithObjects:来对 NSMutableArray 的对象进行初始化。

之后❷处使用 NSMutableArray 的 addObject:方法，在可变数组末尾添加了新增的记录 note4:[noteArray addObject:note4]。

NSMutableArray 提供了多种方法来对一个可变数组进行增加、删除、插入等操作。我们在每种操作的各方法中挑选出最为常用的，见表 4-6。

**表 4-6 NSMutableArray 常用方法**

| 方法 | 介绍 | 示例 |
|---|---|---|
| -(void)addObjectsFromArray:(NSArray *)otherArray | 在调用该方法的 NSMutableArray 对象的末尾添加另一个 NSArray 对象中所包含的元素，来改变该 NSMutableArray 对象 | [array1 addObjectsFromArray:array2]<br>array1: 1 2 → 1 2 3 4 arry1<br>array2: 3 4 |
| -(void)insertObject:(id)anObject atIndex:(NSUInteger)index | 在调用该方法的 NSMutableArray 对象的指定位置 index 处添加一个元素来改变原 NSMutableArray 对象 | [array1 insertObject:newObj atIndex:1]<br>array1: 1 2 → 1 3 2 array1<br>newObj: 3 |

续表

| 方法 | 介绍 | 示例 |
|---|---|---|
| -(void) removeObjectAtIndex:(NSUInteger)index | 移除对象的指定位置 index 处的元素 | [array1 removeObjectAtIndex:1]<br>array1: 1 3 2 → arry1: 1 2 |

以上列举的一些对 NSMutableArray 进行操作的方法，只是最为常用的几个，每种操作除了上面列出的外，还有其他多种形式，如关于移除对象的方法还有-(void)removeObject:(id)anObject 等，在此就不一一列举了。

## 4.4 NSDictionary

NSDictionary 是 Foundation 中另一个非常重要的集合类。每个 NSDictionary 的元素由两部分构成：关键字和值。NSDictionary 在给定的关键字（通常为 NSString 类型的对象）下存储一个相应的值（可以是任意类型的对象），之后我们就可以根据关键字来查找相应的值，正如在现实中的字典一样。

NSDictionary 也是不可变的，如果我们要在程序运行过程中对 NSDictionary 中的元素进行修改，就需要使用它的子类 NSMutableDictionary 来进行这些操作。

NSDictionary 使用的是键查询的优化存储方式，可以立即找出要查询的数据，而无需遍历所有元素进行查找，尤其对于大型数据集来说，使用 NSDictionary 要比使用 NSArray 快很多。

下面我们就使用 NSDictionary 来重新组织一下 Note 程序，将记录保存在 NSDictionary 对象中，用记录的时间作为关键字来进行存取。修改 Note.m 文件如下：

**代码 4.8 Note.m 文件**

```
NSMutableArray *noteArray=[[NSMutableArray alloc]
                              initWithObjects:note1,
                              note2, note3, nil];
    for(int i=0; i<[noteArray count]; i++){
        [[noteArray objectAtIndex:i] showNote];
    }
    NSLog(@"after add a new note:");
    [noteArray addObject:note4];
    for(int i=0; i<[noteArray count]; i++){
        [[noteArray objectAtIndex:i] showNote];
```

```
}
NSDictionary * noteDic=[[NSDictionary alloc]
                          initWithObjectsAndKeys:
                          note1, note1.date,
                          note2, note2.date,          ❶
                          note3, note3.date,
                          note4, note4.date, nil];
NSLog(@"18:00...");
[[noteDic objectForKey:@"18:00"] showNote]; ❷
```

运行程序,得到如图 4-7 所示的输出结果。

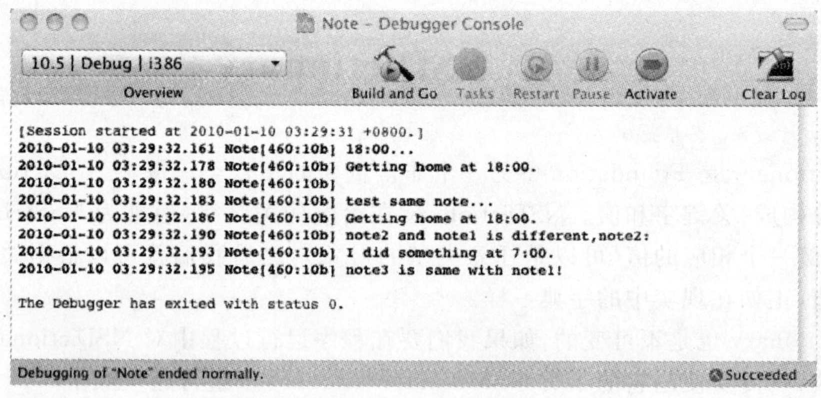


图 4-7 使用 NSDictionary 实现的输出结果

现在,就来分析一下这个使用 NSDictionary 组织记录的程序。

首先来看一下❶处用于创建 NSDictionry 对象的语句:

```
NSDictionary * noteDic =[[NSDictionary alloc]
                  initWithObjectsAndKeys:note1, note1.date,
                  note2, note2.date, note3, note3.date,
                  note4, note4.date, nil];
```

这句代码创建了一个 NSDictionary 的对象 noteDic,并用 NoteClass 类的对象 note1、note2、note3 和 note4 来初始化它,同时又分别使用对应对象的 date 变量作为键值进行存储。这样,就可以通过键值来检索 NSDictionary 中的数据了。NSDictionary 的 initWithObjectsAndKeys:方法声明如下:

```
- (id)initWithObjectsAndKeys:(id)firstObject ,…
```

这个方法与 NSArray 的 initWithObjects:方法类似,后面需要任意个用来初始化 NSDictionary 对象的参数,并在参数列表最后用 nil 表示结束。但要注意的是:在 NSDictionary 中,NSDictionary 的元素包含值与关键字两部分,所以在参数列表当中也需要按照一个值一个对应关键字的次序列出各个参数来初始化 NSDictionary 对象。比如在上面的例子中，note1.date 是 note1 的关键字，note2.date 是 note2 的关键字，它们需要一一对应。

在本例中，我想查看一下自己在 18:00 做了什么事情，所以在程序中的❷处，使用了下面的代码来获取以“18:00”为关键字的记录：

```
[[noteDic objectForKey:@"18:00"] showNote];
```

其中的[noteDic objectForKey:@"18:00"]就表示将关键字为“18:00”的元素的值从 noteDic 中取出，在本程序中也就是 note4 对象。NSDictionary 的 objectForKey:方法声明如下：

```
-(id)objectForKey:(NSString *)key
```

利用这个方法，就可以方便地根据关键字 key 取出对应在 NSDictionary 对象中的值。

NSDictionary 依然是不可变的，如果需要在程序运行过程中改变某些数据的话，我们需要使用 NSDictionary 的子类 NSMutableDictionary 创建对象来组织数据。NSMutableDictionary 与 NSDictionary 的关系同 NSMutableArray 与 NSArray 的关系是类似的。

## 4.5 小结

在本章介绍了一些主要的 Cocoa 类，包括 NSObject、NSString、NSArray 与 NSDictionary。

其中 NSObject 是所有 Cocoa 类的父类，在创建一个自定义类时，要直接或间接地继承 NSObject。

NSString 与 NSMutableString 是 Cocoa 中用来进行字符串操作的类，使用它们可以创建、获取、编辑一个字符串。两者之间不同的是，NSString 类的对象是不可变的，而 NSMutableString 类的对象是可变的。

NSArray、NSMutableArray、NSDictionary 和 NSMutableDictionary 是 Cocoa 中的集合类。NSArray 与 NSMutableArray 相当于其他语言中的数组，但 NSArray 的对象是不可变的，而 NSMutableArray 则是可变的。NSDictionary 与 NSMutableDictionary 中的元素包含两部分——关键字和值，我们通过关键字来对对应元素中的值进行存取。而且 NSDictionary 与 NSMutableDictionary 使用键查询的优化存储方式，可以立即找出要查询的数据，而不需遍历所有元素进行查找，适用于大型数据集。

至此，本章就结束了，下一章中将向你介绍 Objective-C 中类别与协议的有关知识。

# 第5章　类别和协议

本章内容⊙

- 类别的定义与实现
- 类别的使用
- 采用协议
- 自定义协议

类别(Category)与协议(Protocol)是 Objective-C 中既重要又非常有特色的两部分内容。类别允许为已经封装好了的类添加新的行为,是一种对类的功能进行丰富的新手段。协议是一个命名的方法列表,在 Objective-C 中可以通过采用协议来实现某些功能,这有些类似于 Java 中的接口。下面,就来开始本章的学习。

## 5.1　类别

类别是对类的扩展,使用类别,可以不改变已经封装好的类而为它添加新的行为。看到这里,你也许会有疑问:这项工作是不是可以通过创建该类的子类来完成呢? 下面我们就来看两个简单的需要类别的情况:

(1) 当一个类过于庞大时,例如后面章节我们会讲到的用于界面设计的 UIWindow 类,将它放在一个文件中就不太合适,这时便可以使用类别来分散该类的实现。

(2) 有些类如我们前面讲到的 NSString 与 NSArray 等,它们实际上只是一个类簇的前台表示,因而无法用来创建子类。这时如果想为它们添加一些新的功能,就可以通过类别来实现。

现在,假设前面 Note 程序中的 NoteClass 类已经封装好了,不可以对它再进行任何修改,而在 Note.m 文件中,那段用来检测是否有重复记录的代码又过于繁琐,所以决定将检测重复记录这个功能封装为 NoteClass 类的一个方法,这时,便可以使用类别来完成这个功能的添加。

### 5.1.1　类别的声明与实现

使用类别为一个类添加方法,需要先声明该类别,格式与声明一个类类似:

**代码 5.1 类别声明的格式**

```
@interface ClassName (CategoryName)
        addedMethods
    ……
    //不能添加变量
@end
```

(1) 在类别声明中,ClassName 是现有的类,即该类别要为之添加方法的类。CategoryName 是要声明的类别的名称。类别的声明同样要以@end 结束。

(2) 在@interface 与@end 之间是要为 ClassName 类添加的方法,但有一点需要着重注意:在这里只可以添加方法,而不可以添加新变量。

下面,我们就来为 NoteClass 类声明一个类别 NoteCategory,并在该类别中添加用来检测重复记录的方法 testSameNote:。

首先添加两个文件,这一次,选择 other 选项下的 Empty File 模板,如图 5-1 所示。

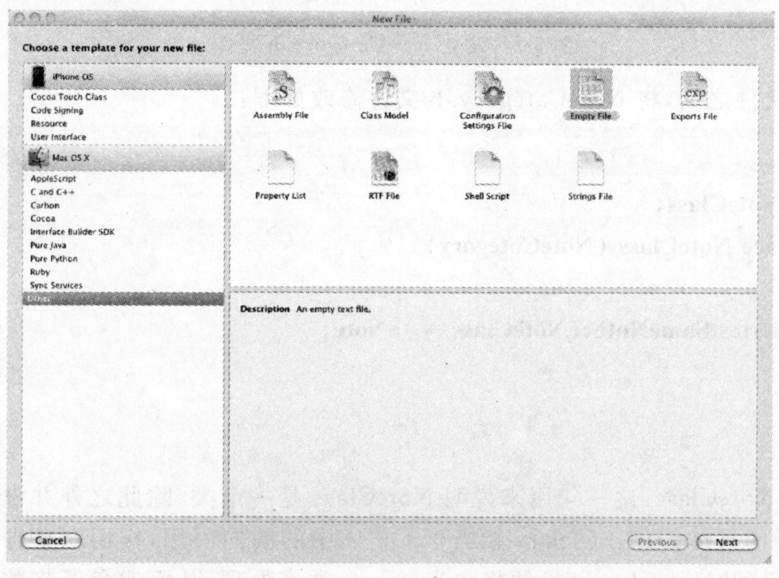


图 5-1 选择 Empty File 文件模板

分别为两个文件命名为 NoteCategory. h 和 NoteCategory. m，如图 5-2 所示。

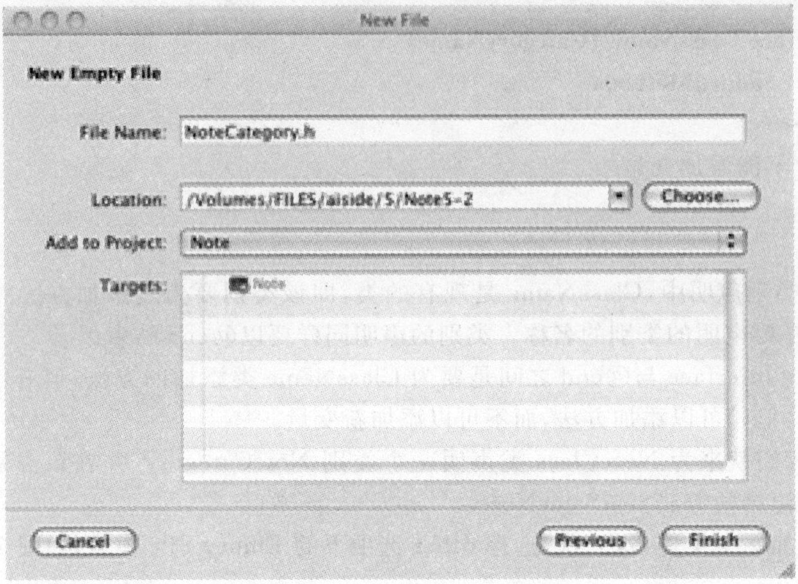

图 5-2　建立 NoteCategory. h 文件

建立好文件之后，将 NoteCategory. h 文件修改如下：

**代码 5.2　NoteCategory. h 文件**

```
@class NoteClass;
@interface NoteClass (NoteCategory)

- (BOOL)testSameNote:(NoteClass * )aNote;

@end
```

@class NoteClass：这一句用来说明 NoteClass 是一个类，除此之外并没有包含任何 NoteClass 类的具体信息。@class 相当于 C++中的前向引用，作用就是告诉下面的程序段如果再见到"NoteClass"时，便将它当做一个类来处理，从而避免了将整个头文件导入进来，造成代码臃肿。

此外，@interface NoteClass(NoteCategory)表示将 NoteCategory 声明为 NoteClass 类的一个类别。接着在该声明之后，为 NoteClass 类添加了一个-（BOOL）testSameNote：(NoteClass * )aNote 方法，该方法用于检测 aNote 与调用该方法的对象是否相同。

这样，类别 NoteCategory 便声明完成了。下面就在 NoteCategory. m 文件中完成对方法-(BOOL)testSameNote：(NoteClass * )aNote 的实现：

**代码 5.3 NoteCategory.m 文件**

```
#import "NoteClass.h"
#import "NoteCategory.h" ❶
@implementation NoteClass (NoteCategory)
-(BOOL)testSameNote:(NoteClass *)aNote {
    if([self.date isEqualTo:aNote.date]
        &&[self.content isEqualTo:aNote.content]){      ❷
        return YES;
    }
    return FALSE;
}
@end
```

文件开头❶处,引入了该类别的头文件:#import "NoteCategory.h"。而且在这里,由于要用到 NoteClass 的变量 date 与 content 等具体信息,所以还需要在文件一开始引进 NoteClass 类的头文件:#import "NoteClass.h"。

与类的实现一样,在@implementation NoteClass (NoteCategory)与@end 两个关键字的中间完成对 NoteCategory 类别中方法的实现。

在❷处,实现了-(BOOL) testSameNote:(NoteClass *)aNote 方法,该方法将检测调用方法的对象与 aNote 对象是否相同,如果相同则返回 YES,不同则返回 NO。

这样,我们便完成了对类别 NoteCategory 的定义。

### 5.1.2 类别的使用

完成了类别的定义,便可以在程序里对类别中新添加的方法进行调用了。调用类别中定义的方法与调用类的一般成员方法形式完全一样。

现在完成了类 NoteClass 的类别 NoteCategory 的定义,我们来修改一下 Note.m 程序。

首先在文件开头引入 NoteCategory 的头文件:#import "NoteCategory.h"。将先前用于检测重复记录的两条 if 语句替换为下面的代码:

**代码 5.4 Note.m 文件的修改**

```
//检查 note1 与 note2 是否是相同记录
if([note1 testSameNote:note2]){ ❶
    NSLog(@"note2 is same with note1!");
}else{
    NSLog(@"note2 is different from note1, note2:");
    [note2 showNote];
```

```
}
//检查 note1 与 note3 是否是相同记录
if([note1 testSameNote:note3]){
    NSLog(@"note3 is same with note1!");
}else{
    NSLog(@"note3 is different from note1, note2:");
    [note3 showNote];
}
```

我们将先前冗余的 if 条件判断语句替换成了在❶处对 testSameNote 方法的调用。运行程序,可以得到如图 5-3 所示的输出结果。

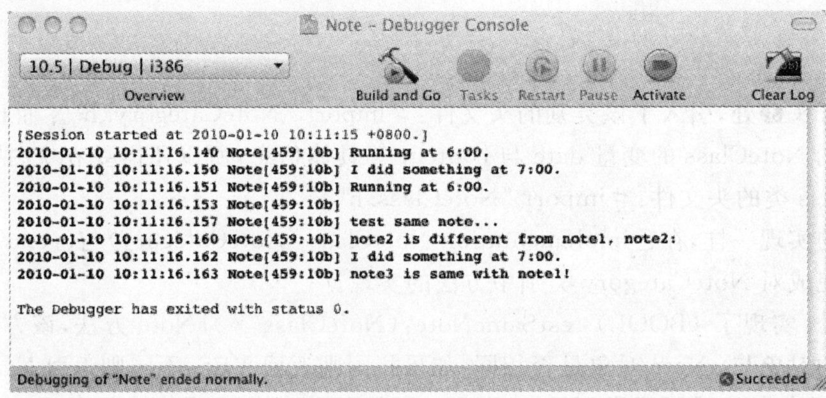

图 5-3　程序输出结果

可以看到,程序的运行结果与先前相同,而程序中对 testSameNote 方法的调用与调用一般成员方法无异:[note1 testSameNote:note2]。

这样,我们便对如何声明、实现一个类别,以及如何使用类别中添加的方法有一定的了解了。在今后的开发过程中,如果遇到需要自己为 NSString 类添加一个自定义方法等情况,便可以使用类别来轻松实现了。

## 5.2　协议

协议是一个命名的方法列表,是 Objective-C 中一个非常重要的语言特性,当一个类采用了某个协议之后,该类通过实现协议中的方法,可以实现一些功能。它类似于 Java 当中的接口,弥补了 Objective-C 不支持多继承的缺陷。

在以前的版本中,如果某个类采用了一个协议,那么这个类必须实现该协议中的所有方法。但在 Objective-C 2.0 中,增加了两个新的协议修饰符:@optional 和@required,这两个新增的修饰符用来修饰协议中的方法,用@optional来修饰则表示该方法可以被选择实现,而用@required修饰的方法则必须实现,@required是默认修饰符。

Cocoa中已经为我们声明好了很多方便好用的协议,如NSCopying协议、NSCoding协议等。通过采用这些协议,可以方便地为自定义的类实现协议所对应的功能。本章,我们将继续修改Note程序。假设有一些重要的记录需要备份,则可以通过采用NSCopying协议来实现复制功能。下面就来学习Objective-C中协议的使用,并为NoteClass实现复制这一功能。

### 5.2.1 深拷贝与浅拷贝

看到我们要采用NSCopying协议来实现记录的备份,你是否有这样的疑惑:使用note2=[note1 retain]可以使note2具有与note1相同的内容,不是一样实现了备份?为什么还要通过采用NSCopying协议呢?这涉及到了深拷贝与浅拷贝的问题。

**1. 浅拷贝**

当我们使用note2=[note1 retain]来为note2赋值时,实现的是note2对note1的浅拷贝。那么什么是浅拷贝呢?

以note2=[note1 retain]为例,该语句执行后,note1与note2虽然是两个对象,但它们指向的是同一块内存区域,也就是说它们只是同一个内容的不同名称。若其中一个对象改变了自己的内容,实际也就改变了另一个对象的内容。这并不是我们对记录进行备份的初衷。实现了浅拷贝的note1与note2的关系如图5-4所示。

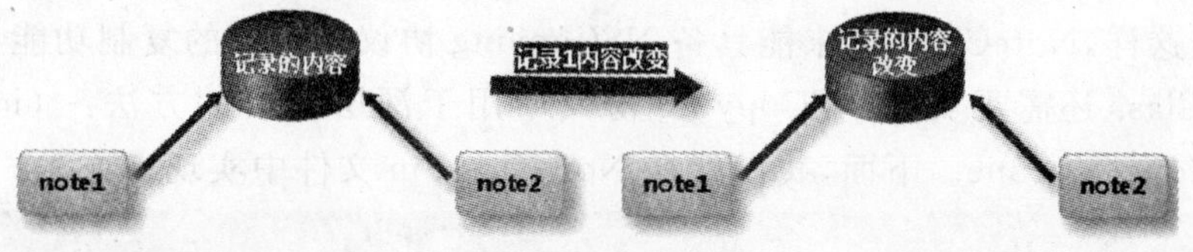


图5-4 浅拷贝示意图

**2. 深拷贝**

如果note2对象对note1对象实现的是深拷贝的话,则会为note2对象重新开辟一块只属于它自己的内存区域,在这块内存区域中存储从note1中拷贝过来的数据。实现了深拷贝后的note1与note2的关系如图5-5所示。

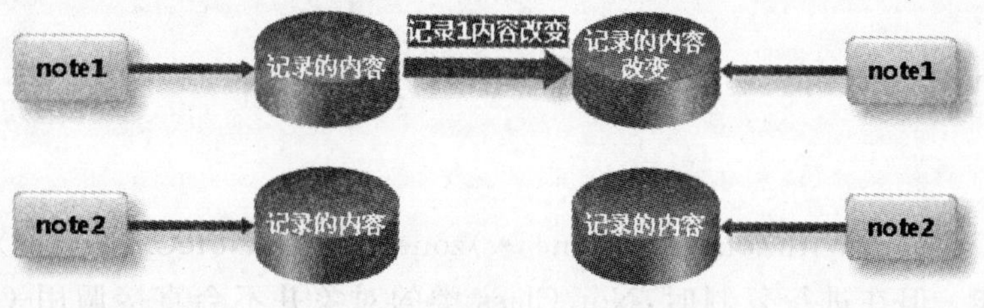


图5-5 深拷贝示意图

此时再修改 note1 的内容时，便不会对 note2 造成任何影响了。下面，具体看一下应该如何通过采用 NSCopying 协议来实现备份的功能。

### 5.2.2　采用协议

在声明一个类的时候，我们要决定该类需要采用哪些协议，并用尖括号将这些协议括起来，格式如下：

```
@interface ClassName : SuperClassName
    <ProtocolName1, ProtocolName2, ……>
```

同一个类在声明时采用的协议可以有多个，中间用","隔开即可，如上面的<ProtocolName1, ProtocolName2, ……>。

现在，就来修改一下 NoteClass 类的声明，使该类采用 NSCopying 协议：

**代码 5.5　在 NoteClass.h 文件中修改 NoteClass 声明**

```
@interface NoteClass : NSObject <NSCopying> { ❶
    NSString * date;
    NSString * content;
}
```

在这里没有做很大的修改，只是在❶处类声明语句最后添加了要采用的 NSCopying 协议：@interface NoteClass : NSObject <NSCopying>。

但仅仅这样，NoteClass 还未能具备 NSCopying 协议所提供的复制功能，要实现这一功能，NoteClass 还需要实现 NSCopying 协议中用于深层拷贝的方法：-(id)copyWithZone:(NSZone *)zone。下面，我们就在 NoteClass.m 文件中实现该方法：

**代码 5.6　-(id)copyWithZone:(NSZone *)zone 的实现**

```
- (id)copyWithZone:(NSZone *)zone{
    NoteClass * noteCopy=[[[self class]
                           allocWithZone:zone] init]; ❶
    noteCopy.date=[self.date copy]; ❷
    noteCopy.content=[self.content copy];

    return noteCopy;
}
```

实现了-(id)copyWithZone:(NSZone *)zone 方法后，NoteClass 类的对象便具有了深拷贝的功能。但在进行复制时，NoteClass 类的对象并不会直接调用-(id)copyWithZone:(NSZone *)zone 方法，而是先调用从 NSObject 继承来的 copy 方法，copy 方法再负责调用 copyWithZone 方法，如图 5-6 所示。

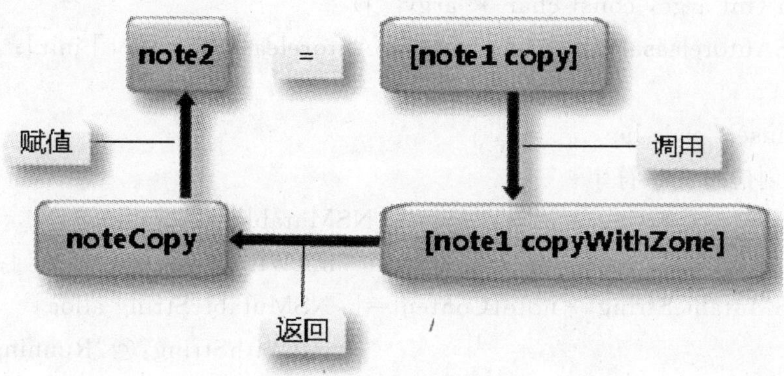


图 5-6 复制过程

下面我们就来分析一下这段代码：

(1) -(id)copyWithZone:(NSZone *)zone:方法会在方法内部复制好对象，并将它作为该方法的返回值。(NSZone *)zone 表示一块可供分配的内存区域，我们就将在这块区域中为新复制的对象开辟内存，并存储复制来的数据。

❶处的 NoteClass *noteCopy=[[[self class] allocWithZone:zone] init]是不是看着较为复杂呢？我们先来介绍一下这当中用到的两个方法：

+(id)allocWithZone:(NSZone *)zone:该方法是一个类方法，它负责在一个给定区域 zone 中创建新对象并将新对象返回。

-(Class)class:它将返回调用该方法的对象所属的类型。

这里使用的[self class]就是为了动态获得要复制的对象所属的类型。假设 NoteClass 还有一个子类 SubNoteClass，它继承了 NoteClass 所采用的协议与 copyWithZone 方法，那么子类 copyWithZone 方法中的[self class]返回的就是 SubNoteClass，而不是 NoteClass 了。这样，我们就动态获取了要复制对象的类型，也避免了子类对 copyWithZone 方法的重复实现。

创建好 noteCopy 对象之后，便可在方法中❷处通过调用 NSString 类的 copy 方法分别实现对 date 变量与 content 变量的深层复制：

```
noteCopy.date=[self.date copy];
noteCopy.content=[self.content copy];
```

这样，便完成了 NoteClass 的深层复制工作，并将复制的数据存放在了 noteCopy 对象对应的内存区域中。最后，我们将 noteCopy 对象作为返回值返回：return noteCopy。

现在已经在 NoteClass 类中通过采用 NSCopying 协议实现了复制功能，下面修改 Note.m 文件，验证一下 NoteClass 是否真的实现了深层复制功能：

**代码 5.7 Note.m 的实现**

```
#import <Foundation/Foundation.h>
#import "NoteClass.h"
```

```
int main (int argc, const char * argv[ ]) {
    NSAutoreleasePool * pool=[[NSAutoreleasePool alloc] init];

    //insert code here...
    //创建可变字符串
    NSMutableString *note1Date=[[NSMutableString alloc]
                                    initWithString:@"6:00"];
    NSMutableString *note1Content=[[NSMutableString alloc]
                                    initWithString:@"Running"];
    //创建记录
    NoteClass *note1=[[NoteClass alloc] initWithDate:note1Date
                                    andContent:note1Content];
    NoteClass *note2=[note1 retain]; ❶
    NoteClass *note3=[note1 copy];
    NSArray *noteArray=[[NSArray alloc ] initWithObjects:
                                    note1, note2, note3, nil];
    for(int i=0; i<[noteArray count]; i++){
        NSLog(@"note%d:", i+1);
        [[noteArray objectAtIndex:i] showNote];
    }
    //改变note1内容
    NSLog(@" ");
    [note1Date setString:@"7:00"]; ❷
    for(int i=0; i<[noteArray count]; i++){
        NSLog(@"note%d:", i+1);
        [[noteArray objectAtIndex:i] showNote];
    }
    [note1Date release];
    [note1Content release];
    [note1 release];
    [note2 release];
    [note3 release];
    [pool drain];
    return 0;
}
@end
```

运行程序来查看结果，如图 5-7 所示。

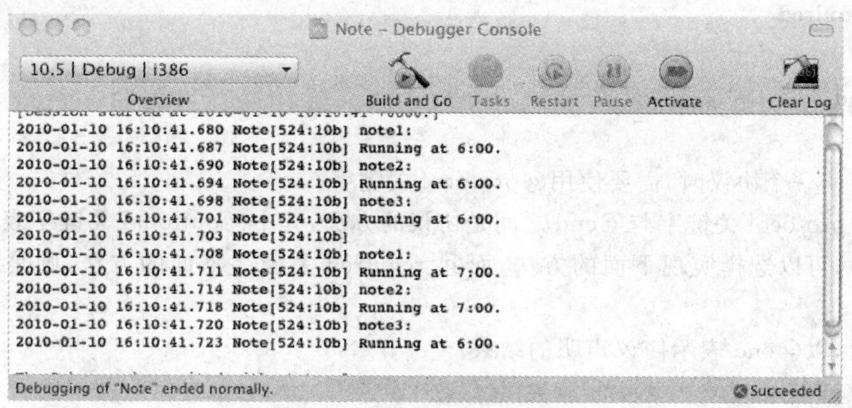


图 5-7　深拷贝输出结果

下面，来分析一下上面的程序代码：

这次在程序中，只初始化了 note1 对象，之后，在代码❶处通过使用 note1 对象分别为 note2 与 note3 赋值来初始化这两个对象。但使用 note1 为两个对象赋值的方式有所不同：

```
NoteClass * note2=[note1 retain];
NoteClass * note3=[note1 copy];
```

对于 note2，使用 retain 方法来为它赋值进行初始化；而对于 note3，则通过调用 copy 方法来为它赋值进行初始化。按照前面所讲的，这样就完成了对 note2 的浅拷贝与对 note3 的深拷贝。

在代码的❷处，对 note1 对象中的时间信息进行了修改，然后再次输出三条记录。按照前面深拷贝与浅拷贝的理论，由于 note2 实现的是对 note1 的浅拷贝，所以在 note1 修改后，note2 的输出内容应该仍然与 note1 相同；而 note3 实现了对 note1 的深拷贝，所以 note1 的任何改变均不会对 note3 造成影响。看看图 5-4 中的输出结果，是不是这样？

现在，我们便通过采用 NSCopying 协议为 NoteClass 类实现了深拷贝的功能。

在 Cocoa 中，还有很多常用且重要的协议，但在这里介绍只会让你感到困惑和乏味。所以关于更多的协议，我们将在后面章节中用到的时候再进行详细讲解。

### 5.2.3　自定义协议

在上一小节，以 NSCopying 为例介绍了如何采用一个 Cocoa 中的协议，我们也可以声明自己的协议，格式如下：

**代码 5.8　自定义协议的格式**

```
@protocol ProtocolName
@optional
```

```
    ……
@required
    ……
@end
```

要声明一个协议时，需要使用@protocol关键字。

在@protocol关键字与@end之间是协议的方法，其中@optional关键字表示在实现该协议时，可以选择实现下面的方法；而@ required关键字下面的方法，则是必须实现的。

最后，以@end表示协议声明的结束。

# 5.3　小结

这一章我们介绍了Objective-C中两个非常重要的语言特性：类别和协议。

使用类别，可以为特定的类添加新的功能，从而可以对已经封装好的类进行扩展，或者将过于庞大的类进行分解实现。在5.1节，讲解了一个类别从声明到实现一直到使用的过程。在使用类别对类进行扩展时，我们只可以为类添加新的方法，而不可以对类添加新的变量。

在5.2节中，讲解了如何采用一个Cocoa协议以及如何自定义一个协议，并通过让NoteClass采用NSCopying协议，使本类的对象具有了复制（深拷贝）的能力。

至此，本书关于Objective-C语言的介绍便结束了，从下一章开始，我们将开始iPhone开发的学习！

# 第3篇　核心篇

本篇通过大量的实例，讲解了 iPhone 开发中的基本概念和基础知识，是读者进一步学习本书扩展篇和丛书其他分册的基础。

核心篇包括7章：

第6章视图和控件，讲解了 iPhone 开发中视图和控件的基本概念和使用方法，介绍了如何在 iPhone 程序中使用自定义视图。

第7章视图控制器，通过三个小程序介绍了 iPhone 开发中视图控制器的使用方法，包括自定义视图控制器、导航控制器和标签栏控制器，本章还介绍了视图间转换动画的两种实现方法。

第8章表视图，系统而详尽地讲解了表视图的概念和常用操作，包括创建表、填充数据、显示数据及操作表中的数据等。

第9章数据持久性存储，介绍了应用程序沙盒的概念，系统地介绍了持久保存数据的四个方法：属性列表、归档、SQLite3 和 CoreData。

第10章用户设置，通过 NoteSetting 程序介绍了如何在 Settings 中和程序内部添加设置选项，本章还讲解了表行上添加控件的方法。

第11章触摸、手势和事件，介绍了 iPhone 上触摸的底层机制，讲解了触摸的各种常用手势的实现方法，包括轻击、拖拽、轻扫、捏合等。

第12章国际化和本地化，介绍了国际化和本地化的概念，通过对一个应用程序进行国际化和本地化操作，详细讲述了应用程序如何实现多语言版本。

# 第 6 章　视图和控件

**本章内容**

- 视图的概念和使用方法
- 理解 IBOutlet 和 IBAction
- 如何使用各种简单控件
- 如何实现与控件交互
- 如何创建一个自定义视图

你一定为 App Store 上丰富多彩的应用程序而心动过吧！从这一章开始，就进入了本书的核心部分：用 Xcode 和 Interface Builder 编写 iPhone 应用程序。

如果想进行 iPhone 程序开发，本章的内容是必不可少的，这是我们开发的应用程序通向 App Store 的必经之路。

在这一章中，首先介绍了 iPhone 开发中最基本的视图的概念，以及各个视图之间的继承和层次关系；然后介绍了程序中要用到的一些基本控件，通过一个名为"Controls"的程序解释了如何利用 Xcode 和 Interface Builder 编写 iPhone 应用程序；最后我们对"Controls"程序进行了优化，并用自定义视图方式实现了它。这一章涵盖了 iPhone 开发所需的关键概念，学习完本章，相信你一定能开发出能在 iPhone/iPod touch 上完美运行的程序。

现在，就让我们开始吧！

## 6.1　视图概述

在 iPhone 程序开发中，视图是最基础的部分。一般来说，几乎所有显示在 iPhone 屏幕上的内容都可以被称为视图，比如通常见到的按钮、开关、图片等。我们可以设置视图的大小和位置，以及视图之间的层次关系。如图 6-1 所示，屏幕上显示的文字、图形等都是视图。

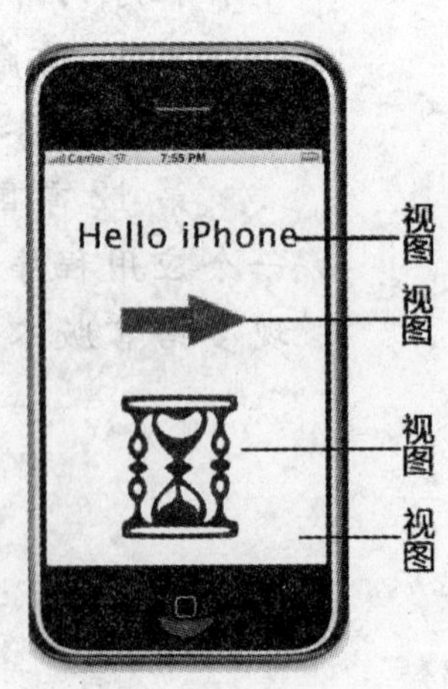


图 6-1　视图

### 6.1.1　视图和窗口

视图是显示可见元素的图形用户界面(GUI)，而窗口则是存放视图的容器，它们都是构建 iPhone 程序界面的基本组成部分。具体来说，窗口提供了一个显示具体内容的平台，而视图则承担了大部分的绘制界面和用户响应的工作。iPhone 界面的设计通常采用“单窗口－多视图”的模式，即程序只有一个窗口却可以有多个视图，打个比方，如图 6-2 所示，把窗口比作显示器，视图就如同显示出来的丰富多彩的内容。显示器里面的内容是不断更新和变化的，而显示器却是不变的。它不像 Windows IE 那样可以是多个窗口同时浏览，如图 6-3 所示。

图 6-2 iPhone 的“单窗口－多视图”模式

图 6-3 Window IE 可以打开多个窗口同时进行浏览

窗口，即窗口类(UIWindow)的一个实例。它用来定义一个定位和管理应用程序界面的对象。窗口有层次地保存所有的视图，并且位于所有层的根部。从根本上说，窗口就是一种特殊的视图，它对于桌面系统非常重要，但对 iPhone 应用程序就没有那么重要了，因为 iPhone 应用程序一般只有一个窗口，毕竟 iPhone 的屏幕太过小巧，这也是 iPhone 程序采用“单窗口－多视图”模式的直接原因。

视图，即视图类(UIView)的一个实例，它通常用于定义屏幕上的一个矩形区域，可以用来显示各种控件、图像等可视元素，这些元素都是以子视图的方式加载在视图上的，这就是我们稍后要讲解的视图层次关系。

### 6.1.2 视图的继承

在 iPhone 开发中，UIView 的继承是一部分重要的内容，具体来说，视图一般拥有一个父视图和多个子视图。图 6-4 中列出了 UIView 的这种继承关系，从中我们可以很清楚地看出 UIView 的父类是 UIResponder，而它的子类是 UIControl、UIWindow 等。UIControl 拥有很多子类，其中包括 UIButton、UITextField、UISwitch、UISlider 等等，这些基本上就是本章所要讲解的内容。

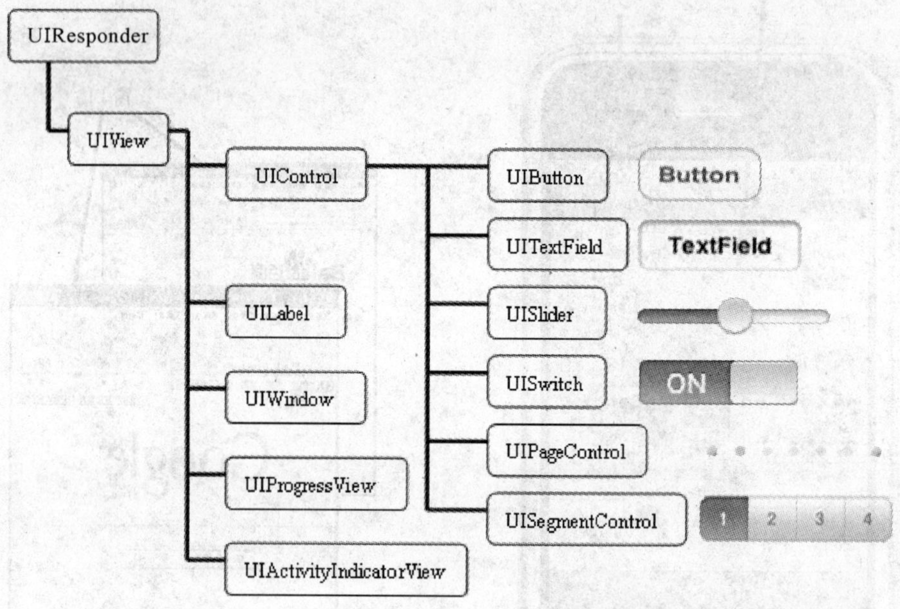


图 6-4 视图的继承

这里简单介绍一下 UIControl。UIControl 继承自 UIView，从图 6-4 中可以清楚地看出这种关系。UIControl 的主要作用是实现用户和程序的交互：当用户操作某个控件时，将触发相应的事件来完成一定的功能或过程；UIControl 还可以控制控件的状态，例如设置它们的 enabled 或者 highlighted 属性。

### 6.1.3 视图的层次结构

程序中每个视图和显示该视图的窗口是相互关联的，一个窗口中的所有视图可以根据视图层次连在一起。iPhone 按照视图加入的先后顺序，由后向前显示，这说明了视图层次是一种空间上的叠加关系。所有视图都可以有子视图，一个视图可能拥有很多子视图，也可能没有子视图。

从图 6-5 World Clock 视图层次中可以看出，包含状态栏的是程序窗口，显示的是在 iPhone 上运行的程序的界面。标号为 2 的是去除状态栏的程序“视图”，3 是程序“视图”的导航栏部分，4 是标签栏部分，5 是位于导航栏和标签栏之间的自定义视图部分。2、3、

4、5部分叠加在一起形成程序视图1。

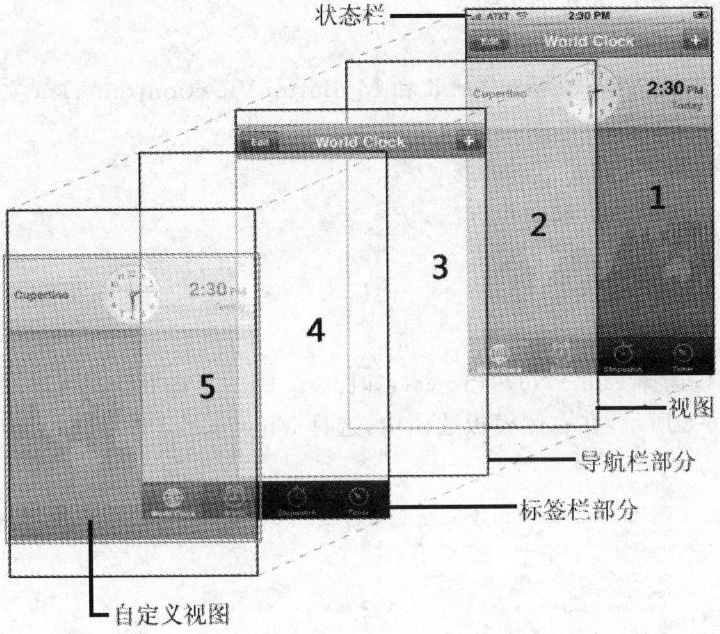

图 6-5 World Clock 视图层次

视图层次，就是视图在空间上的相互位置关系，它可以帮助我们从整体上理解视图，也可以为接下来学习控件的知识做铺垫。

## 6.2 基本控件介绍和使用

了解了有关视图和窗口的知识后，接下来我们学习基本控件的功能和使用方法，这些控件是以后构建 iPhone 应用程序的主要元素。

### 6.2.1 UILabel 和 UIButton

在实现本章开头部分提到的 Controls 程序之前，我们先来实现一个简单的 MyButton 程序。如果你学习过其他的程序设计语言，那一定还记得经典的“Hello World”程序。接下来我们将完成一个类似的程序，程序的最终效果如图 6-6 所示，它只包含一个标签和一个按钮，点击按钮，标签中显示的是“Hello iPhone”，通过这个程序，我们会逐渐熟悉 Xcode 和 Interface Builder 开发环境，以及

图 6-6 MyButton 程序效果图

iPhone 程序开发的大致过程。

此程序大概需要以下几个步骤：

- 新建项目 MyButton；
- 在 MyButtonViewController.h 和 MyButtonViewController.m 文件中添加程序代码；
- 构建程序界面；
- 为控件连接输出口和操作方法；
- 编译并运行。

**1. 新建项目 MyButton**

打开 Xcode，选择 File→New Project，如图 6-7 所示。在 iPhone OS 层级下选择 Application，如图 6-8 所示，在右侧模板选项中，选择 View-based Application(基于视图的应用程序)。

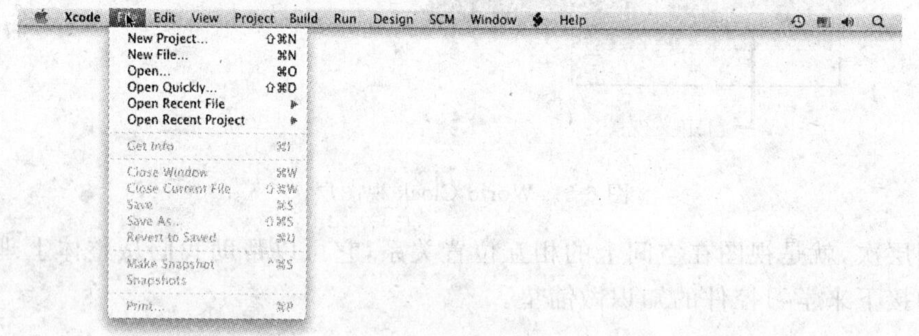


图 6-7　创建新项目

图 6-8　选择项目模板

点击 Choose，将项目命名为 MyButton，如图 6-9 所示，同时可以为程序指定一个储存路径，用于存放我们以后编写的所有示例。

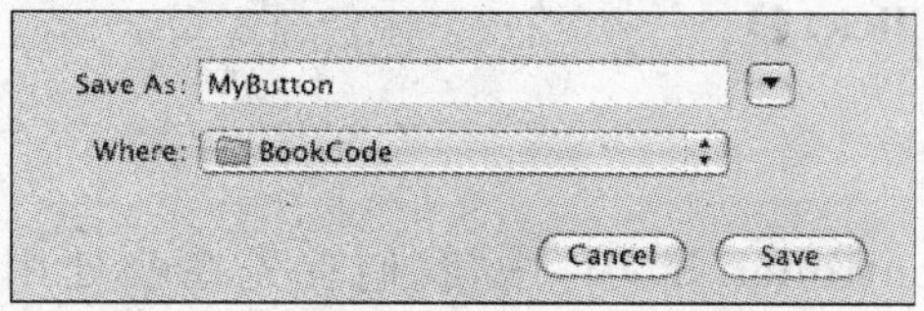


图 6-9 给项目命名并指定存储路径

点击 Save 之后，展现在我们眼前的应该是一个完整的 Xcode 框架，你一定注意到了 Xcode 左侧的 Groups & Files 窗格吧，展开 Classes、Resources 和 Frameworks 文件夹，将看到如图 6-10 所示的项目视图，它展示的是这个项目的结构。

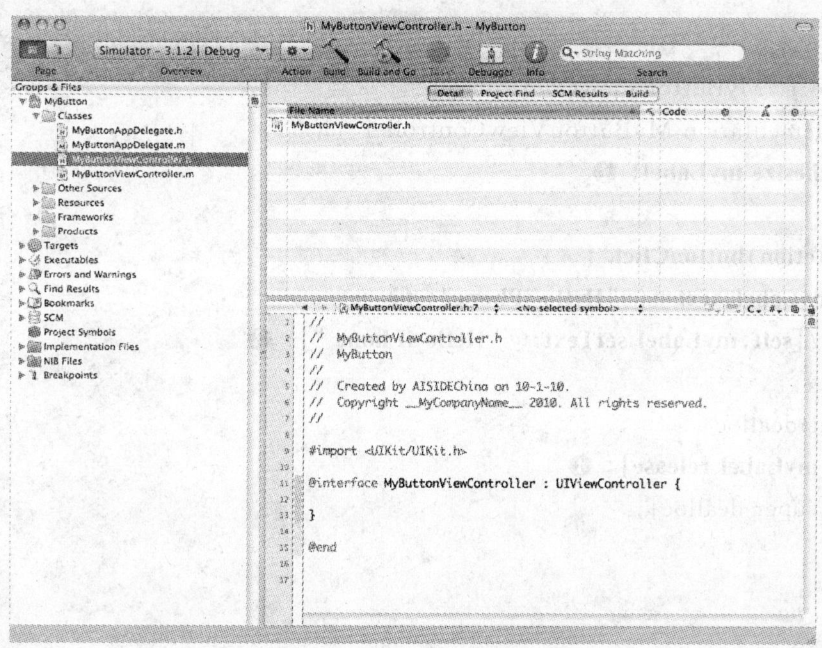


图 6-10 Xcode 下的项目视图

**2. 在 MyButtonViewController.h 和 MyButtonViewController.m 文件中添加程序代码**

选中 MyButtonViewController.h 文件，在里面定义一个 UILabel 类型的输出口 myLabel 和一个 buttonClick 操作方法。具体代码如下：

**代码 6.1 MyButtonViewController.h 文件**

```
#import <UIKit/UIKit.h>
@interface MyButtonViewController : UIViewController {
    IBOutlet UILabel *myLabel; ❶
```

```
}
@property (nonatomic, retain) IBOutlet UILabel * myLabel; ❷
- (IBAction) buttonClick;❸
@end
```

在 MyButtonViewController. h 文件中，我们首先在❶处用 IBOutlet 关键字声明一个 UILabel 类型的输出口；在❷处用 property 通知编译器编译时如何创建该类的成员；在❸处用 IBAction 关键字声明了一个按钮单击时的操作方法；关于 IBOutlet 和 IBAction 的问题我们稍后将进行讲解。

接下来，选中 MyButtonViewController. m 文件，实现定义的 buttonClick 操作方法。它用来实现单击按钮时在标签上显示“Hello iPhone”的功能，具体代码如下：

**代码 6.2 MyButtonViewController. m 文件**

```
#import "MyButtonViewController.h"
@implementation MyButtonViewController
@synthesize myLabel; ❶

- (IBAction)buttonClick {
    myLabel.text=@"Hello iPhone"; ❷
    //[self.myLabel setText:@"Hello iPhone "]; ❸
}
- (void)dealloc {
    [myLabel release]; ❹
    [super dealloc];
}
@end
```

在 MyButtonViewController. m 文件中，首先在❶处用 synthesize 关键字说明编译器该如何设置这个成员变量；接着在 buttonClick 操作方法中，使用了两种方法显示字符串：在❷处是通过设置标签 text 属性的方法来实现的，而❸处是通过调用 setText: 方法来实现的，两种方法具有相同的效果；每当程序结束时都要调用❹处的 dealloc 方法，它是用来释放对象所占的内存资源的。

至此，程序代码编写工作完成了，接下来的工作是构建程序界面。

**3. 构建程序界面**

本程序的界面比较简单，只需要添加一个标签和一个按钮。具体步骤如下：

(1) 展开 Resources 文件夹，双击 MyButtonViewController. xib 文件，进入 Interface Builder 中，如图 6-11 所示，从左到右分别表示 Xib 窗口、View 窗口和 Inspector 窗口。

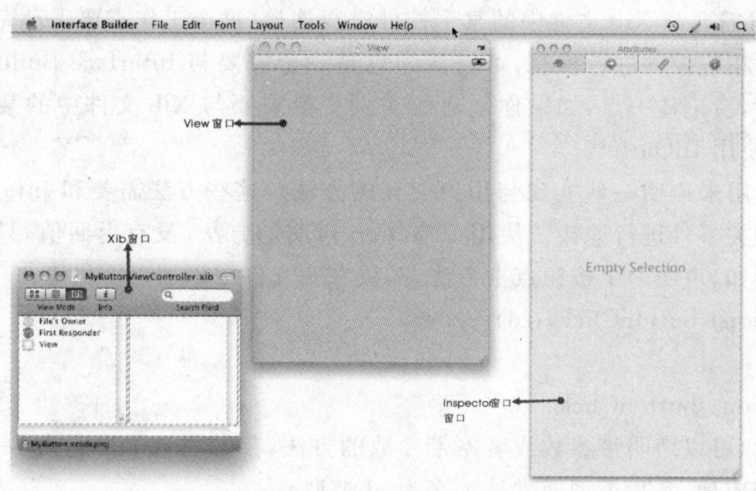


图 6-11 Interface Builder 界面

(2) 选择 Tools→Library(或使用快捷键⌘+Shift+L),打开 Library 窗口(如图6-12所示),并从 Library 窗口中拖出一个 Label 控件和 Round Rect Button 控件到 View 窗口中(如果当前没有显示 View,可以双击 Xib 窗口中的 View 图标)。调整 Label 大小,将 Button 的标题设为"Click Me"。这样用户界面就创建好了。

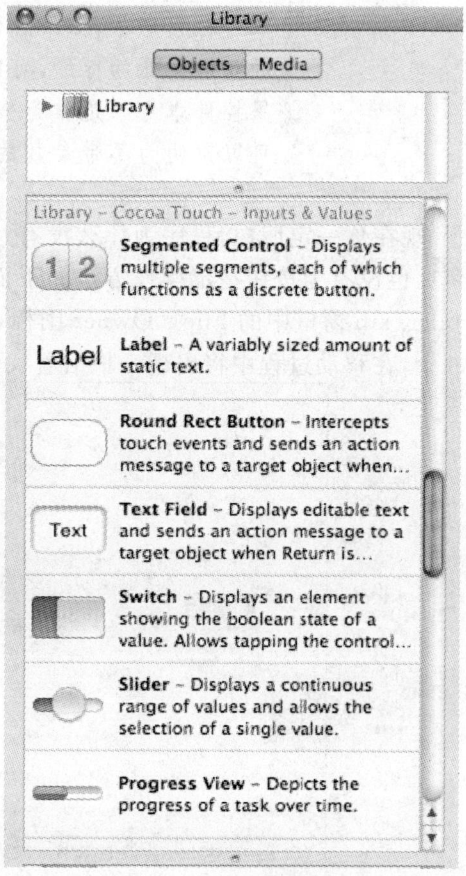


图 6-12 Library 窗口

**4. 为控件连接输出口和操作方法**

在为控件连接输出口和操作方法之前,我们先来运行一下程序。点击 Xcode 中 Build and Go 图标(或使用快捷键⌘+R)运行程序,弹出 iPhone 模拟器,点击 Click Me 按钮,并没有出现我们希望的效果。为什么会这样呢?这就是程序中存在 IBOutlet 和 IBAction 的缘故。

**IBOutlet** 经常被称为输出口,可以理解为程序中用来动态输出数据的对象。它主要用于在头文件(.h 文件)中声明一个实例变量。例如:要声明一个 UILabel 类型的实例变量 myLabel,只需在头文件中添加如下代码:

```
IBOutlet UILabel *myLabel;
```

它用来传递一个消息给 Interface Builder:该实例变量需要与 Xib 文件中的某个控件进行连接(这个工作需要手动完成),从而实现在程序中对相应控件进行控制,可以认为它是一个需要与 Interface Builder 进行连接的标识。因而,请记住一点:如果你创建的实例变量需要与 Xib 文件中的某个对象关联起来,就必须使用 IBOutlet。

**IBAction** 用来声明一些实现与用户交互的方法。这些方法需要和 Interface Builder 中的控件的相关事件进行连接。使用 IBAction 声明的方法,没有返回值,只是用来执行一些操作。例如,声明一个按钮点击方法,需要在头文件中添加以下代码:

```
- (IBAction) buttonClick:(id)sender;
```

或:

```
- (IBAction) buttonClick;
```

如上所示,可以声明带参数或者不带参数的方法,参数名通常指定为 sender,即事件的发送者。相应地,参数类型通常被定义为 id 类型。

**Tips**

在程序中,如果我们希望获取控件相关属性的值,如获取按钮的标题,可以声明为带参数的操作方法;如果控件本身只是作为事件的发起者,与本身属性无关,可以声明为不带参数的操作方法。

现在你一定明白,程序为什么没有达到预期的效果了,因为我们没有为标签连接输出口,也没有为按钮指定操作方法。下面就来完成这些工作。选中 MyButtonViewController.xib 窗口中的 File's Owner 图标,同时按住鼠标右键,拖动鼠标到 View 中的 Label 上,在移动过程中将出现一根沿着我们移动方向的蓝色引导线,如图 6-13 所示。

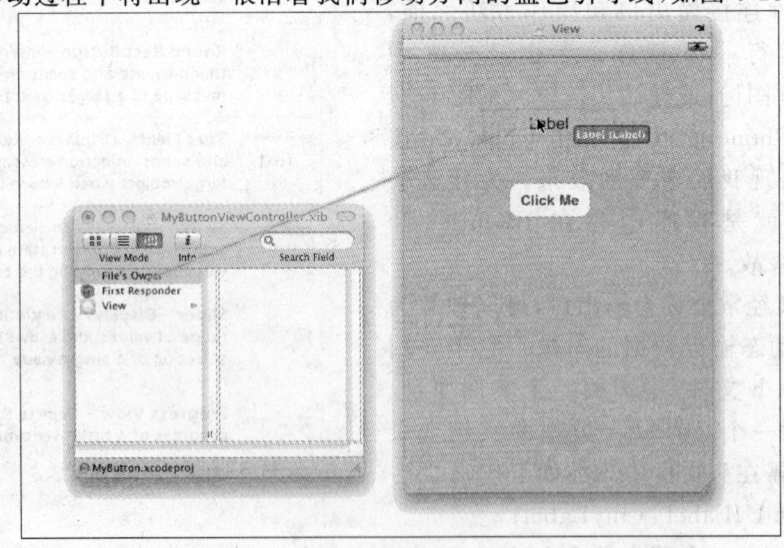


图 6-13 连接输出口

释放鼠标右键,此时会出现如图 6-14 所示的 Outlets 选项框,里面有一个我们之前定义的输出口 myLabel,选中它就可以了。

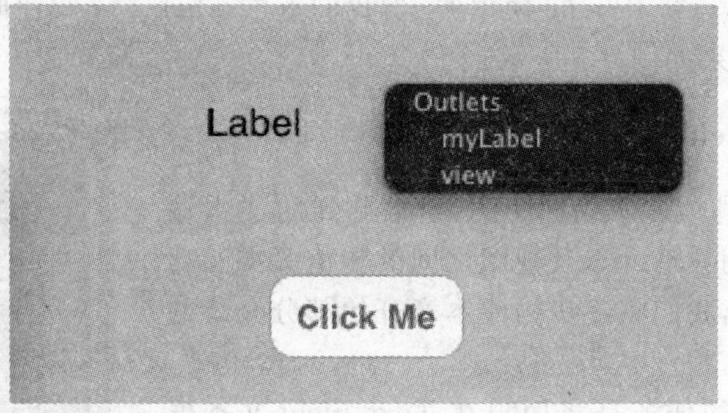


图 6-14 选择输出口 myLabel

接下来要完成另外一个连接:为按钮的单击事件指定 buttonClick 操作方法。选中 View 窗口下的 Click Me 按钮,同时按下鼠标右键,拖动鼠标到 Xib 窗口中的 File's Owner 图标上,注意,这次移动方向和连接输出口时相反;释放鼠标右键,此时会出现如图 6-15 所示的 Events(事件)选项框,同样里面有我们定义的 buttonClick 方法,选中它就可以了。

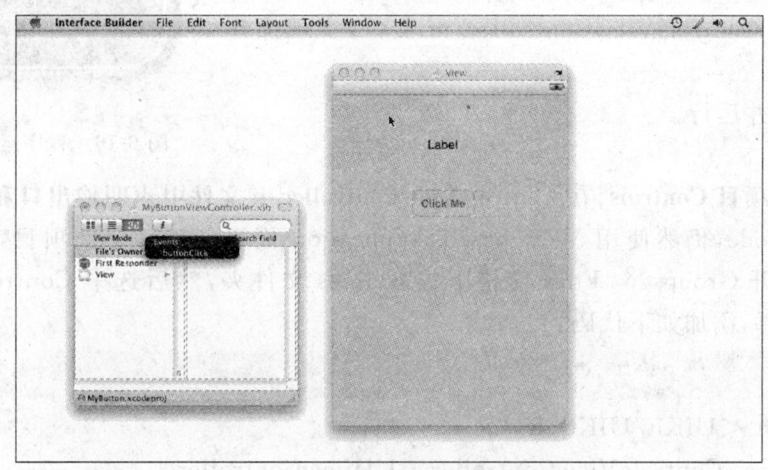


图 6-15 为控件连接操作方法

**Tips**

除了上述连接操作方法的方式以外,我们还可以使用快捷键⌘+2 调出 Connections 窗口,在 Events 层级下选择 Touch Up Inside 事件,连接到 File's Owner 图标;同时在弹出的 Events 选项框下选择相应的操作方法即可。

**5. 编译并运行**

至此，这个小程序才算真正完成了，现在再次运行一下看看效果。点击 Xcode 中的 Build and Go 图标，运行 iPhone 模拟器，当我们点击 Click Me 按钮时，将在标签上显示“Hello iPhone”，效果如图 6-6 所示。

### 6.2.2 UITextField

Text Field 控件就是实际应用中经常见到的文本输入框，它主要提供用户输入以及视图之间的传值等交互功能。从本小节开始，我们将逐步构建本章开始部分介绍的 Controls 程序。

这一节要完成的任务是将用户在 Text Field 文本框中输入的信息显示出来。程序的最终效果如图 6-16 所示。

要完成此程序需要做以下工作：

- 新建项目 Controls，在 ControlsViewController.h 文件中声明输出口和操作方法；
- 构建程序界面，并连接输出口和操作方法；
- 修改 ControlsViewController.m 文件，实现操作方法；
- 编译并运行。

图 6-16 程序运行效果图

**1. 新建项目 Controls，在 ControlsViewController.h 文件中声明输出口和操作方法**

打开 Xcode，仍然使用 View-based Application 模板创建一个新项目，将其命名为 Controls；展开 Groups & Files 窗格下的 Classes 文件夹，然后选中 ControlsViewController.h 文件，添加如下代码：

**代码 6.3 ControlsViewController.h 文件**

```
#import <UIKit/UIKit.h>
@interface ControlsViewController : UIViewController {
    IBOutlet UITextField *topic; ❶
    IBOutlet UITextField *form; ❷
}
@property (nonatomic, retain) UITextField *topic;
@property (nonatomic, retain) UITextField *form;
- (IBAction)buttonClick; ❸
@end
```

在❶和❷处声明了两个 UITextField 类型的输出口 topic 和 form，分别用来接收用户输入的数据；此外，在❸处还声明了一个 buttonClick 操作方法，该方法用来响应用户点击按钮的操作。

**2. 构建程序界面，并连接输出口和操作方法**

双击打开 ControlsViewController. xib 文件，将出现 Interface Builder 界面，从 Library 窗口中拖出两个 Label、两个 Text Field 和一个 Round Rect Button 控件到 View 窗口中，设置各个控件的大小、位置和标题，如图 6-17 所示。

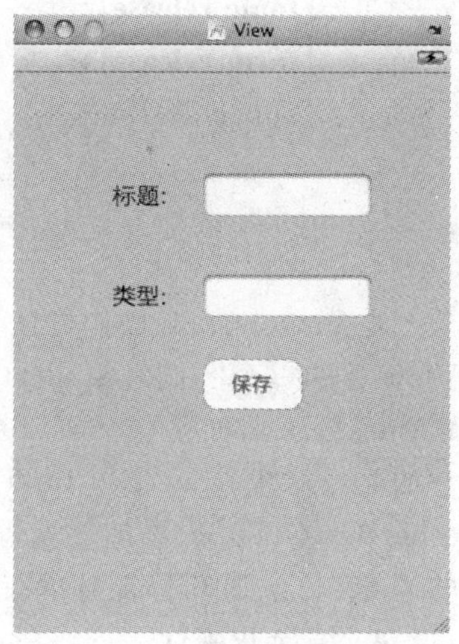


图 6-17　Controls 程序界面

最后，我们需要将 ControlsViewController. h 文件中声明的两个输出口分别与界面上的 Text Field 控件连接；同时，还需要为按钮的 Touch Up Inside 事件指定一个操作方法，即 buttonClick 方法。具体的连接方法和上一小节完全相同。

**3. 修改 ControlsViewController. m 文件，实现操作方法**

为了能够显示用户输入的信息，需要实现 buttonClick 操作方法，选中 ControlsViewController. m 文件，添加如下黑体字所示的代码：

**代码 6.4　ControlsViewController. m 文件**

```
#import "ControlsViewController.h"
@implementation ControlsViewController
@synthesize topic;
@synthesize form;

- (IBAction) buttonClick {
    //获得两个 TextField 中的内容
    NSString *stringTopic=self.topic.text;
    NSString *stringForm=self.form.text;
    UIAlertView *alert=[[UIAlertView alloc]
                initWithTitle: @"显示信息"
                message:[NSString stringWithFormat:@"标题:
                ↳%@\n 类型: %@",stringTopic,stringForm]    ❶
                delegate:nil
                cancelButtonTitle:@"确定"
                otherButtonTitles:nil];
```

```
        [alert show]; ❷
        [alert release]; ❸
    }
    - (void)dealloc {
        [topic release];
        [form release];
        [super dealloc];
    }
    @end
```

在 buttonClick 操作方法中，我们主要看一下❶处的代码。首先，创建了 UIAlertView 类的一个实例对象 alert，其中 initWithTitle：@"显示信息"，用于设置警告框的标题；message 后面是一个 NSString 对象，表示标题下的警告信息；delegate 后面用于设置它的委托对象；cancelButtonTitle 用于设置警告框上默认的按钮标题，我们还可以通过 otherButtonTitles 添加并设置其他按钮的标题；其次，创建完成对象之后，需要调用❷处的 show 方法通知警告显示在屏幕上；最后别忘了调用❸处的 release 方法将它释放掉。

**4. 编译并运行**

点击 Build and Go 图标，编译并运行程序，在文本框中输入相关信息，点击"Click Me"按钮，程序将弹出一个警告框来显示用户输入的信息。具体效果如图 6-16 所示。

### 6.2.3 UISwitch

上一小节已经完成了 Controls 程序的一部分功能，在这里将再加入一个 Switch(开关)控件、一个 Label 和一个 Text Field，其中添加的开关用来控制 Label 和 Text Field 控件是否可见。程序运行时，初始界面如图 6-18 所示，开关处于 ON 状态，新添加的两个控件显示在视图上；当点击开关处于 OFF 状态时，新添加的标签和文本框将被隐藏，如图 6-19 所示。

图 6-18　Switch 处于打开状态

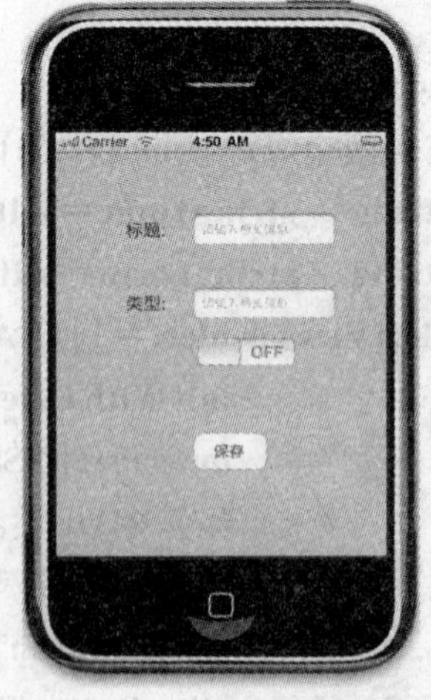


图 6-19　Switch 处于关闭状态

用户输入完成之后，点击按钮将弹出警告提示框。若开关打开，警告框中将显示所有用户输入信息，如图 6-20 所示；若开关关闭，警告框中没有“内容”的相关信息，如图6-21所示。

图 6-20　开关打开时的提示信息

图 6-21　开关关闭时的提示信息

程序的步骤大致如下：

- 修改 ControlsViewController.h 文件，添加输出口和操作方法；
- 修改程序界面，并连接输出口和操作方法；
- 修改 ControlsViewController.m 文件，实现操作方法；
- 编译并运行。

**1. 修改 ControlsViewController.h 文件，添加输出口和操作方法**

使用 Xcode 打开 Controls 程序，选中 ControlsViewController.h 文件，并添加如下黑体字所示的代码：

**代码 6.5　ControlsViewController.h 文件**

```
#import <UIKit/UIKit.h>

@interface ControlsViewController : UIViewController {
    IBOutlet UITextField *topic;
    IBOutlet UITextField *form;
```

```
        IBOutlet UILabel  * contentLabel;
        IBOutlet UITextField  * contentField;
        IBOutlet UISwitch  * contentSwitch;
    }
    @property (nonatomic, retain) UITextField  * topic;                    ❶
    @property (nonatomic, retain) UITextField  * form;
    @property (nonatomic, retain) UILabel  * contentLabel;
    @property (nonatomic, retain) UITextField  * contentField;
    @property (nonatomic, retain) UISwitch  * contentSwitch;
    - (IBAction)hidden; ❷
    - (IBAction)buttonClick; ❸
    - (IBAction)textFieldDoneEditing:(id)sender; ❹
    @end
```

在 ControlsViewController.h 文件❶处增加了三个输出口：contentLabel、contentField 和 contentSwitch。在❷处声明的 hidden 方法用来控制新添加的控件是否可见；在❸处声明的 buttonClick 方法用来显示用户输入信息；在程序❹处声明的 textFieldDoneEditing:方法，用来撤销输入过程中弹出的键盘。

**2. 修改程序界面，并连接输出口和操作方法**

接着我们需要对 Controls 程序原有界面做简单修改。在已有的标签和文本框的基础上再往视图上添加一个开关，按下⌘+1，打开开关的 Attributes 窗口，如图 6-22 所示，将初始状态设为 On。然后再添加一个标签和一个文本框，适当地调整它们的大小和位置。最后效果如图 6-23 所示。

我们还需要为相应的控件连接输出口，方法同上一小节；此外还需要为控件连接操作方法，选择开关控件，打开 Connections 窗口，点击鼠标左键选择 Value Changed 事件，把它拖到 File's Owner 上，选择 hidden 方法；选择按钮，打开 Connections 窗口，将其 Touch Up Inside 事件连接到 File's Owner 上并选择 buttonClick 方法；最后将三个文本框的 Did End on Exit 事件，均连接到 File's Owner 上并选择 textFieldDoneEditing:方法。

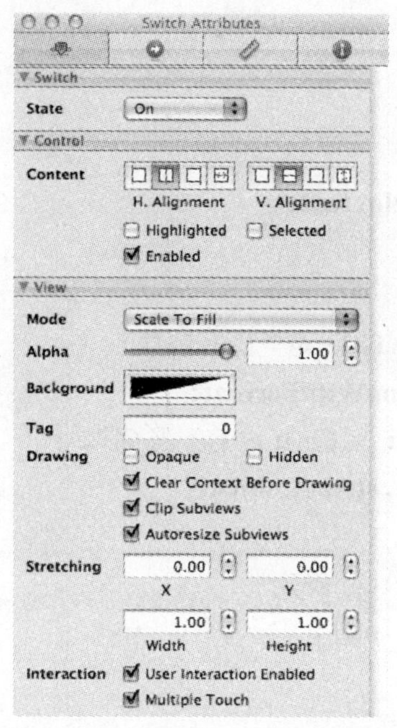


图 6-22 Switch 的 Attributes 窗口

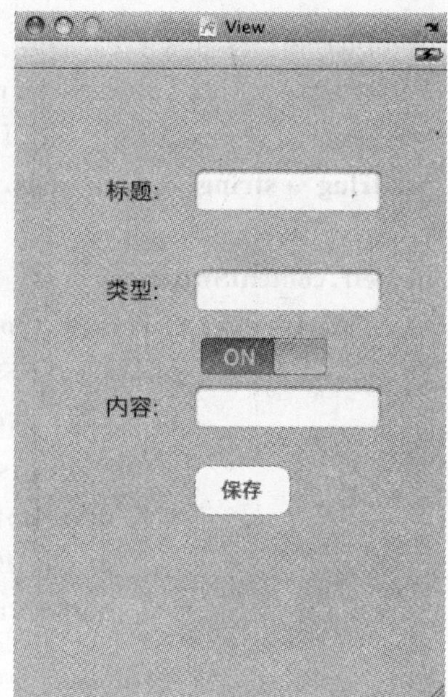


图 6-23 添加新控件后的界面

**3. 修改 ControlsViewController.m 文件,实现操作方法**

保存上述操作并返回 Xcode,选中 ControlsViewController.m 文件,添加如下黑体字所示的代码:

**代码 6.6 ControlsViewController.m 文件**

```
#import "ControlsViewController.h"
@implementation ControlsViewController
@synthesize topic;
@synthesize form;
@synthesize contentLabel;
@synthesize contentField;
@synthesize contentSwitch;
- (IBAction)hidden {
    if (self.contentSwitch.isOn){
        [self.contentLabel setHidden:NO];
        [self.contentField setHidden: NO];
    }                                              ❶
    else{
        [self.contentLabel setHidden: YES];
        [self.contentField setHidden: YES];
    }
```

```
}
- (IBAction)buttonClick {
    NSString * stringTopic=self.topic.text;
    NSString * stringForm=self.form.text;
    NSString * stringContent=self.contentField.text;
    UIAlertView * alert;
    if (self.contentSwitch.isOn){
        alert=[[UIAlertView alloc]initWithTitle:@"显示信息"
                    message:[NSString stringWithFormat:
                    @"标题: %@\n 类型: %@\n 内容: %@",
                    stringTopic,stringForm,stringContent]
                        delegate:nil
                        dcancelButtonTitle:@"确定"
                        dotherButtonTitles:nil];
    }
    else {
        alert=[[UIAlertView alloc]initWithTitle: @"显示信息"
                    message:[NSString stringWithFormat:
                    @"标题: %@\n 类型: %@",
                    stringTopic,stringForm]
                    delegate:nil
                    cancelButtonTitle:@"确定"
                        otherButtonTitles:nil];
    }
        [alert show];
        [alert release];
}                                                                  ❷

- (IBAction)textFieldDoneEditing:(id)sender{
    [sender resignFirstResponder];                                 ❸
}

- (void)viewDidLoad {
    if (!self.contentSwitch.isOn){
        [self.contentLabel setHidden:YES];
        [self.contentField setHidden:YES];
    }
```

```
    self.topic.placeholder=@"请输入相关信息";
    self.form.placeholder=@"请输入相关信息";
    self.contentField.placeholder=@"请输入相关信息";
    [super viewDidLoad];
}
- (void)dealloc {
    [topic release];
    [form release];
    [contentLabel release];
    [contentField release];
    [contentSwitch release];
    [super dealloc];
}
@end
```

在此文件中,❶处的 hidden 方法是根据开关的状态来控制相应的控件是否显示的;我们对 buttonClick 方法做了简单修改,如❷处代码段所示,当开关处于 ON 状态,contentLabel 和 contentField 显示,可以显示用户输入的全部信息;当开关处于 OFF 状态,contentLabel 和 contentField 将被隐藏。最后,还需要调用 show 方法来显示警告框,并用 release 方法释放资源。

运行程序时,点击文本输入框将推出一个键盘提供用户输入,输入完成之后我们需要将它隐藏起来,这就需要调用❸处的 textFieldDoneEditing:方法,程序运行时,点击键盘上 Return 键,就能隐藏键盘了。

**4. 编译并运行**

点击 Build and Go 按钮,编译并运行程序,将会出现图 6-18 所示的界面,输入信息并点击按钮,将弹出图 6-20 所示的警告框;当点击开关转换为 OFF 状态时,将出现图 6-19 所示的界面,此时输入信息之后,点击按钮,将弹出图 6-21 所示的警告框。

## 6.2.4 用代码创建按钮控件

在上两小节中,我们都是利用 Interface Builder 设计程序界面,要在视图中添加一个控件非常简单,但上述方式需要在 Xcode 和 Interface Builder 之间来回切换,操作比较繁琐。本小节我们将介绍构建界面的另一种方式——用代码实现,下面的 ControlWithCode 程序,就是使用代码在界面上添加一个 Button。

本程序的主要步骤如下:

- 创建项目,修改 ControlWithCodeViewController.h 文件;
- 修改 ControlWithCodeViewController.m 文件,在主视图上创建一个按钮;
- 编译并运行。

**1. 创建项目,修改 ControlWithCodeViewController.h 文件**

打开 Xcode,使用 View-based Application 模板创建一个新项目,并将其命名为 ControlWithCode。选中 ControlWithCodeViewController.h 文件,在文件中添加如下黑体字所示的代码:

**代码 6.7　ControlWithCodeViewController.h 文件**

```
#import <UIKit/UIKit.h>
@interface ControlWithCodeViewController: UIViewController {
    UIButton *myButton;  ❶
}
@property (nonatomic, retain) UIButton *myButton;
@end
```

注意:在❶处我们声明了一个 UIButton 类型的变量 myButton,但是并没有使用关键字 IBOutlet,因为这里不会用到 Interface Builder。

**2. 修改 ControlWithCodeViewController.m 文件,在主视图上创建一个按钮**

选中 ControlWithCodeViewController.m 文件,需要在文件中用代码创建一个按钮,同时为它指定类型和标题,具体代码如下:

**代码 6.8　ControlWithCodeViewController.m 文件**

```
#import "ControlWithCodeViewController.h"
@implementation ControlWithCodeViewController
@synthesize myButton;
- (void) viewDidLoad {
    CGRect frame=CGRectMake (105.0f, 150.0f, 100.0f, 50.0f);❶
    myButton=[UIButton buttonWithType:
                        UIButtonTypeRoundedRect]; ❷
    [myButton setTitle:@"Click Me"forState:UIControlStateNormal]; ❸
    myButton.frame=frame; ❹
    [self.view addSubview: self.myButton]; ❺
    [super viewDidLoad];
}

- (void) dealloc {
    [myButton release];
    [super dealloc];
}
@end
```

在 viewDidLoad 方法中，我们首先在❶处创建一个矩形区域 frame，用来定位按钮的位置和大小，其中的参数分别为矩形框的横坐标、纵坐标、宽和高，在❹处，将 frame 赋给了按钮；在❷处为文件中声明的按钮实例指定类型，Type 属性是 UIButton 的一个重要属性，在 Xcode 中，当这条语句写到 UIButtonType 时，按下 Esc 键，将会出现如图 6-24 所示的提示框，你可以根据需要选择不同的按钮类型；在❸处设置了按钮的标题和状态，同样，这条语句写到 UIControlState 时，按下 Esc 键，会出现如图 6-25 的提示框，里面包含了按钮的多种状态，这里我们设置了按钮正常状态下显示的标题；最后，一定不要忘了使用❺处的 addSubview：方法将 myButton 加载到视图中。

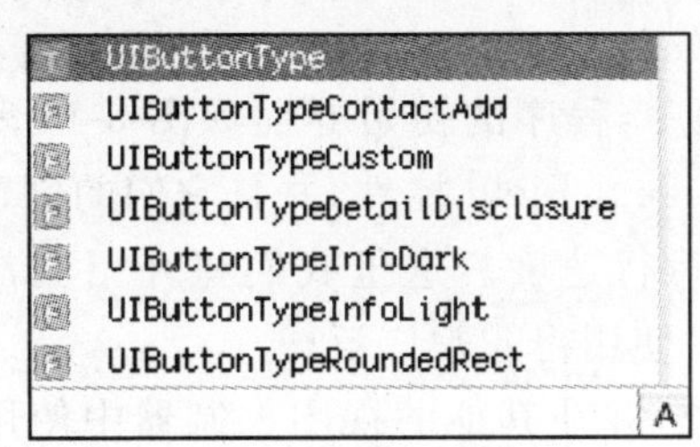


图 6-24 Button 类型

UIControlState
UIControlStateApplication
UIControlStateDisabled
UIControlStateHighlighted
UIControlStateNormal
UIControlStateReserved
UIControlStateSelected
A

图 6-25 Button 状态

**3. 编译并运行**

到目前为止，用代码创建按钮程序就算完成了。点击 Build and Go 图标，编译并运行程序，将出现如图 6-26 所示的效果。

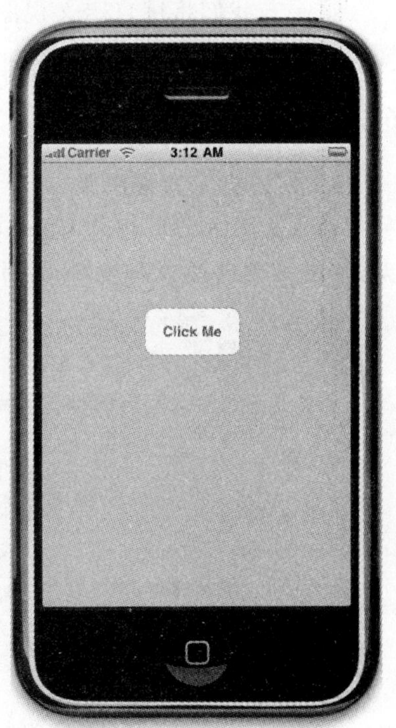


图 6-26 用代码创建按钮程序的运行效果

从图中看到，创建的按钮和 MyButton 实例中的按钮是一样的，但是点击按钮时，却不会有任何效果，那是因为没有为按钮制定任何操作方法。如果想让程序变得更丰富点，可以尝试着用代码再创建一个标签，并添加相应的操作方法，就能达到前面小节程序的效果了，想想看该怎么实现。

### 6.2.5 其他控件

前面已经详细介绍了 Round Rect Button、Text Field、Switch 等控件，再次打开 Interface Builder，选择 Tools→Library，调出 Library 窗口（或使用快捷键⌘＋Shift＋L），可以看到 Library 窗口提供了各种各样的控件，这些控件主要来自 iPhone UIKit 框架（iPhone UIKit 是用于创建应用程序用户界面的框架）。以下列举了部分控件和它的作用：

**Segmented Control**，分段控件，由多个 segment 组成，各个部分可看成相对独立的按钮。

**Slider**，滑块控件，主要用于标识某个对象的变化。我们会在本书第 10 章用户设置中介绍这个控件的使用。

**Progress View**，进度条，用来描述任务的进度，值在0.0与1.0之间变化。

**Activity Indicator View**，活性指示器，旨在将当前任务的进度反馈给用户。

**Page Control**，页面控制器，iPhone应用程序可能有若干页，在页与页之间切换时，Page Control控件使用一个高亮的dot指示当前页。

# 6.3 自定义视图

在上一节中，我们创建了Controls应用程序，程序的初始界面如图6-18所示。仔细观察可以发现，界面上有三组对应的Label和Text Field控件，并且它们的创建方式也相同，这就造成了代码上的冗余，将影响程序的运行速度。这里我们将介绍更常用的创建方法——自定义视图，它在iPhone应用程序中使用得非常广泛。

自定义视图就是创建一个自定义的视图类，并在其他的视图控制器中使用它的实例对象来加载视图，从而实现代码重用。

这一节我们将用自定义视图方法，创建一个自定义的视图，并用它实现Controls程序。

## 6.3.1 创建自定义视图

### 1. 创建自定义视图类

（1）使用Xcode打开Controls程序，我们需要添加几个新的文件，展开Classes文件夹，点击右键选择Add→New File，如图6-27所示。

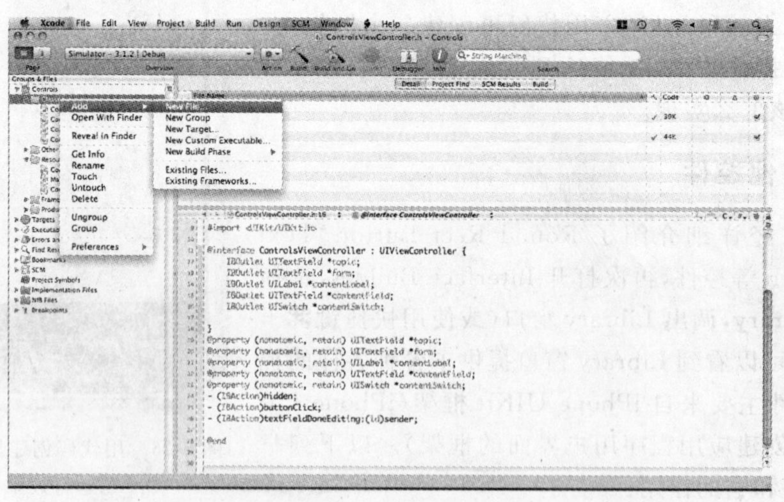

图6-27 添加新文件

（2）在弹出的对话框中，选择Cocoa Touch Class层级下的Objective－C class模板，

并确保 Subclass of 选择为 UIView，如图 6-28 所示。点击 Next 按钮，将文件命名为 CustomView。这时 Classes 文件夹下多了两个新的文件：CustomView. h 和 CustomView. m。

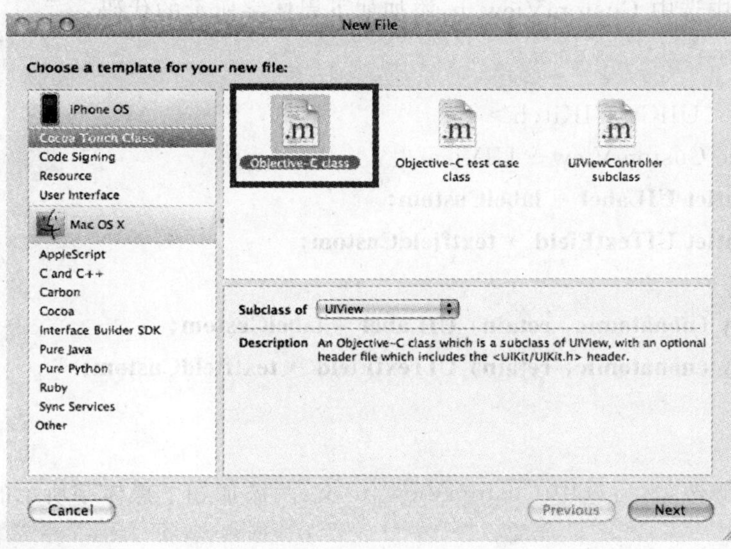


图 6-28 创建自定义视图类

（3）接着在 Resources 文件夹下新建一个 Xib 文件。展开 Resources 文件夹，点击右键选择 Add→New File，在 New File 对话框中选择 User Interface 层级下的 View XIB，如图 6-29 所示，同样命名为 CustomView。

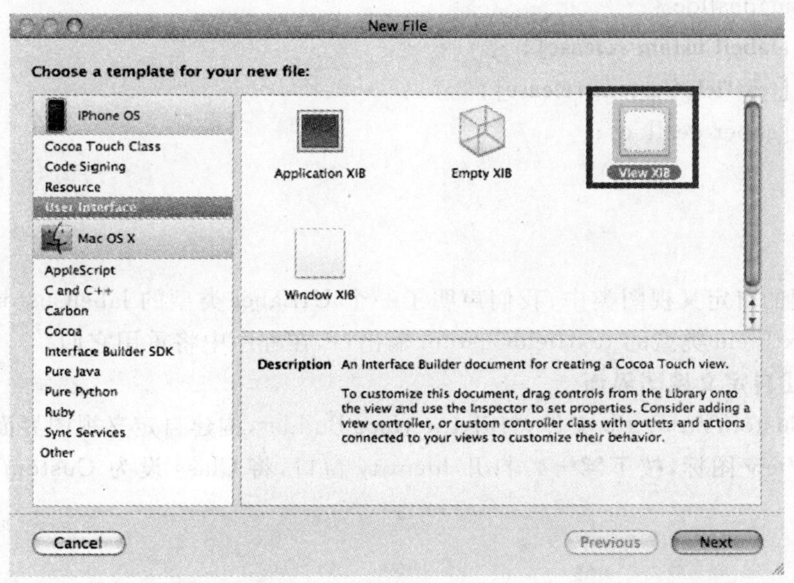


图 6-29 创建 Xib 文件

这样，文件就添加完毕了。其中 Classes 文件下的 CustomView 是我们要实现的自定义视图类，而 Resources 下的 CustomView. xib 是对应的视图。

**2. 修改自定义视图类代码**

在 Xcode 中选中 CustomView. h，添加如下黑体字所示的代码：

**代码 6.9　CustomView. h 文件**

```
#import <UIKit/UIKit.h>
@interface CustomView : UIView {
    IBOutlet UILabel  *labelCustom;
    IBOutlet UITextField  *textfieldCustom;
}
@property (nonatomic, retain) UILabel  *labelCustom;
@property (nonatomic, retain) UITextField  *textfieldCustom;
@end
```

点击保存修改，然后选中 CustomView. m 文件，添加如下黑体字所示的代码：

**代码 6.10　CustomView. m 文件**

```
#import "CustomView.h"
@implementation CustomView
@synthesize labelCustom;
@synthesize textfieldCustom;
- (void)dealloc {
    [labelCustom release];
    [textfieldCustom release];
    [super dealloc];
}
@end
```

在创建的自定义视图类中，我们声明了一个 UILabel 类型的 labelCustom 输出口和一个 UITextField 类型的 textfieldCustom 输出口，在程序中将重用它们。

**3. 构建自定义视图界面**

双击 CustomView. xib 文件，打开 Interface Builder，构建自定义视图界面。选择 Xib 窗口下的 View 图标，按下⌘+4，打开 Identity 窗口，将 Class 设为 CustomView，如图 6-30所示。

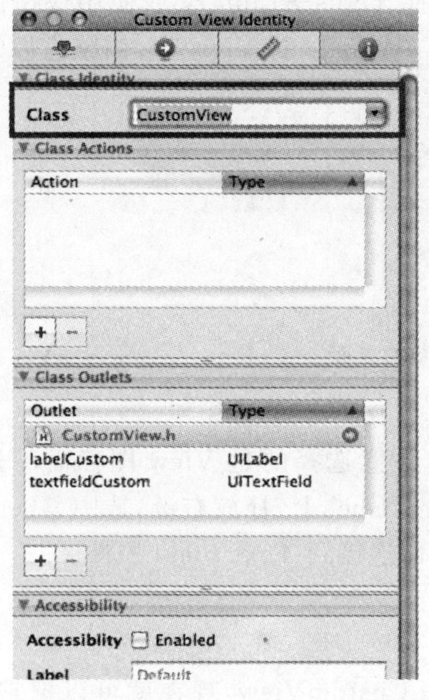


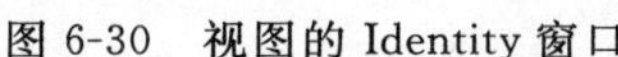
图 6-30 视图的 Identity 窗口

图 6-31 视图的 Attributes 窗口

按下⌘+1，打开 View 的 Attributes 窗口，并将 Status Bar 设置为 None，如图 6-31 所示。因为程序的主视图已经默认加载了 Status Bar，所以在自定义视图中不需要显示状态栏；按下⌘+3 键，打开 Size 窗口，给 CustomView 设置适当的尺寸，宽度设为 300，高度设为 49，如图 6-32 所示。

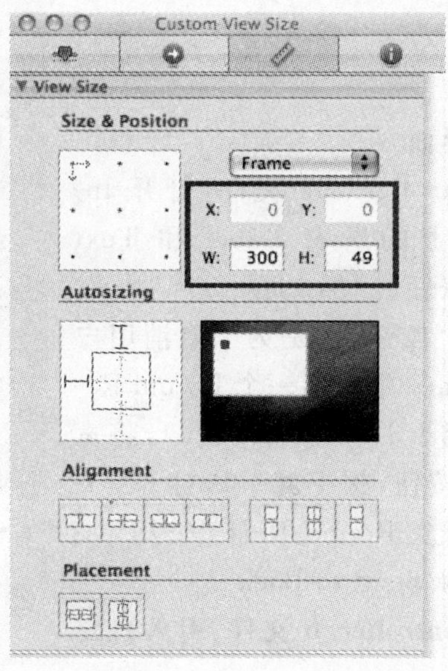


图 6-32 视图的 Size 窗口

接下来，需要在自定义视图中添加控件。选择 Tools→Library 或使用快捷键（⌘+Shift+L），从 Library 中拖出一个 Label 和一个 Text Field 到 View 中，调整它们的大小、位置，如图 6-33 所示。

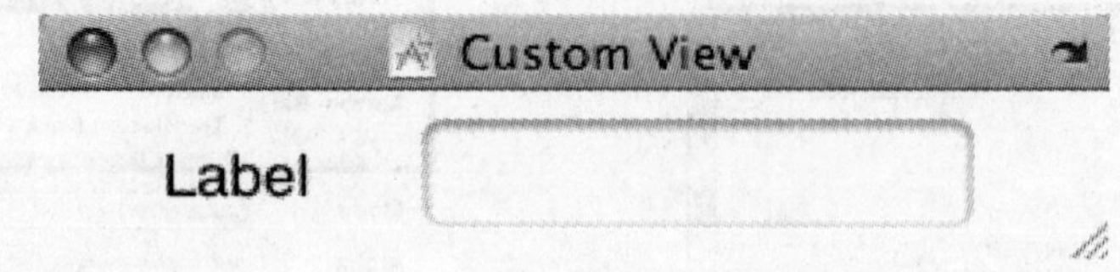


图 6-33　自定义视图的界面

**4. 连接输出口**

选中 CustomView. xib 窗口中的 View 图标（注意选择的是 View 图标而不是 **File's Owner** 图标），按住鼠标右键，拖动鼠标到视图中的 Label 上，释放右键，选中 labelCustom 输出口；使用同样的方法将 textfieldCustom 输出口连接到 Text Field 控件。

### 6.3.2　使用自定义视图

上一小节我们已经创建了一个自定义的视图 CustomView，现在就可以把它应用到 Controls 程序中了。

程序的大致步骤如下：

· 修改程序界面；

· 修改 ControlsViewController. h 文件，引入自定义视图；

· 修改 ControlsViewController. m 文件，使用自定义视图；

· 编译并运行程序。

**1. 修改 Controls 程序界面**

双击 ControlsViewController. xib 文件，打开 Interface Builder。删除视图上的所有 Label 和 Text Field 控件，只留下开关和按钮控件，如图 6-34 所示。此外，我们还要另外添加一个按钮，它是为了取消用户输入时弹出的键盘。从 Library 中拖出一个 Round Rect Button 到 View 中，按下⌘+3，打开 Size 窗口，将宽设为 320，高设为 460；选中按钮，按下⌘+1，打开 Attributes 窗口，将按钮的类型从 Round Rect 改为 Custom；再从 Layout 菜单中选择 Send to Back。

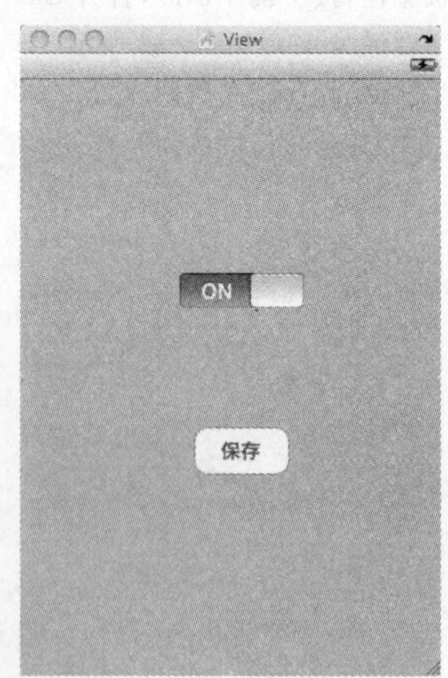


图 6-34　调整后的界面

**2. 修改 ControlsViewController. h 文件，引入自定义视图**

选中 ControlsViewController. h 文件，我们需要对此文件做以下修改：

**代码 6.11 ControlsViewController.h 文件**

```
#import <UIKit/UIKit.h>
#import "CustomView.h" ❶
@interface ControlsViewController : UIViewController {
    IBOutlet UITextField * topic;
    IBOutlet UITextField * form;
    IBOutlet UILabel * contentLabel;
    IBOutlet UITextField * contentField;
    IBOutlet UISwitch * contentSwitch;
    CustomView * topicView;
    CustomView * formView;      ❷
    CustomView * contentView;
}
@property (nonatomic, retain) UITextField * topic;
@property (nonatomic, retain) UITextField * form;
@property (nonatomic, retain) UILabel * contentLabel;
@property (nonatomic, retain) UITextField * contentField;
@property (nonatomic, retain) UISwitch * contentSwitch;
@property (nonatomic, retain) CustomView * topicView;
@property (nonatomic, retain) CustomView * formView;
@property (nonatomic, retain) CustomView * contentView;
- (IBAction)hidden;
- (IBAction)buttonClick;
- (IBAction)textFieldDoneEditing:(id)sender;
- (IBAction)backgroundpressed:(id)sender; ❸
@end
```

首先，在❶处我们引入了自定义视图类头文件 CustomView.h；其次，在❷处声明了三个 CustomView 的实例对象，对应着原来的三组 Label 和 Text Field 控件；最后，我们删除 textFieldDoneEditing:方法，添加了一个取消弹出键盘的操作方法❸。

**3. 修改 ControlsViewController.m 文件，使用自定义视图**

选中 ControlsViewController.m 文件，对它做以下修改：

**代码 6.12 ControlsViewController.m 文件**

```
#import "ControlsViewController.h"
@implementation ControlsViewController
```

```
@synthesize topic;
@synthesize form;
@synthesize contentLabel;
@synthesize contentField;
@synthesize contentSwitch;
@synthesize topicView;
@synthesize formView;
@synthesize contentView;
- (IBAction)hidden {
    if(self.contentSwitch.isOn) {
        [self.contentLabel setHidden:NO];
        [self.contentField setHidden:NO];
        CGRect contentFrame = CGRectMake(0.0f, 200.0f,
            320.0f, 49.0f);
        NSArray *thirdView = [[NSBundle mainBundle]
                              loadNibNamed:@"CustomView"
                              owner:self
                              options:nil];
        self.contentView = [thirdView objectAtIndex:0];
        self.contentView.labelCustom .text = @"内容:";              ❶
        self.contentView.frame = contentFrame;
        self.contentView.background = [UIColor clearColor];
        [self.view addSubview:self.contentView];
    }
    else {
        [self.contentLabel setHidden:YES];
        [self.contentField setHidden:YES];
        [self.contentView removeFromSuperview];
    }
}
- (IBAction)buttonClick {
    NSString *stringTopic = self.topic.text;
    NSString *stringForm = self.form.text;
    NSString *stringContent = self.contentField.text;
```

```
    NSString * stringTopic = self.topicView.textfieldCustom.text;
    NSString * stringForm = self.formView.textfieldCustom.text;          ❷
    NSString * stringContent=self.contentView.textfieldCustom.text;
    UIAlertView * alert;
        if (self.contentSwitch.isOn){
            alert = [[UIAlertView alloc] initWithTitle:@"显示信息"
                        message:[NSString
                        stringWithFormat:@"标题: %@\n
                        ↳类型: %@\n 内容: %@",
                        stringTopic, stringForm, stringContent]
                    delegate:nil
                        cancelButtonTitle:@"确定"
                        otherButtonTitles:nil];
        }
        else{
            alert = [[UIAlertView alloc] initWithTitle:@"显示信息"
                        message:[NSString
                        stringWithFormat:@"标题: %@\n
                        ↳类型: %@", stringTopic, stringForm]
                        delegate:nil
                        cancelButtonTitle:@"确定"
                        otherButtonTitles:nil];
    }
    [alert show];
    [alert release];
}
- (IBAction)textFieldDoneEditing:(id)sender{
    [sender resignFirstResponder];
}
- (IB Action)backgroundpressed:(id)sender{
    [self.topicView.textfieldCustom resignFirstResponder];
    [self.formView.textfieldCustom resignFirstResponder];          ❸
    [self.contentView.textfieldCustom resignFirstResponder];
}
- (void)viewDidLoad {
    if (!self.contentSwitch.isOn){
```

```
            [self.contentLabel setHidden:YES];
            [self.contentField setHidden:YES];
        }
        self.topic.placeholder = @"请输入相关信息";
        self.form.placeholder = @"请输入相关信息";
        self.contentField.placeholder = @"请输入相关信息";
        CGRect topicFrame = CGRectMake(0.0f, 25.0f, 320.0f, 49.0f);
        NSArray *firstView = [[NSBundle mainBundle]
                  loadNibNamed:@"CustomView"
                  owner:self
                  options:nil];
        self.topicView = [firstView objectAtIndex:0];
        self.topicView.labelCustom.text = @"标题:";
        self.topicView.frame = topicFrame;
        self.topicView.background = [UIColor clearColor];
        [self.view addSubview:self.topicView];                      ❶
        CGRect formFrame = CGRectMake(0.0f, 95.0f, 320.0f, 49.0f);
        NSArray *secondView = [[NSBundle mainBundle]
                  loadNibNamed:@"CustomView"
                  owner:self
                  options:nil];
        self.formView = [secondView objectAtIndex:0];
        self.formView.labelCustom.text = @"类型:";
        self.formView.frame = formFrame;
        self.formView.background = [UIColor clearColor];
        [self.view addSubview:self.formView];
        [super viewDidLoad];
    }
    - (void)dealloc {
        [topic release];
        [form release];
        [contentLabel release];
        [contentField release];
        [contentSwitch release];
        [topicView release];
        [formView release];
```

```
        [contentView release];
        [super dealloc];
    }
    @end
```

下面我们来看一下 ControlsViewController.m 文件中变动的地方。

变动之一：❶处的 hidden 方法，if 语句的判断条件没有发生改变，仍然是开关的状态，只不过添加控件的方式改变了，原来是通过控制控件的可见性来改变视图，现在通过控制自定义视图的加载来改变视图。当开关状态为 ON 时，使用如下黑体代码创建一个自定义视图对象，并使用 addSubview：方法加载自定义视图，设置视图 frame 属性、background 属性以及 labelCustom 的 text 属性；当开关状态为 OFF 时，调用 removeFromSuperview 方法移除自定义视图。

```
NSArray *thirdView = [[NSBundle mainBundle]
                            loadNibNamed:@"CustomView"
                            owner:self
                            options:nil];
self.contentView = [thirdView objectAtIndex:0];
```

变动之二：❷处的代码表示通过自定义视图对象来获取用户输入的相关信息。

变动之三：我们删除了取消键盘的 textFieldDoneEditing：方法，在❸处添加了一个 backgroundpressed：方法，这是我们介绍的又一种取消键盘的方法。程序运行时，点击视图中没有控件的区域都将触发这个操作，从而关闭键盘。

变动之四：在 viewDidLoad 方法中，在代码段❹中重用了两次自定义视图，首先创建一个 frame 用于设定视图的位置和大小，然后加载了可重用的视图单元，并将它赋给了自定义视图实例对象，最后使用 addSubview：方法在主视图上添加子视图。

**4. 连接操作方法**

代码添加完成之后，还需要返回 Interface Builder，将控件和相应的方法连接起来。选择开关控件，按下⌘+2，打开 Connections 窗口，选择 Value Changed 事件，按住鼠标左键拖动至 File's Owner 上，选中 hidden 方法即可；同时将"保存"按钮和背景按钮的 Touch Up Inside 事件，分别与 buttonClick 操作方法和 backgroundpressed：操作方法连接。

**5. 编译并运行程序**

一切工作完成之后，点击 Build and Go 图标，编译并运行程序，效果如图 6-18 所示，点击开关，便会出现如图 6-19 所示的效果，用户输入信息之后，点击按钮，效果如图 6-20 和图 6-21 所示。

## 6.4　小结

到目前为止,你应该对 iPhone 开发有了进一步的了解了。

通过本章的学习,我们应当可以解决以下问题:

(1) 如何创建一个简单的 iPhone 程序,并掌握控件的使用;

(2) 理解 IBOutlet 和 IBAction 的作用,如何为控件连接输出口和操作方法;

(3) 如何使用代码创建简单的控件;

(4) 如何用自定义视图实现代码的重用,自定义视图为我们提供了一种不同的视图创建方法,也提供了另一种编程体验。

本章介绍的是 iPhone 开发的基础知识,需要我们认真地弄懂每一个知识点,为后面章节的学习打好基础。

# 第 7 章 视图控制器

**本章内容**

- 自定义视图控制器
- 丰富多彩的动画效果
- 导航控制器
- 标签栏控制器
- 表视图控制器

前面章节中曾提到过 iPhone 应用程序界面通常采用“单窗口－多视图”的设计模式。iPhone 中音乐播放器是这个模式的一个很好的运用(如图 7-1 所示)。支持多视图让 iPhone 能够拥有更加丰富的功能,而单窗口的限制又使得程序要不断地在视图之间进行转换。对于这样一个拥有多种功能的软件来说,视图间的转换是如何实现的?这就需要用到本章将要讲解的内容了。

图 7-1 音乐播放器视图

## 7.1 视图控制器概述

在 iPhone 开发中,视图控制器起着举足轻重的作用。任何应用程序都至少有一个视图控制器,对于多视图应用程序,每个视图都有唯一的视图控制器,而一个视图控制器可以控制一个或同时控制多个视图。这些视图控制器共同搭建起应用程序的框架,在界面中显示出一个视图或实现多个视图间的转换。

根据视图控制器所控制的视图的数量,将其分为单视图控制器和多视图控制器两种。

### 7.1.1 单视图控制器

顾名思义,单视图控制器只控制一个视图,它不仅出现在单视图应用程序中,多视图应用程序中也经常用到。

第 6 章中所构建的都是单视图应用程序,例如在应用程序 Controls 中,使用的是

View-based Application 模板，该模板已经创建了一个视图控制器 ControlsViewController，控制着有多个控件的视图。

单视图控制器在多视图应用程序中更为常见，它通常是根视图控制器的一个子控制器。在本章 7.4 节所构建的应用程序中，"关于"视图就是受单视图控制器控制着，它是标签栏控制器的一个子控制器。

### 7.1.2 多视图控制器

在多视图应用程序中，程序由多个内容视图组成，每个视图都有自己的控制器、输出口和动作。视图之间不断切换，必须要有统一的控制器来控制它们的显示。

视图控制器是 UIViewController 类的一个实例，UIViewController 类在 UIKit 框架中定义，如图 7-2 所示。

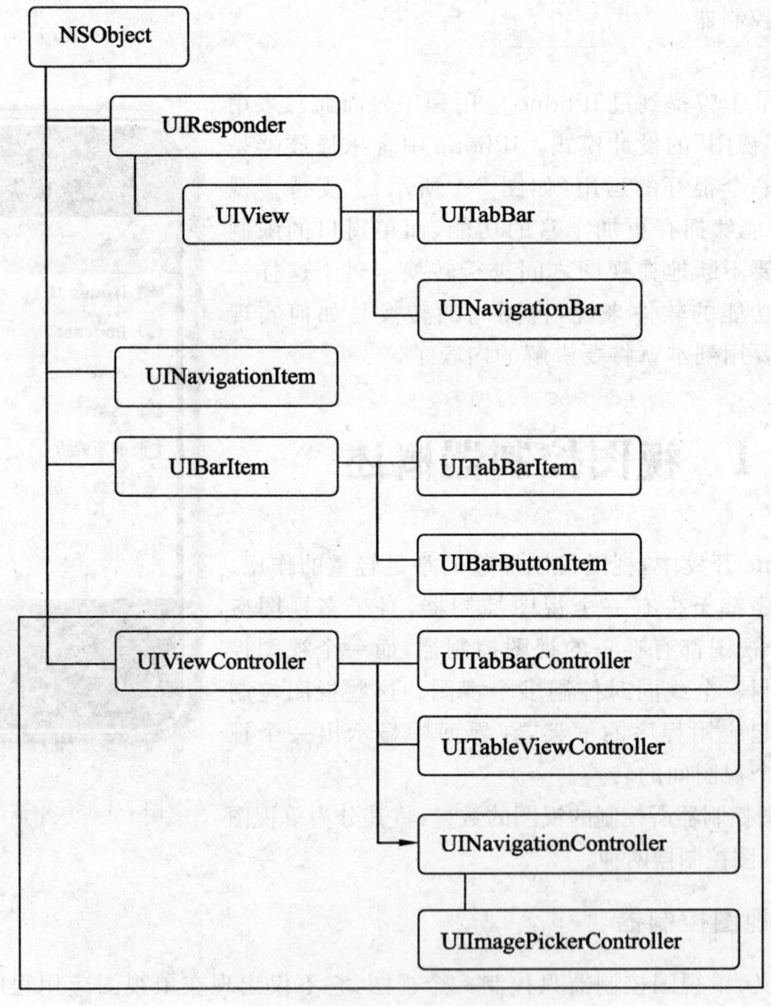


图 7-2 UIViewController 类

从图 7-2 可以看出,UIViewController 的父类为 NSObject;它有三个子类:UINavigationController 、UITabBarController、UITableViewController,分别对应着不同类型的视图控制器:

(1) UINavigationController 是导航控制器类,控制多个具有层次关系的视图,是构建分层应用程序的主要工具;

(2) UITabBarController 是标签栏控制器类,它用于管理多个具有相对独立关系的视图,用户可以在不同视图之间切换,每一个视图代表着不同的运行模块;

(3) UITableViewController 是表视图控制器类,它用于显示数据列表,可以被配置成所需的任何形式,是用来显示数据的最常用的机制。

上述三种视图控制器类是 UIKit 框架提供的视图控制器类,我们还可以直接使用其父类 UIViewController 定义自己需要的视图控制器,即自定义视图控制器。

本章中将用三个应用程序分别介绍自定义视图控制器、导航控制器和标签栏控制器。前两个程序都是完成视图的转换和数据传递的功能,只是实现方式不同;第三个程序中我们把导航控制器加在标签栏控制器中,作为其中的一个独立功能模块。最后简单介绍了一下表视图控制器,它作为最常用的视图控制器,我们将在第 8 章中予以详细介绍。

## 7.2 ViewController

在这一节中,我们构建一个基于窗口的应用程序,亲自创建三个视图控制器。其中一个控制器作为根视图控制器,控制另外两个视图之间的转换及数据的传递。

Window-based Application 模板是所有项目模板中功能最简单的模板,它仅提供一个窗口和一个应用程序委托。使用此模板构建应用程序需要自己定义视图或控制器,手动创建对象间的连接。由于操作的繁琐性,我们很少使用这个模板创建项目。不过现在,对于让我们更加深刻地了解视图控制器的工作原理,这确实是一个不错的方法。

### 7.2.1 构建基于 Window 的应用程序

下面开始构建这个名为 NoteController 的多视图应用程序。程序的初始界面如图 7-3 所示。屏幕的底部是一个工具栏(Toolbar),该工具栏右侧有一个按钮。在该视图(以下简称输入视图)上方三个 Text Field 中输入相关信息,单击“转换”按钮,视图转换为显示相关信息的视图(以下简称显示视图),如图 7-4 所示,输入的信息在对应的三个 Label 中被成功读取出来。

图 7-3 程序运行时的输入视图　　图 7-4 程序运行时的显示视图

此程序操作比较复杂，需要以下 7 步：

· 新建基于窗口的项目 NoteController；

· 新建文件 RootViewController. m、SaveViewController. m 及 ShowViewController. m 及相应的头文件；

· 新建文件 SaveViewController. xib 和 ShowViewController. xib；

· 修改 NoteControllerAppDelegate. h 和 NoteControllerAppDelegate. m 及 MainWindow. xib，创建根视图控制器及 Toolbar；

· 修改 SaveViewController. h、SaveViewController. m 及 SaveViewController. xib，设计输入视图；

· 修改 ShowViewController. h、ShowViewController. m 及 ShowViewController. xib，设计显示视图；

· 修改 RootViewController. h 和 RootViewController. m，实现两视图间的转换。

**1. 新建基于窗口的项目 NoteController**

打开 Xcode，选择 File→ New Project，在 New Project 窗口 iPhone OS 层级下选择 Application 中的 Window-based Application，如图 7-5 所示，点击 Choose 按钮后输入项目名称 NoteController。

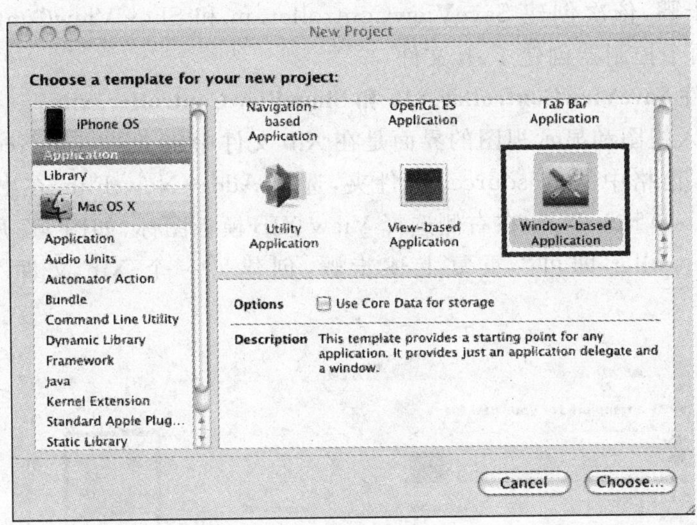


图 7-5　新建基于窗口的应用程序

**2. 新建文件 RootViewController. m、SaveViewController. m 及 ShowViewController. m 及相应的头文件**

接下来,开始创建所需要的三个视图控制器。单击 Groups & Files 窗格中的 Classes 文件夹,单击右键选择 Add→New File 之后,出现 New File 窗口。在该窗口左侧选择 Cocoa Touch Class,可以看到大量用于 Cocoa Touch 类的模板,如图 7-6 所示。选择 UIViewController subclass,不要勾选 With XIB for user interface。单击 Next,输入文件名 RootViewController. m,并且确保选中 Also create "RootViewController. h"复选框,单击 Finish。查看 Xcode 中 Classes 文件夹,这时已经增加了两个新文件:RootViewController. h 文件和 RootViewController. m 文件。

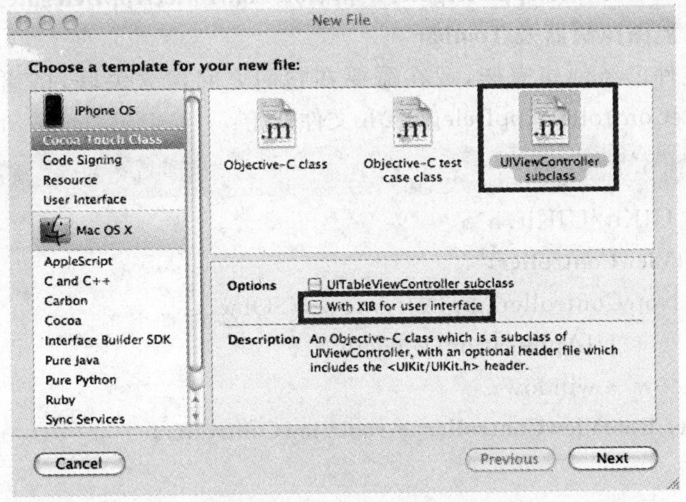


图 7-6　新建一个 UIViewController 类

重复上述步骤，依次创建 SaveViewController. m 和 ShowViewController. m 文件，并确保也为这两个控制器创建了.h 文件。

**3. 新建文件 SaveViewController. xib 和 ShowViewController. xib**

本例中，输入视图和显示视图的界面是在 Xib 文件中定义的。在 Xcode 中右键单击 Groups & Files 窗格中的 Resources 文件夹，选择 Add→New File，在出现的 New File 窗口左侧选择 User Interface，在右侧选择 View XIB 模板图标，如图 7-7 所示。输入文件名 SaveViewController 即可。重复上述步骤，创建另一个 Xib 文件 ShowViewController. xib。

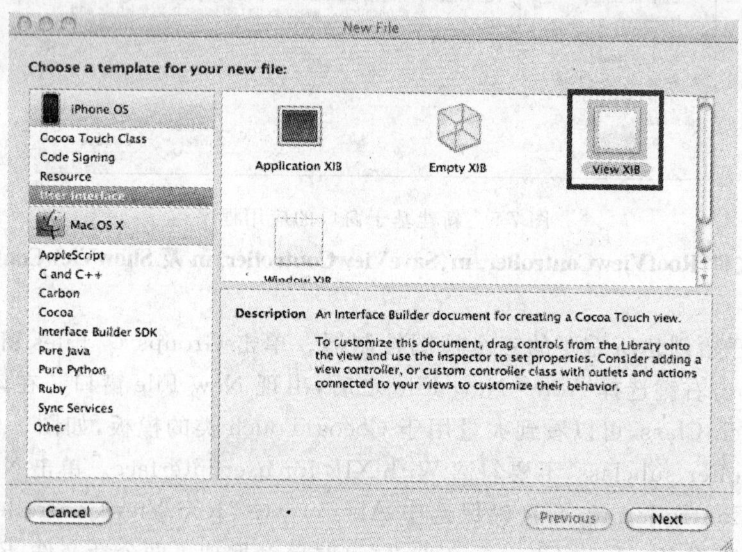


图 7-7 新建一个 Xib 文件

**4. 修改 NoteControllerAppDelegate. h 和 NoteControllerAppDelegate. m 及 MainWindow. xib，创建根视图控制器及 Toolbar**

所需要的文件已经创建完毕，现在需要在它们之间建立关联，搭建好程序的框架。我们从修改 NoteControllerAppDelegate. h 文件开始：

**代码 7.1 NoteControllerAppDelegate. h 文件**

```
#import <UIKit/UIKit.h>
@class RootViewController;
@interface NoteControllerAppDelegate : NSObject
            <UIApplicationDelegate> {
    UIWindow * window;
    IBOutlet RootViewController * rootViewController;
}
@property (nonatomic, retain) IBOutlet UIWindow * window;
@property (nonatomic, retain)
```

```
    IBOutlet RootViewController * rootViewController;
@end
```

接下来，在 NoteAppDelegate.m 中添加如下黑体字所示的代码：

**代码 7.2 NoteControllerAppDelegate.m 文件**

```
#import "NoteControllerAppDelegate.h"
#import "RootViewController.h"
@implementation NoteControllerAppDelegate
@synthesize window;
@synthesize rootViewController;
- (void)applicationDidFinishLaunching:
              (UIApplication *)application {
    // Override point for customization after application launch
    [window addSubview:rootViewController.view];
    [window makeKeyAndVisible];
}
- (void)dealloc {
    [window release];
    [rootViewController release];
    [super dealloc];
}
@end
```

我们在应用程序的委托中声明了 RootViewController 类的一个实例对象 rootViewController，并将根视图控制器的视图添加到应用程序的窗口，作为它的子视图。

在 Resources 文件夹下选择 MainWindow.xib 文件，在 Interface Builder 中将其打开。其中有四个图标：File's Owner、First Responder、NoteController App Delegate、Window。由于没有视图控制器，需要我们自己创建。按下⌘＋Shift＋L，打开 Library 窗口，如图 7-8 所示。在 Library 窗口中，Cocoa Touch-Controllers 子项是专门存放控制器的库，前面讲过的几类视图控制器都在其中，在本章后面会陆续用到它们。

图 7-8 Library 中的各类控制器

选中 View Controller，把它拖进 MainWindow. xib 中，当鼠标旁显示一个绿色的小加号时，放开鼠标，ViewController 便添加到了 MainWindow. xib，结果如图 7-9 所示。

选择 ViewController，按下⌘＋4，打开 Identity 窗口，在 Class 属性的下拉列表中选择 RootViewController，此时 MainWindow. xib 窗口中视图控制器的名称变为 RootViewController。双击打开它，是一个灰色的、中间显示一个 View 字样的窗口，这表示该视图控制器现在处于缺少 View 的状态。再次打开 Library，拖动一个 View 到 RootViewController 窗口，View 自动填充整个屏幕。右键点击 Note Controller App Delegate，拖拽至 RootViewController，连接输出口 rootViewController。

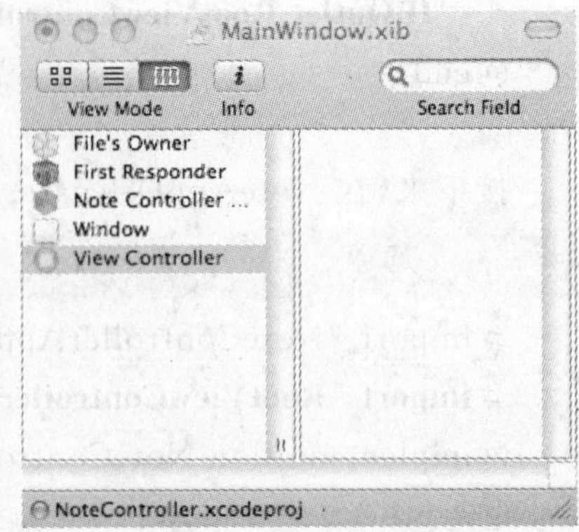


图 7-9 视图控制器被添加到 Xib 文件

程序中有一个工具栏，为了能够控制输入视图和显示视图间的切换，需要将它放置在根视图控制器的视图中。因此，从 Library 中找到 Toolbar，拖到刚才加入的视图的底部。默认状态下，工具条中的按钮是在左边放置的，有一个工具可以将按钮调整到工具条的右边：Flexible Space Bar Button Item，将它从 Library 中拖至按钮前面。最后双击按钮，更改按钮标题为“转换”，如图 7-10 所示。

图 7-10 RootViewController 设计界面

图 7-11 ToolBar 界面中的显示

返回 Xcode,编译并运行程序。界面下方带按钮的工具条已经出现了,工具条上方的屏幕却是白的,如图 7-11 所示,这是由于我们还未定义程序的初始界面,因此我们应先设计输入视图和显示视图的界面。

**5. 修改 SaveViewController. h、SaveViewController. m 及 SaveViewController. xib,设计输入视图**

打开 SaveViewController. h 文件,添加如下黑体字所示的代码:

**代码 7.3 SaveViewController. h 文件**

```
#import <UIKit/UIKit.h>
@interface SaveViewController : UIViewController {
    IBOutlet UITextField * saveTopic;
    IBOutlet UITextField * saveForm;
    IBOutlet UITextField * saveContent;
}
@property (nonatomic, retain) IBOutlet UITextField * saveTopic;
@property (nonatomic, retain) IBOutlet UITextField * saveForm;
@property (nonatomic, retain) IBOutlet UITextField * saveContent;
- (IBAction) textFieldDoneEditing:(id)sender;
@end
```

我们声明了三个 UITextField 类型的输出口,用来输入消息内容。textFieldDoneEditing:方法在第 6 章中已经用过,当完成输入时通过点击键盘的 Return 键取消键盘。

在 SaveViewController. m 中添加如下黑体字所示的代码:

**代码 7.4 SaveViewController. m 文件**

```
#import "SaveViewController.h"
@implementation SaveViewController
@synthesize saveTopic;
@synthesize saveForm;
@synthesize saveContent;
- (IBAction) textFieldDoneEditing:(id)sender {
    [sender resignFirstResponder];
}
……
- (void)dealloc {
    [saveTopic release];
    [saveForm release];
    [saveContent release];
```

```
    [super dealloc];
}
@end
```

完成了代码的添加，接下来，设计输入视图的界面。

(1) 将所需要的图片导入 Resources 文件夹下。选中 Resources 文件夹，在 Xcode 上方菜单栏中选择 Project→Add to Project，在打开的提示框中选择后缀为.png 的两张图片 saveImage.png 和 showImage.png，单击 Add，Resources 文件夹下显示了添加的图像。

(2) 双击 SaveViewController.xib 文件，打开 Interface Builder。由于根视图控制器的视图中有状态条(高度为 20)和 Toolbar(高度为 44)，因此我们需要调整 SaveViewController 视图的大小为 416(480－20－44)。按下⌘＋1，调出 Attributes 窗口，将 Status Bar 设为 None 后，按下⌘＋3，调出 Size 窗口，将该视图高度设为 416。注意，只有没有状态条的视图才可以更改其大小。

(3) 选择 File's Owner，调出 Attributes 窗口，更改其 Class 属性为 SaveViewController。如果忘记了这一步，程序在加载 Xib 文件时将无法连接 SaveViewController 类。此时若在 File's Owner 上点击右键，发现 view 输出口并未连接，选中 File's Owner，按住鼠标左键，将其拖到 View 图标上，出现输出口 view，选中即可。

(4) 接下来，在界面中加入背景图片。按下⌘＋Shift＋L，在 Library 中找到 Image View，拖入 View 窗口，打开 Attributes 窗口，在其 Image 属性的下拉列表中选择 saveImage.png。此时，Image View 中就有了图像。

(5) 从 Library 中拖入三个 Text Field，将它们放到视图中合适的位置，参见图 7-3。首先，将三个 Text Field 分别连接相应的输出口 saveTopic、saveForm、saveContent；然后按下⌘＋2，调出 Connections 窗口，将 Did End On Exit 事件连接 textFieldDoneEditing:方法。

完成连接工作后保存，关闭 Interface Builder，返回 Xcode。同样地，对 ShowViewController 做相似的操作。

**6. 修改 ShowViewController.h、ShowViewController.m 及 ShowViewController.xib，设计显示视图**

在 ShowViewController.h 中添加如下黑体字所示的代码：

**代码 7.5 ShowViewController.h 文件**

```
#import <UIKit/UIKit.h>
@interface ShowViewController : UIViewController {
IBOutlet UILabel *showTopic;
IBOutlet UILabel *showForm;
IBOutlet UILabel *showContent;
```

```
}
@property (nonatomic, retain) IBOutlet UILabel * showTopic;
@property (nonatomic, retain) IBOutlet UILabel * showForm;
@property (nonatomic, retain) IBOutlet UILabel * showContent;
@end
```

对应地，在 ShowViewController. m 中修改代码：

**代码 7.6 ShowViewController. m 文件**

```
#import "ShowViewController.h"
@implementation ShowViewController
@synthesize showTopic;
@synthesize showForm;
@synthesize showContent;
……

- (void)dealloc {
    [showTopic release];
    [showForm release];
    [showContent release];
    [super dealloc];
}
@end
```

在 ShowViewController. h 文件和 ShowViewController. m 文件中，我们仅声明了三个 UILabel 类型的输出口，用于显示已经存储的数据。

打开 Interface Builder，在 ShowViewController. xib 中，选择 File's Owner，更改其 Class 属性为 ShowViewController。将三个 Label 拖入视图中，具体设计可参见图 7-4。为三个 Label 分别连接输出口 showTopic、showForm、showContent。

根视图控制器控制的两个视图设计好后，便可以在 RootViewController 中定义两个视图转换及数据传递的操作了。选择 RootViewController. h 文件，添加如下黑体字所示的代码：

**代码 7.7 RootViewController. h 文件**

```
#import <UIKit/UIKit.h>
@class SaveViewController;
@class ShowViewController;
@interface RootViewController : UIViewController {
    SaveViewController * saveViewController;
```

```
    ShowViewController * showViewController;
}
@property (nonatomic, retain)
        SaveViewController * saveViewController;
@property (nonatomic, retain)
        ShowViewController * showViewController;
- (IBAction)switchViews:(id)sender;
@end
```

代码中引入了两个视图的视图控制器类及一个 switchViews:方法。该方法在点击 Toolbar 上按钮时被调用。因此,在 MainWindow. xib 中打开 RootViewController 窗口,为按钮连接 Touch Up Inside 事件所触发的操作 switchViews:方法。

**7. 修改 RootViewController. h 和 RootViewController. m,实现两视图间转换**

保存并关闭 Interface Builder,在 RootViewController. m 中添加如下黑体字所示的代码:

**代码 7.8　RootViewController. m 文件**

```
#import "RootViewController.h"
#import "SaveViewController.h"
#import "ShowViewController.h"
@implementation RootViewController
@synthesize saveViewController;
@synthesize showViewController;
……
- (void)viewDidLoad {
        SaveViewController * saveMessage = [[SaveViewController alloc]
                initWithNibName:@"SaveViewController"
                        bundle:nil];
        self.saveViewController = saveMessage;                    ❶
        [self.view addSubview: saveViewController.view];
        [saveMessage release];
        [super viewDidLoad];
}
- (IBAction)switchViews:(id)sender {
    if (self.showViewController.view.superview == nil) {
        if (self.showViewController == nil) {
                ShowViewController * showMessage =
```

```
                [[ShowViewController alloc ]
            initWithNibName:@"ShowViewController"
                      bundle:nil ];
            self.showViewController = showMessage;
            [self.view addSubView:showViewController.view ]
            [showMessage release ];
        }
    showViewController.showTopic.text=
                    saveViewController.saveTopic.text;
      showViewController.showForm.text=
                    saveViewController.saveForm.text;        ❷
      showViewController.showContent.text=
                    saveViewController.saveContent.text;
    [saveViewController.view removeFromSuperview];❸
    [self.view insertSubview:showViewController.view atIndex:0];❹
    }
    else {
        [showViewController.view removeFromSuperview];
        [self.view insertSubview:
                saveViewController.view atIndex:0];
        }
}
- (void)dealloc {
    [saveViewController release];
    [showViewController release];
    [super dealloc];
}
@end
```

在 viewDidLoad 方法中❶处，将 SaveViewController 视图作为根视图控制器的视图出现后第一次载入的视图。switchViews:方法中的逻辑判断并不复杂，不做过多解释。真正起到视图间转换作用的是以下两个方法(通常成对出现)：❸处 removeFromSuperview 方法用于将视图从父视图移除，然后在❹处使用 insertSubview：atIndex:方法载入另一视图。

当显示为 SaveViewController 的视图时点击按钮，switchViews:方法会将 Text Field 中输入的值传递到 ShowViewController 中 Label 的值并显示出来，两视图间的传值在❷处简单地实现了。

返回 Xcode，编译并运行程序。视图之间实现了切换，在输入视图中输入的数据在显示视图中被读取出来。

## 7.2.2 丰富多彩的动画效果

### 1. 加入动画效果

至此,几乎所有的工作都完成了,视图之间成功实现了转换。不过,并没有看到明显的转换效果。如果在转换时加入动画效果,程序会变得美观很多,而这并不需要我们做太多的工作。在 RootViewController.m 文件中添加如下黑体字所示的代码:

**代码 7.9 RootViewController.m 文件**

```
- (IBAction)switchViews:(id)sender {
    [UIView beginAnimations:@" Curve"context:nil];
    [UIView setAnimationDuration:1.25];
    [UIView setAnimationCurve:UIViewAnimationCurveEaseInOut]; ❶
    if (self.showViewController.view.superview == nil) {
        if (self.showViewController == nil) {
            ShowViewController *showMessage =
                    [[ShowViewController alloc]
                initWithNibName:@"ShowViewController"
                        bundle:nil];
            self.showViewController = showMessage;
            [showMessage release];
        }
        self.showViewController.showTopic.text = saveTopic.text;
        self.showViewController.showForm.text = saveForm.text;
        self.showViewController.showContent.text=saveContent.text;
        [UIView setAnimationTransition:
                UIViewAnimationTransitionCurlUp
                    forView:self.view
                    cache:YES];                        ❷
        [saveViewController.view removeFromSuperview];
        [self.view insertSubview:showViewController.view atIndex:0];
    }
    else {
        [UIView setAnimationTransition:
                UIViewAnimationTransitionCurlDown
                    forView:self.view
            cache:YES];                                ❸
        [showViewController.view removeFromSuperview];
```

```
        [self.view insertSubview:
                saveViewController.view atIndex:0];
        }
    [UIView commitAnimations];
}
```

UIView 类的几个方法用来修改动画的属性，见表 7-1。

**表 7-1 UIView 类方法及相应更改的属性**

| UIView 类中的方法 | 动画的属性 |
|---|---|
| beginAnimations: context: | 开始位置 |
| setAnimationDuration: | 持续时间 |
| setAnimationCurve: | 动画速度 |
| setAnimationTransition: forView: cache: | 转换类型 |
| commitAnimations | 结束位置 |

我们想营造一种前后翻页的视觉效果，因此设置转换动画时，在❷处由输入视图转化到显示视图时，使用了向上卷起的动画，而在❸处由显示视图转换到输入视图时，使用了向下卷放的动画。为使动画显得更加自然，在❶处设置动画速度时，其返回类型为 UIViewAnimationCurveEaseInOut，使动画在开始和结束的时候速度较慢，中间过程较快。运行程序看看效果，如图 7-12 所示。

图 7-12 翻页动画效果

除了 UIViewAnimationCurveEaseInOut，setAnimationCurve:方法还有以下几种返回类型：

（1）UIViewAnimationCurve:匀速；

（2）UIViewAnimationCurveEaseIn:开始慢，后来快；

（3）UIViewAnimationCurveEaseOut:开始快，后来慢；

（4）UIViewAnimationCurveLiear:线性动画效果，平坦运行。

在这组动画转换效果中，起决定性作用的就是 setAnimationTransition:方法，它除了支持 UIViewAnimationTransitionCurlUp/Down 这组上下翻页的动画外，还支持 UIViewAnimationTransitionFlipFromLeft/Right 这组左右翻转动画。

效果同样很明显，不妨试一下。不过，对于想把程序变得更加美观的我们，这些动画还远远不足。下面我们换一种方式实现更加丰富的动画效果。

**2. 更加丰富的动画效果**

这次要用到 QuartzCore 框架中 CATransition 类。该程序所使用的 Window-based Application 模板中没有引入 QuartzCore. framework，需要手动把它加入 Frameworks 文件夹下。右键选中左侧 Groups & Files 窗格中 Targets 项下的 NoteController，在菜单中选择 Get Info，打开 Info 窗口。点击 General 选项卡下 Linked Libraries 项左下方的加号，在出现的 Frameworks 列表中选择 QuartzCore. framework，点击 Add，如图 7-13 所示。

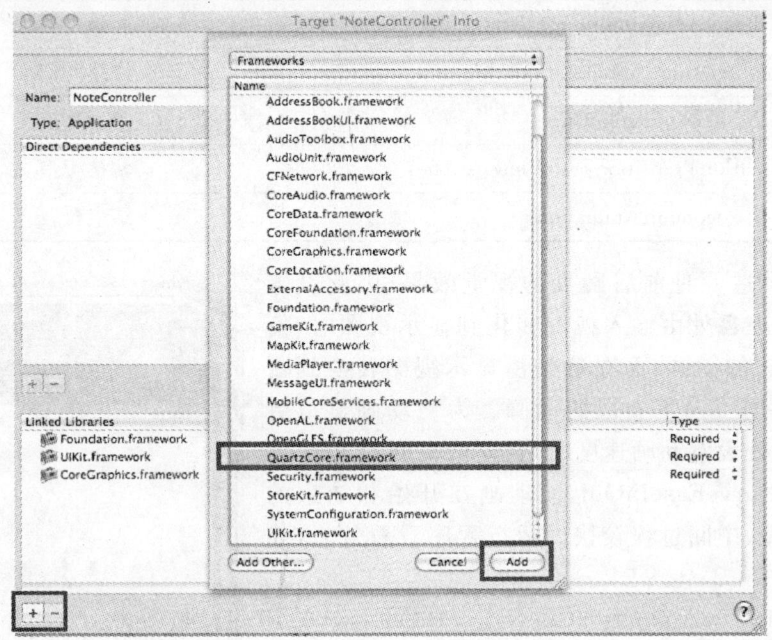


图 7-13 添加 QuartzCore. framework

关闭 Info 窗口，返回 Xcode，你会发现 Frameworks 文件夹中已经增加了 QuartzCore. framework。接下来，在 RootViewController. m 中做以下更改：

**代码 7.10 RootViewController. m 文件**

- (IBAction)switchViews:(id)sender {

~~[UIView beginAnimations:@" Curve"context:nil];~~

~~[UIView setAnimationDuration:1.25];~~

~~[UIView setAnimationCurve:UIViewAnimationCurveEaseInOut];~~

**CATransition * animation = [CATransition animation];**

**[animation setDuration:1.25f];**

```
    [animation setTimingFunction: [CAMediaTimingFunction
    functionWithName:kCAMediaTimingFunctionEaseIn]]; ❶
    if (self.showViewController.view.superview == nil) {
        if (self.showViewController == nil) {
                ShowViewController * showMessage =
                                        [[ShowViewController alloc]
            initWithNibName:@"ShowViewController"
                    bundle:nil];
        self.showViewController = showMessage;
        [showMessage release];
    }
self.showViewController.showTopic.text = saveTopic.text;
self.showViewController.showForm.text = saveForm.text;
self.showViewController.showContent.text=saveContent.text;
[UIView setAnimationTransition:
            UIViewAnimationTransitionCurlUp
                forView:self.view
            cache:YES];
[animation setType:kCATransitionReveal]; ❷
[animation setSubtype: kCATransitionFromBottom]; ❸
[self.view.layer addAnimation:animation forKey:@"Reveal"]; ❹
[saveViewController.view removeFromSuperview];
[self.view insertSubview:showViewController.view atIndex:0];
}
else {
    [UIView setAnimationTransition:
            UIViewAnimationTransitionCurlDown
                forView:self.view
            cache:YES];
    [animation setType:@"suckEffect"]; ❺
    [self.view.layer addAnimation:animation
                    forKey:@"suckEffect"]; ❻
[showViewController.view removeFromSuperview];
[self.view insertSubview:
        saveViewController.view atIndex:0];
}
```

```
}
```

CATransition 类能够遍历默认的动画属性,它也有自己的一些修改动画属性的方法,见表 7-2。

**表 7-2 设置动画属性的 CATransition 类方法**

| CATransition 类方法 | 动画属性 |
|---|---|
| setDuration: | 持续时间 |
| setTimingFunction: | 动画速度 |
| setType: | 动画类型 |
| setSubtype: | 动画子类型 |

❶处的 setTimingFunction:方法与 setAnimationCurve:方法的作用完全相同,都是设置动画的速度,即何时快何时慢,虽然返回类型有所不同,但同样仅支持上面提及的五种动画速度样式。

也许你已经注意到了,两视图间的转换在❷处和❸处设置动画类型的方式是不同的,在由输入视图转换到显示视图时,我们使用了标准的设置动画类型的方式。setType:方法支持以下四种动画类型:

(1) kCATransitionFade:淡入淡出;

(2) kCATransitionMoveIn:新视图滑入,覆盖原视图;

(3) kCATransitionPush:推入;

(4) kCATransitionReveal:原视图滑出,新视图显现(效果如图 7-14 所示)。

图 7-14 转换到显示视图时的动画效果

还可以设置转换动画的方向,即在❸处使用的 setSubtype:方法,称为动画的子类型。它也有四种方式:

(1) kCATransitionFromRight;

(2) kCATransitionFromLeft;

(3) kCATransitionFromTop;

(4) kCATransitionFromBottom。

在❺处由显示视图转换到输入视图时,我们使用了官方文档中未定义的动画类型,iPhone 同样可以支持。这样的方式丰富了可用的动画类型,但未经官方发布,因此并不稳定,随时会被删除或更改。以下是其中的几种:

(1) suckEffect:三角(效果如图 7-15 所示);

(2) rippleEffect:水波抖动;

(3) pageCurl:上翻页(效果同图 7-12);

(4) pageUnCurl ：下翻页；

(5) oglFlip：上下翻转。

最后，❹处 addAnimation：forKey：方法将刚刚定义的动画加到视图的层上。请注意，与 UIView 不同，CATransition 作用于层，而不是视图本身。layer 是视图的属性之一，一个视图上允许很多个层的叠加。比起视图，层可以有更复杂的转换，比如旋转、倾斜、放大、缩小等。

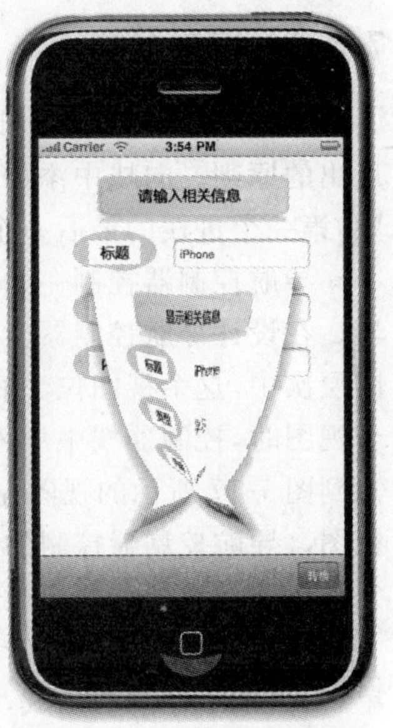


图 7-15　转换到输入视图的动画效果

## 7.3 NavigationController

上一节，我们认识了多视图应用程序的结构，并自定义了一个多视图控制器。在很多 iPhone 应用程序中，屏幕上方都会有一个导航条，如图 7-16 所示。点击右上角的导航按钮，切换到“播放列表”视图，如图 7-17 所示。这就是多视图间转换的另一种重要的实现方式：导航控制器。

NavigationController 是用于构建分层应用程序的主要工具，使程序能够在具有层次关系的视图间不断切换。像音乐播放器这样的多视图应用程序，一部分视图会具有某种层次关系，由 NavigationController 直接管理，另一部分受控于导航控制器的子视图控制器或根视图控制器。

图 7-16　音乐播放器中的导航控制器

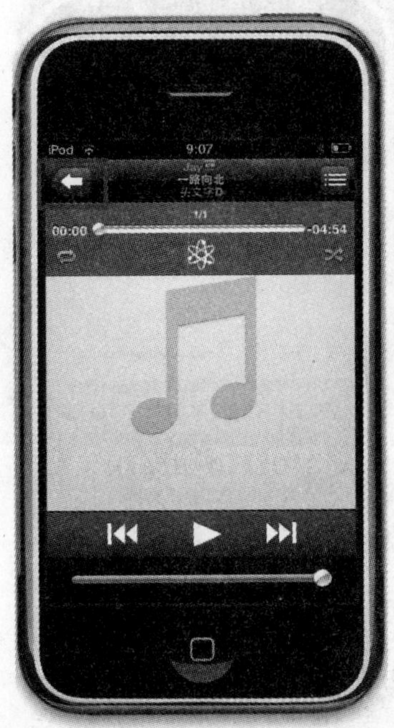


图 7-17　转换到“正在播放”视图

### 7.3.1 控制器栈

UINavigationController 是作为栈来实现的。栈是一种常用的数据结构，采用先进后出的原则。向栈中添加对象称为入栈，从栈中删除对象称为出栈。根据先进后出的原则，第一个进栈的最后一个出栈，最后一个进栈的第一个出栈。

导航控制器控制一个视图控制器栈，在这个栈中存放着应用程序所有的视图控制器。在设计导航控制器时，你需要指定应用程序运行时的第一个视图，在程序的整个视图层次中，这个视图位于底层，而它对应的控制器就是根控制器。当用户选择查看下一个视图时，控制器栈中加入一个新的对象，这些新的对象称为子控制器。相信你已经注意到图 7-17 所示的视图左上方的按钮了，这个按钮用于由子控制器视图切换到父控制器视图。导航控制器控制的每个子视图中，都有相应的返回按钮，返回按钮可以让程序在父视图和子视图之间切换。随着程序的运行，栈中的控制器对象不断进栈出栈。

我们将要构建的这个程序，实现的效果还是数据的传递和视图的切换，程序比较简单，是由两个控制器构成，一个充当根控制器，另一个充当子控制器。

### 7.3.2 构建应用程序 NoteNav

该程序仍然实现 NoteController 程序中的功能，只不过这次用导航控制器的方式实现，最终效果如图 7-18 所示。在输入视图中输入相关信息，点击屏幕右上方的“提交”按钮，进入下级的显示视图，如图 7-19 所示，之前输入的文字在该视图中显示出来。

图 7-18 程序运行时的根视图

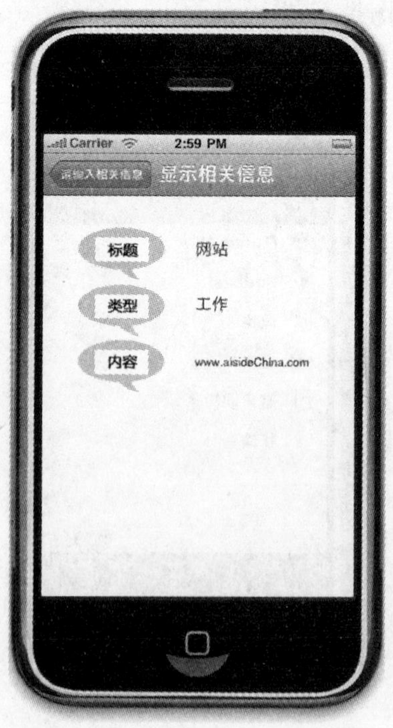


图 7-19 程序运行时的子视图

Xcode 为创建基于导航控制器的应用程序提供了一个很好的模板，该模板已经建立了一个导航控制器，无需我们自己创建。不过，手动创建一个导航控制器更有助于我们理解其工作机制，因此本例中我们将继续使用基于窗口的应用程序模板。

打开 Xcode，新建一个 Window-based Application，将它命名为 NoteNav。与 NoteController 程序结构不同的是，我们将 SaveViewController 作为根控制器，将 ShowViewController 作为子控制器。

首先，创建根控制器，右键点击 Classes 选择 Add→New File，在出现的向导中选择 Cocoa Touch Class 下的 UIViewController subclass。本例中我们选择自动创建相应 Xib 文件，因此应选中 With XIB for user interface 复选框，单击 Next 将此文件命名为 SaveViewController.m，确保选中创建头文件的复选框。此时，Groups & Files 窗格中 Classes 文件夹下新增了三个以 SaveViewController 命名的不同类型的文件。用同样的方式创建子视图控制器 ShowViewController 的三个文件。

接下来，我们为程序添加导航控制器。选择 NoteNavAppDelegate.h 文件，添加如下黑体字所示的代码：

**代码 7.11　NoteNavAppDelegate.h 文件**

```
#import <UIKit/UIKit.h>
@interface NoteNavAppDelegate : NSObject
        <UIApplicationDelegate> {
    UIWindow * window;
    IBOutlet UINavigationController * navController;
}
@property (nonatomic, retain) IBOutlet UIWindow * window;
@property (nonatomic, retain)
                IBOutlet UINavigationController * navController;
@end
```

在 NoteNavAppDelegate.m 文件中，添加如下黑体字所示的代码：

**代码 7.12　NoteNavAppDelegate.m 文件**

```
#import "NoteNavAppDelegate.h"
@implementation NoteNavAppDelegate
@synthesize window;
@synthesize navController;
- (void)applicationDidFinishLaunching:
        (UIApplication *)application {
    [window addSubview:navController.view];
    [window makeKeyAndVisible];
```

```
    }
    - (void)dealloc {
                [navController release];
                [window release];
                [super dealloc];
    }
```

我们定义了一个 UINavigationController 类型的输出口，并将导航控制器的视图添加到应用程序的窗口，作为它的子视图。

双击 Resources 下的 MainWindow. xib 文件，与添加 ViewController 相似，在库中拖入一个 Navigation Controller。我们要将这个导航控制器的实例和刚才声明的 navController 连接起来。用鼠标右键点击 Note Nav App Delegate，拖拽至 Navigation Controller 上，释放鼠标会出现一个输出口 navController，将其选中。

导航控制器必须要指定根控制器，在 MainWindow. xib 窗口中展开 Navigation Controller，可以看到两个子项：Navigation Bar 和 View Controller。选中 View Controller 图标，按下⌘＋4，打开 Identity 窗口，将其 Class 属性更改为 SaveViewController，此时 ViewController 变为 Save View Controller。打开 Attributes 窗口，在 NIB Name 属性下拉列表中选择 SaveViewController，在本例中这一操作是可选的，因为 SaveViewController 的 Xib 文件是自动生成的，已指定其类为 SaveViewController。SaveViewController 有一个子项 Navigation Item，打开 Attributes 窗口，将 Title 属性设为“请输入相关信息”。

保存并关闭 Interface Builder，如果此时运行程序，一个标题为“请输入相关信息”的导航条会出现在界面的上方。这个视图就是根控制器 SaveViewController 的视图。由于还未设计输入视图的界面，因此显示为空。

在 Xcode 中打开 SaveViewController. h 文件，添加如下黑体字所示的代码：

**代码 7.13　SaveViewController. h 文件**

```
#import <UIKit/UIKit.h>
@class ShowViewController;
@interface SaveViewController : UIViewController {
    IBOutlet UITextField * saveTopic;
    IBOutlet UITextField * saveForm;
    IBOutlet UITextField * saveContent;
    ShowViewController * showViewController;
}
@property (nonatomic, retain) IBOutlet UITextField * saveTopic;
@property (nonatomic, retain) IBOutlet UITextField * saveForm;
```

```
@property (nonatomic, retain) IBOutlet UITextField * saveContent;
@property (nonatomic, retain)
        ShowViewController * showViewController;
- (IBAction) textFieldDoneEditing:(id)sender;
- (IBAction) saveMessage;
@end
```

上述文件中，我们创建了显示视图的一个实例 showViewController，并声明了一个方法 saveMessage，在点击导航控制器右侧按钮时被调用。

打开 SaveViewController.xib，界面设计如图 7-20 所示。

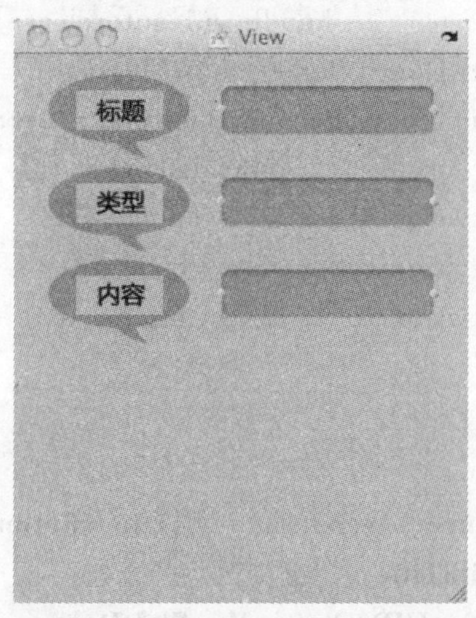


图 7-20 SaveViewController.xib 界面设计

界面设计的具体操作步骤可参见 NoteController 程序中的描述，这里不再重述。也许你已经发现，本例中的背景图片改变了，没有了上方的提示信息。这个提示信息显示在了导航条的标题上。因此，在引入图片时应选择 messageImage.png 文件，在两个视图中重用。

分别打开三个 Text Field 的 Connections 窗口，点击 Did End On Exit 事件后的小圆圈拖拽到 File's Owner，连接 textFieldDoneEditing:方法。此事件在按下键盘右下角 Return 键时触发。

接下来，修改 SaveViewController.m 文件：

**代码 7.14 SaveViewController.m 文件**

```
#import "SaveViewController.h"
#import "ShowViewController.h"
@implementation SaveViewController
```

```
@synthesize saveTopic;
@synthesize saveForm;
@synthesize saveContent;
@synthesize showViewController;
- (IBAction) textFieldDoneEditing:(id)sender {
    [sender resignFirstResponder];
}
- (IBAction) saveMessage {
    ShowViewController * showMessage = [[[ShowViewController alloc]
                        initWithNibName:@"ShowViewController"
                            bundle:nil] autorelease];
      self.showViewController = showMessage;
      [self.navigationController pushViewController:showViewController
                  animated:YES]; ❶
      self.showViewController.showTopic.text = saveTopic.text;
      self.showViewController.showForm.text = saveForm.text;
      self.showViewController.showContent.text = saveContent.text;
}
……
- (void)viewDidLoad {
     UIBarButtonItem * saveButton = [[UIBarButtonItem alloc]
          init WithTitle:@"提交"
               style:UIBarButtonItemStyleDone
                  target:self
                  action:@selector(saveMessage)];        ❷
     self.navigationItem.rightBarButtonItem = saveButton;
     [super viewDidLoad];
}
- (void)dealloc {
     [saveTopic release];
     [saveForm release];
     [saveContent release];
     [showViewController release];
     [super dealloc];
}
@end
```

程序中，在❷处 viewDidLoad 方法创建了导航条上右侧的按钮，使用 initWithTitle: style: target: action:方法定义了按钮的标题、类型、对象及事件。

点击提交按钮时调用 saveMessage 方法，在该方法中，首先初始化了 ShowViewController，并在❶处使用 pushViewController: animated:方法将 ShowViewController 子控制器推入控制器栈，从而进入子视图。最后将 TextField 中输入的信息显示在 Label 中。

打开 ShowViewController.h 文件，添加以下代码：

**代码 7.15 ShowViewController.h 文件**

```
#import <UIKit/UIKit.h>
@interface ShowViewController : UIViewController {
    IBOutlet UILabel *showTopic;
    IBOutlet UILabel *showForm;
    IBOutlet UILabel *showContent;
}
@property (nonatomic, retain) IBOutlet UILabel *showTopic;
@property (nonatomic, retain) IBOutlet UILabel *showForm;
@property (nonatomic, retain) IBOutlet UILabel *showContent;
@end
```

在 Interface Builder 中打开 ShowViewController.xib 文件，设计子视图界面，如图7-21所示，不要忘了为三个 Label 连接输出口。

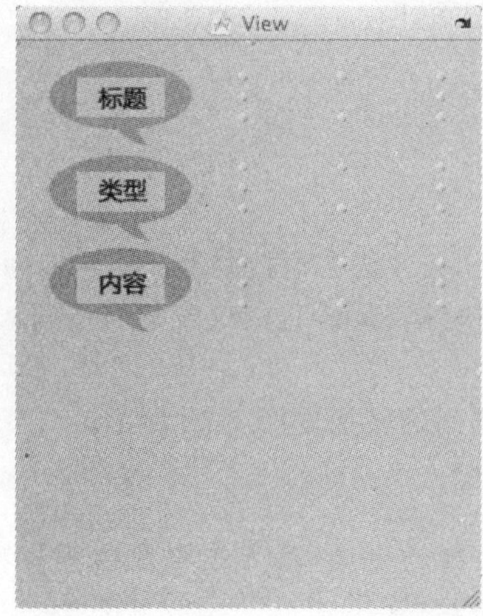


图 7-21 ShowViewController 界面设计

在 ShowViewController.m 中添加如下黑体字所示的代码：

**代码 7.16 ShowViewController.m 文件**

```
#import "ShowViewController.h"
@implementation ShowViewController
@synthesize showTopic;
@synthesize showForm;
@synthesize showContent;
- (id)initWithNibName:(NSString *)nibNameOrNil
              bundle:(NSBundle *)nibBundleOrNil {
    if (self = [super initWithNibName:nibNameOrNil
                               bundle:nibBundleOrNil]) {
        // Custom initialization
        self.title = @"显示相关信息";
    }
    return self;
}
……
- (void)dealloc {
    [showTopic release];
    [showForm release];
    [showContent release];
    [super dealloc];
}
@end
```

initWithNibName：bundle:方法在根控制器 saveMessage:方法中初始化显示视图时被调用，我们仅仅设置了子视图中导航条的标题。

运行程序看看效果，视图间的转换如图 7-22 所示。

图 7-22 视图间转换效果

> **Tips**
>
> 在此应谨记一点：必须先将视图控制推入栈才可以成功传递数据。

# 7.4 Tab Bar Controller

前面我们分别用自定义视图控制器和导航控制器的方式实现了两个视图之间的切换和数据传递，这一节我们再介绍一种新的方式：Tab Bar Controller(标签栏控制器)。

## 7.4.1 Tab Bar Controller 概述

iPhone 上的应用程序很多都使用了标签栏，Tab Bar Controller 作为根控制器来控制各个子控制器，各子控制器之间是相互独立的，实现不同的功能。如果一个应用程序有若干功能独立的模块，Tab Bar Controller 是个非常好的选择。

如图 7-23 所示，在音乐播放器应用程序中，播放列表、歌曲、表演者、专辑等这些相互独立的功能，以标签的形式放置在屏幕的下方，点击标签实现视图间的切换，如图 7-24 所示。

图 7-23 音乐播放器中的标签栏控制器

图 7-24 转换到“播放列表”视图

同音乐播放器一样，在 iPhone 应用程序中，Tab Bar Controller 经常和 Navigation Controller 结合起来使用，使程序更加具有层次性。图 7-25 展示了标签栏控制器和导航控制器结合使用时的层次关系。本节的程序 NoteTab，就是用这种方式构建的。

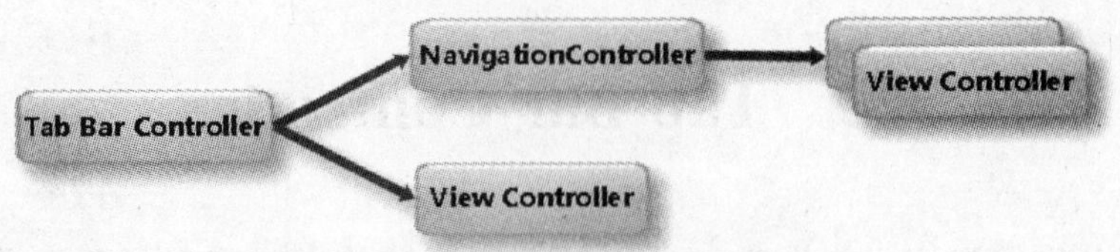


图 7-25 标签栏控制器与导航控制器结合使用

### 7.4.2 构建应用程序 NoteTab

Xcode 没有提供基于标签栏的应用程序模板，所以我们必须自己搭建程序的主框架，实现两个相互独立的功能："记录"功能和"关于"功能。我们把上节 NoteNav 程序实现的功能完全移植到了本程序中的"记录"功能模块，如图 7-26 和图 7-27 所示。"关于"功能提供该软件的介绍信息，如图 7-28 所示。虽然仅显示了一张图片，但不要小看它，几乎所有在 AppStore 上发布的软件的版本信息、版权声明等都会选择用这种方式表达出来。

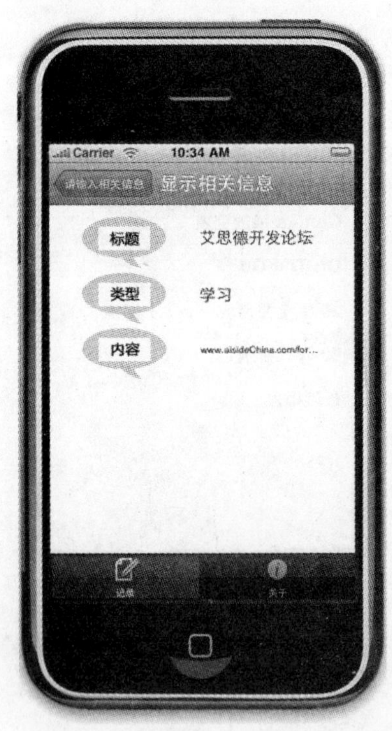


图 7-26 程序运行时的输入视图

图 7-27 程序运行时的显示视图

打开 Xcode，使用 Window-based Application 模板创建一个新项目，将其命名为 NoteTab。通过前面两个程序的学习，也许你已经轻车熟路，知道我们下一步要做些什么了。打开 NoteTabAppDelegate.h 文件，添加如下黑体字所示的代码：

图 7-28 程序运行时的"关于"视图

**代码 7.17 NoteTabAppDelegate.h 文件**

```
#import <UIKit/UIKit.h>
@interface NoteTabAppDelegate :
    NSObject <UIApplicationDelegate> {
    UIWindow *window;
    IBOutlet UITabBarController *tabController;
}
@property (nonatomic, retain) IBOutlet UIWin-
dow *window;
@property (nonatomic, retain)
    IBOutlet UITabBarController *tabController;
@end
```

该文件中声明了一个 UITabBarController 类型的输出口，命名为 tabController。在 NoteTabAppDelegate.m 文件中，添加如下黑体字所示的代码：

**代码 7.18 NoteTabAppDelegate.m 文件**

```
#import "NoteTabAppDelegate.h"
@implementation NoteTabAppDelegate
@synthesize window;
@synthesize tabController;
- (void)applicationDidFinishLaunching:
        (UIApplication *)application {
    [window addSubview:tabController.view];
    [window makeKeyAndVisible];
}
- (void)dealloc {
    [tabController release];
    [window release];
    [super dealloc];
}
@end
```

首先，在程序中把标签栏控制器视图加载到程序的主窗口中，与前面 ViewController 和 NavigationController 作为窗口的子视图采用了相同的方法。

接下来，要在 Interface Builder 中创建一个标签栏的实例。在 Interface Builder 中打

开 MainWindow. xib，从 Library 窗口中找到 Tab Bar Controller，拖入 MainWindow. xib 窗口中。右键选择 Note Tab App Delegate，把它拖到 Tab Bar Controller 上，连接 tabController 输出口。

双击图标打开 Tab Bar Controller，如图 7-29 所示。可以看到，视图下方的标签栏，默认标签有两项。保存并关闭 Interface Builder，返回 Xcode 中。

我们要新建四个视图控制器的子类，分别命名为 NavViewController、SaveViewController、ShowViewController 和 AboutViewController。作为记录功能的根控制器，NavViewController 不需创建 Xib 文件，其余三个视图控制器的子类在创建时应勾选 With XIB for user interface 复选框，以便同时创建相应的 Xib 文件。

在 Classes 文件夹下增加了 11 个文件后，还需要将本程序所需的四个图片加入 Resources 文件夹下：messageImage. png 作为“记录”功能视图中的背景图片；aboutImage. png 在“关于”功能中会用到；record. png 和 about. png 是“记录”和“关于”功能对应的两个标签上的图像。

图 7-29 UITabBarController 的 view 视图

重新打开 MainWindow. xib 文件，在 Tab Bar Controller 窗口中选择第一个视图控制器（注意：单击鼠标第一次选中的是控制器，如果在其图标上再次点击一下，选中的是 Item1 这个标签）。调出 Identity 窗口，更改其 Class 属性为 NavViewController，同时在 Attributes 窗口中你也可以为其指定一个 Xib 文件，这里我们保留这个属性为空。选中第二个视图控制器，更改其 Class 属性为 AboutViewController。

接下来两次单击（间隔 1 秒）第一个标签，调出 Tab Bar Item 的 Attributes 窗口，将它的标题设为“记录”，在 Image 属性的下拉列表中选择 record. png 作为图像，同样地，更改第二个标签的属性，将其标题设为“关于”，图像为 about. png。

调整之后的标签栏控制器视图如图 7-30 所示。

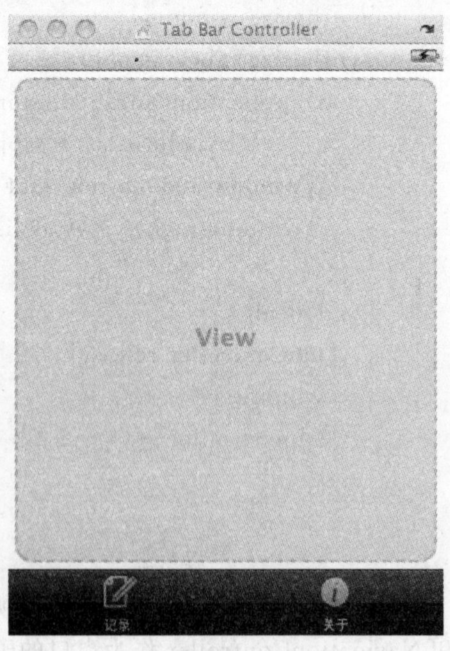


图 7-30 添加标题和图片后的标签栏

这一次我们选择用代码在“记录”功能的根控

制器中添加导航条。在 NavigationController. h 中添加如下黑体字所示的代码：

**代码 7.19 NavViewController. h 文件**

```
#import <UIKit/UIKit.h>
@interface NavViewController : UIViewController {
    UINavigationController * navController;
}
@property (nonatomic, retain)
        UINavigationController * navController;
@end
```

在该文件中声明了一个 NavigationController 的实例，而且不需要输出口。接下来，在 NavViewController. m 文件中添加如下黑体字所示的代码：

**代码 7.20 NavViewController. m 文件**

```
#import "NavViewController.h"
import "SaveViewController.h"
@implementation NavViewController
@synthesize navController;
……
- (void)viewDidLoad {
    //自定义 Navigation
    SaveViewController * saveController =
        [[SaveViewController alloc]
        initWithNibName:@"SaveViewController"
            bundle:nil];
     UINavigationController * nav = [[UINavigationController alloc]
             initWithRootViewController:saveController];
     self.navController = nav;
     [self.view addSubview:self.navController.view];
     [super viewDidLoad];
}
……
- (void)dealloc {
    [navController release];
[super dealloc];
}
@end
```

首先，定义了一个 SaveViewController 的实例 saveController，然后使用 initWithRootViewController：方法将它作为导航控制器的根控制器，最后将这个创建的导航控制器的视图加载到根视图。这样，在“记录”功能中便有了一个导航控制器，并且它的根控制器为 SaveViewController，主视图为输入视图。

SaveViewController 和 ShowViewController 的头文件和执行文件中代码添加与 NoteNav 程序几乎完全相同，你可以将代码复制并粘贴到此程序中，这是提高效率的一个不错的方法。唯一不同的是，主视图中导航控制器的标题不是在 Attributes 窗口中更改的，而是在 SaveViewController. m 文件中，因此应添加如下黑体字所示的语句到 SaveViewController. m 文件：

**代码 7.21 SaveViewController. m 文件**

```
- (id)initWithNibName:(NSString *)nibNameOrNil
              bundle:(NSBundle *)nibBundleOrNil {
    if (self = [super initWithNibName:nibNameOrNil
                              bundle:nibBundleOrNil]) {
        self.title = @"请输入相关信息";
    }
    return self;
}
```

输入视图和显示视图的设计与 NoteNav 程序中操作完全相同，“关于”视图的设计就更加简单了。打开 AboutViewController. xib，去掉视图状态条，调整高度为 416，然后只需要拖入一个 Image View 控件，选择图片 aboutImage. png 即可。由于只在“关于”功能中显示一个图片，因此无需在 AboutViewController 的头文件和执行文件中添加任何代码。

至此，NoteTab 应用程序构建完成，编译并运行一下程序吧！可见，除去与上节程序中重复的操作，实现基于 TabBarViewController 的应用程序并不复杂。

## 7.5 TableViewController

除了前面介绍的视图控制器外，还有一种非常重要的视图控制器：表视图控制器（TableView Controller）。同 UINavigationController 和 UITabBarController 一样，UITableViewController 也是 UIViewController 的子类。表视图控制器将在本书第 8 章详细介绍。

在使用表视图控制器时，我们会大量地使用数据源和数据委托。应用程序将一些工作分配给它的委托，委托的方法在指定时被调用。数据源的工作原理和委托类似，通常情况下,使用数据源获得表单元数,使用委托方法进行有关表的操作。

表视图控制器和导航控制器是密不可分的,并且它们两者也可以和标签栏控制器结合起来使用,如图 7-31 所示。点击列表第三行,转换到"风格"视图,如图 7-32 所示。

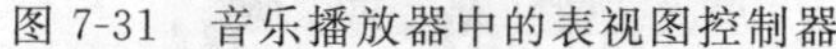

图 7-31 音乐播放器中的表视图控制器

图 7-32 转换到"风格"视图

# 7.6 小结

这一章,我们使用三种不同的方式实现了一个视图切换和数据保存的应用程序,从中学习了视图控制器及其子类导航控制器和标签栏控制器的工作原理。

TableViewController 是一个非常重要的视图控制器,本章中只是做了简单的介绍,接下来我们会使用一章的内容详细学习。将表视图控制器与导航控制器结合起来,是非常好的一种视图转换的方式,值得你耐心地深入研究。

# 第8章　表视图

## 本章内容

- 表视图的概述与分类
- 数据源与委托
- 如何创建表视图
- 表视图单元的各种操作
- 添加搜索功能
- 自定义表视图单元
- 表视图的美化

通过前面的学习我们已经对视图和视图控制器有了一定的了解，本章将重点介绍表视图的有关知识。

表视图是 iPhone 上显示数据的元素。如图 8-1 和图 8-2 所示，iPhone 中电子邮件、设置、时钟、浏览器、书签等软件都使用了表视图，只是它们的外观和显示的内容不尽相同。

图 8-1　表视图(一)

图 8-2　表视图(二)

# 8.1 表视图概述

在 iPhone 应用程序中，表视图应用非常广泛，它最主要的功能是以列表形式向用户显示数据。在 iPhone 中表视图并没有对行的数量进行限制，用户可以通过垂直滚动的方式导航到一个表视图的任意行上，并可以自定义每一行数据的显示方式，需要注意的是 iPhone 中每个表视图只能有一列。

## 8.1.1 表视图简介

表视图是用来显示列表数据的视图对象，从本质上讲，表视图是 UITableView 类的一个实例；从结构组成上讲，表视图是由多个行组成的，其中每一行都是由表视图单元 UITableViewCell 类来实现的。

在表视图中显示数据是通过实现两个协议 UITableViewDelegate 和 UITableViewDataSource 中的方法来完成的。

虽然每个表视图只能有一列，但是图 8-3 所示的表视图却不止"一列"，这是通过创建自定义表视图单元来实现的。

图 8-3 表视图的多列"假象"

## 8.1.2 分组表和索引表

表视图有两种基本样式：分组表和索引表。

图 8-4 所示的 iPhone 中的设置就采用了分组表的样式。分组表可以包含多组，当然也可以只有一个组，每个组都由嵌入在圆角矩形中的多个行共同构成。

图 8-5 所示的索引表视图也称无格式表。如果数据源提供了必要的信息，我们可以在索引表右侧添加索引来导航视图，图中的索引是 26 个英文字母。

图 8-4 分组表

图 8-5 索引表

### 8.1.3 表视图的结构

图 8-6 展示了分组表的结构。其中的每一单独部分称为分区(Section),而每个分区又由多个行构成,此外每个分区都拥有一个头(Header)和尾(Footer)。例如,在分组表中,一个分组就是一个分区。

图 8-7 展示了索引表的结构。在索引表中,每个字母段都是一个分区,如上面图 8-5 联系人列表中的“A”分区、“B”分区等。

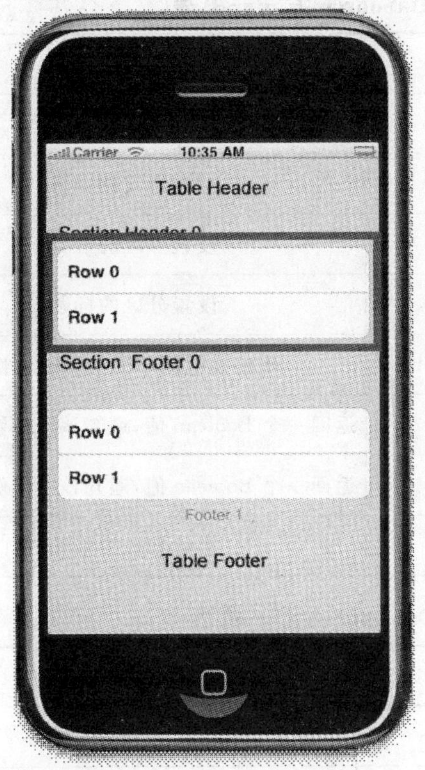


图 8-6 分组表的结构

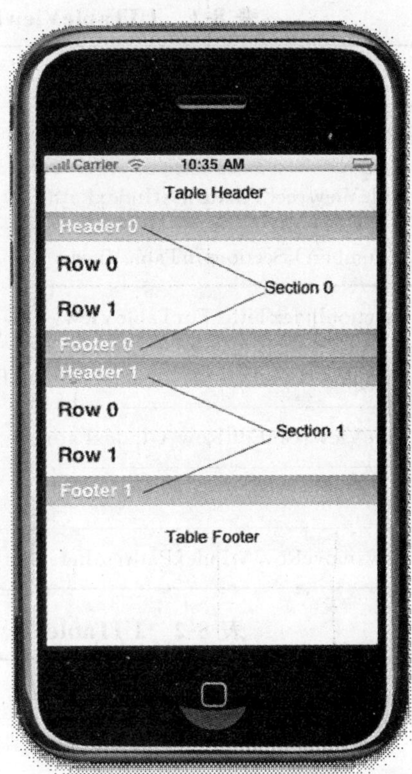


图 8-7 索引表的结构

### 8.1.4 UITableView 和 UITableViewController

一般每一个表视图都有表视图控制器(UITableViewController),它作为 UITableView 的视图控制类,负责管理 UITableView 并控制 UITableView 的生命周期。我们可以把 UITableViewController 作为表视图的数据源和委托,然后利用相对应的协议的方法来定义表视图不同的显示风格,甚至可以自定义表视图单元的每一个元素。

### 8.1.5 数据源和委托

要实现一个表视图必须为其指定一个数据源和一个委托,其中数据源为表视图提供显示的内容,委托用来处理用户对表行的操作。

在程序中我们需要添加 UITableViewDataSource 和 UITableViewDelegate 协议,通过实现这两个协议中声明的方法来构建表视图。

表 8-1 和表 8-2 列举了 UITableViewDataSource 和 UITableViewDelegate 协议的一些方法,在后面的章节中我们会陆续用到。

表 8-1 UITableViewDataSource 方法一览表

| UITableViewDataSource 方法 | 方法描述 |
|---|---|
| tableView:numberOfRowsInSection: | 设置分区里的行数 |
| tableView:cellForRowAtIndexPath: | 绘制特定的一行 |
| numberOfSectionsInTableView: | 设置分区的数目 |
| sectionIndexTitlesForTableView: | 设置分区的标题 |
| tableView:commitEditingStyle:forRowAtIndexPath: | 确定表视图单元的可编辑性 |
| tableView:canEditRowAtIndexPath: | 返回一个 Boolean 值,通知该行能够被编辑 |
| tableView:canMoveRowAtIndexPath: | 返回一个 Boolean 值,通知该行能够被移动 |
| tableView:moveRowAtIndexPath:toIndexPath: | 移动表视图单元 |

表 8-2 UITableViewDelegate 方法一览表

| UITableViewDelegate 方法 | 方法描述 |
|---|---|
| tableView:heightForRowAtIndexPath: | 定义表的行高 |
| tableView:accessoryButtonTappedForRowWithIndexPath: | 添加细节扩展按钮 |
| tableView:willSelectRowAtIndexPath: | 处理将被选中的行 |
| tableView:didSelectRowAtIndexPath: | 处理已被选中的行 |
| tableView:editingStyleForRowAtIndexPath: | 返回行的编辑类型 |

# 8.2 实现一个简单的表

至此,我们对表视图已经有一定的了解了,在这一节里将真正地实现一个简单的表。图 8-8 是这个简单表的最终效果。

下面是完成这个项目的具体步骤：

- 修改 NoteScanViewController. h 文件，声明遵守表视图协议；
- 修改 NoteScanViewController. xib 文件，构建表视图；
- 修改 NoteScanViewController. m 文件，实现协议方法。

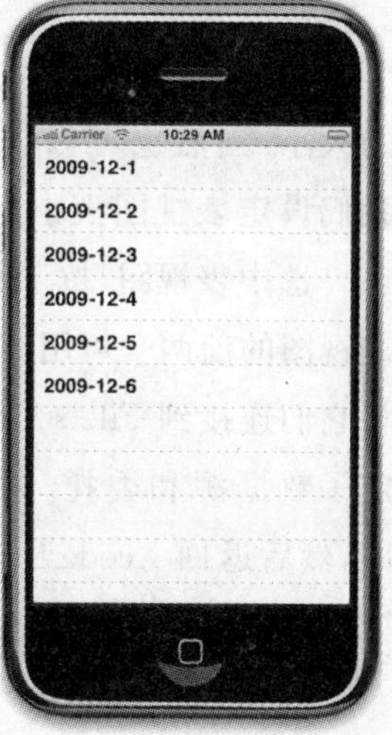


图 8-8　简单表

**1. 修改 NoteScanViewController. h 文件，声明遵守表视图协议**

首先打开 Xcode，并使用 View-based Application 模板创建一个新项目，命名为 NoteScan，打开 Classes 和 Resources 文件夹。

选中 NoteScanViewController. h 文件，添加如下黑体字所示的代码：

**代码 8.1　NoteScanViewController. h 文件**

```
#import <UIKit/UIKit.h>
@interface NoteScanViewController : UIViewController
    <UITableViewDelegate, UITableViewDataSource>{ ❶
        NSArray * notelist; ❷
}
@ property (nonatomic, retain) NSArray *
noteList;
@end
```

在❶处添加了 UITableViewDelegate 和 UITableViewDataSource 协议，实际上是让表视图控制器类使用这两个协议来充当表视图的委托和数据源。

在❷处声明了一个数组，用来存放将要显示在表中的数据。接下来需要对 NoteScanViewController. xib 文件进行操作，构建表视图。

**2. 修改 NoteScanViewController. xib 文件，构建表视图**

双击打开 NoteScanViewController. xib 文件，打开 View 窗口，接下来的操作很简单：只需将 Table View（如图 8-9 所示）从 Library 窗口中拖到 View 窗口里。

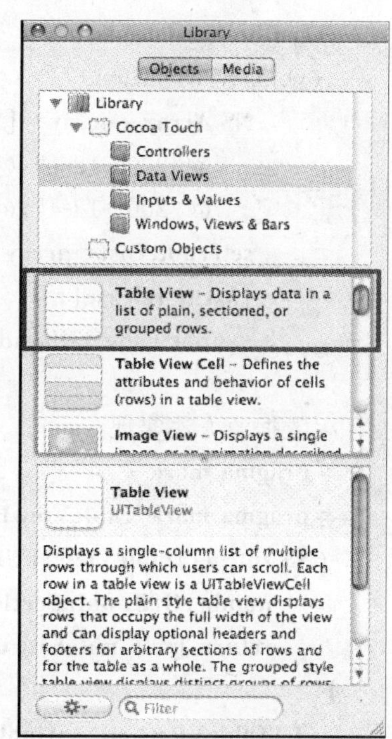


图 8-9　Table View

表视图会自动调整高度和宽度以适应 View 窗口的大小，当然也可以自己定制表视图的大小。在这里我们设定表视图大小为 320 * 460，如图 8-10 所示。

选中表视图，按下⌘+2 键，调出 Connections 窗口，表视图的前两个可用连接是 datasource 和 delegate，分别将它们连接到 File's Owner 图标上，这样就为表视图指定了数据源和委托。完成上面的这些操作，保存并关闭，然后返回 Xcode 中，修改 NoteScanViewController. m 文件，以实现协议方法。

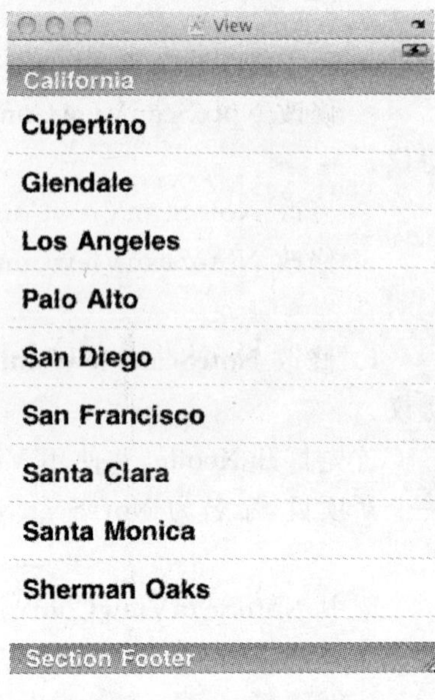


图 8-10 表视图

**3. 修改 NoteScanViewController. m 文件，实现协议方法**

选中 NoteScanViewController. m 文件，添加如下黑体字所示的代码：

**代码 8.2 NoteScanViewController. m 文件**

```
#import "NoteScanViewController.h"
@implementation NoteScanViewController
@synthesize noteList;
- (void)viewDidLoad {
        NS Array * array=[[NSArray alloc] initWithObjects:
            @"2009-12-1",@"2009-12-2",@"2009-12-3",
            @"2009-12-4",@"2009-12-5",@"2009-12-6",nil];    ❶
        self.noteList=array;
        [array release];
        [super viewDidLoad];
}
//数据源方法实现
#pragma mark -
#pragma mark TableViewDataSource Methods
-(NS Integer)tableView:(UITableView  *)tableView
      numberOfRowsInSection:(NSInteger)section{    ❷
      return [self.noteList count];
}
- (UITableViewCell  *)tableView:(UITableView  *)tableView
      cellForRowAtIndexPath:(NSIndexPath  *)indexPath {    ❸
```

```
    static NSString *NoteScanIdentifier=@"NoteScanIdentifier";❹
    UITableViewCell *cell=[tableView
    dequeueReusableCellWithIdentifier: NoteScanIdentifier];❺
    if(ce ll==nil){
        cell=[[[UITableViewCell alloc]initWithStyle:
              U ITableViewCellStyleDefault                  ❻
                reuseIdentifier:NoteScanIdentifier];
     }
     NSUInteger row=[indexPath row];
     cell.textLabel.text=[noteList objectAtIndex:row];  ❼
     return cell;
}
@end
```

在 ViewDidLoad 方法里进行了初始化操作，通过❶处的代码将要在表视图中显示的数据存储在数组中。

接下来，我们重点说明如何在程序中添加 TableView 数据源方法。

❷处的方法用于设定指定的分区中有多少行。默认的分区数量为 1。在这里用它来返回组成文本列表分区的行数。

❸处是程序绘制表行的方法。第一个参数 tableView 是 UITabelView 类型的实例对象，第二个参数 IndexPath 是 NSIndexPath 实例变量，用来确定表行的位置。

❹处声明了一个静态的字符串实例，作为表示表视图单元的键。在这个表里，我们只使用一种表视图单元，所以只定义了一种标识符。

❺处是使用 NoteScanIdentifer 类型的可重用单元。

❻处的代码检查一下单元是否为空(nil)。如果是，就要使用前面所提到的标识符字符串来创建一个新的表视图单元。

到现在为止，我们已经构建了表视图框架和一个可以重用的表视图单元，接下来需通过❼处的代码把要显示的数据放置到表视图单元里。

编译并运行程序，就会得到图 8-8 所示的效果了。

## 8.3 表的简单操作

接下来我们将介绍对表格的一些基本操作，并结合上一章导航控制器的内容，构建一个项目，来实现行的移动、删除操作。

## 8.3.1 构建项目框架

本项目将要构建一个包含表视图和一个导航控制器的框架，如图 8-11 所示。

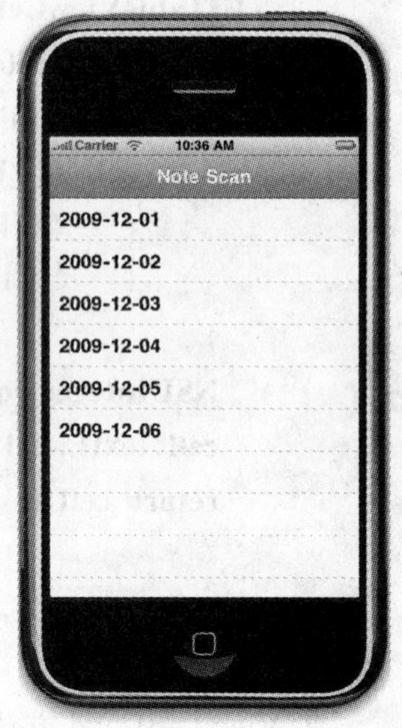


图 8-11 项目框架

接下来，我们将通过以下步骤完成此项目：

- 创建项目；
- 修改 RootViewController. h 文件；
- 修改 RootViewController. m 文件。

### 1. 创建项目

打开 Xcode，使用 Navigation-based Application 模板创建一个项目，并把它命名为 NoteNav，如图 8-12 所示。

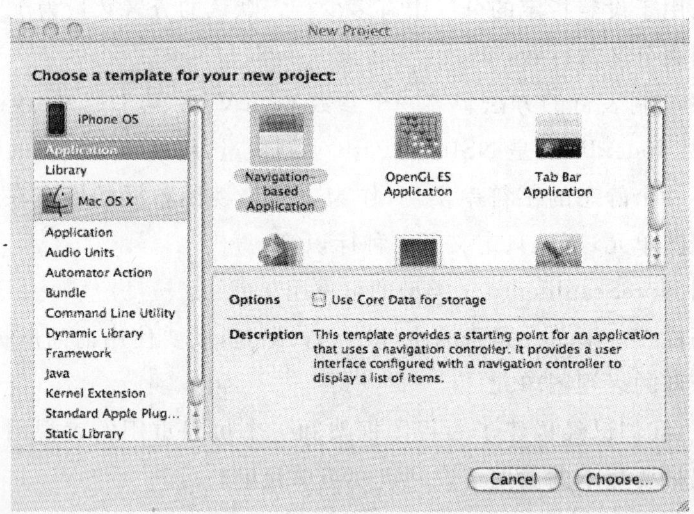


图 8-12 新建项目

这与之前创建项目时所选择的模式有所不同，这一节我们的主要目的在于表的操作，所以直接使用自带导航控制器和表视图的模板来构建项目，当然也可以利用前面所学的知识创建一个带有导航控制器和表视图的框架，这对于前面知识的巩固很有帮助。

打开 Classes 和 Resources 文件夹，共有四个文件。RootViewController.h 和 RootViewController.m 两个文件是模板的根视图控制器文件，也是本项目进行代码编辑的“主战场”。

编译并运行一下，看看这个新模板产生的视图效果，以便对它有更加直观的了解，如图 8-13 所示。

图 8-13 模板效果

**2. 修改 RootViewController.h 文件**

这里要实现的效果和上一节 NoteScan 项目是一样的，只是这里的表视图要在根视图里显示，所以接下来的变量和方法声明都会在根控制器文件内添加。

选中 RootViewController.h 文件，添加如下黑体字所示的代码：

**代码 8.3 RootViewController.h 文件**

```
@interface RootViewController : UITableViewController
    <UITableViewDataSource, UITableViewDelegate>{
    NSMutableArray *noteList; ❶
}
@property (nonatomic, retain)NSMutableArray *noteList;
@end
```

在❶处声明一个可变数组 MutableArray，而不是简单的数组 Array，这是为了后面进行移动行和删除行的操作而特意准备的。

**3. 修改 RootViewController.m 文件**

选中 RootViewController.m 文件，添加如下黑体字所示的代码：

**代码 8.4 RootViewController.m 文件**

```
#import "RootViewController.h"
@implementation RootViewController
@synthesize noteList;
- (void)viewDidLoad {
    self.title=@"Note Scan";❶
    NSMutableArray *array=
    [[NSMutableArray alloc]initWithObjects:
            @"2008-12-01",@"2008-12-02",@"2008-12-03",
            @"2008-12-04",@"2008-12-05",@"2008-12-06",nil];
    self.noteList=array;
```

```
        [array release];
        [super viewDidLoad];
}
- (void)didReceiveMemoryWarning {
        [super didReceiveMemoryWarning];
}
- (void)viewDidUnload {
}
//数据源方法实现
#pragma mark Tableview DataSource Methods
- (NSInteger)numberOfSectionsInTableView:(UITableView *)tableView{
      return 1;
}
- (NSInteger)tableView:(UITableView *)
tableView numberOfRowsInSection:(NSInteger)section {
      return [self.noteList count];
}
-(UITableViewCell *)tableView:(UITableView *)
 tableView cellForRowAtIndexPath:(NSIndexPath *)indexPath {
     static NSString *CellIdentifier = @"Cell";
     UITableViewCell *cell =
[tableView dequeueReusableCellWithIdentifier:CellIdentifier];
      if (cell == nil) {
            cell = [[[UITableViewCell alloc]
            initWithStyle:UITableViewCellStyleDefault
            reuseIdentifier:CellIdentifier] autorelease];
        }
        NSUInteger row=[indexPath row];
        cell.textLabel.text=[noteList objectAtIndex:row];
        return cell;
}
- (void)dealloc {
        [noteList release];
        [super dealloc];
}
@end
```

❶处添加的代码 self.title=@"Note Scan"，用来设置导航控制器的标题。

到现在为止我们已经把前面 NoteScan 项目里面的表格“移植”到新项目里来了，现在编译并运行程序，就得到图 8-11 所示的效果。

### 8.3.2 移动表视图单元

我们将在前面的基础上添加行的移动操作，这种操作对于应用程序来说是很有帮助的，如图 8-14 所示。

图 8-14 可移动表视图

这里我们需要完成以下工作：

- 修改 RootViewController.h 文件，添加移动方法；
- 修改 RootViewController.m 文件，实现移动操作；
- 编译并运行。

**1. 修改 RootViewController.h 文件，添加移动方法**

选中 RootViewController.h 文件，添加如下黑体字所示的代码：

**代码 8.5 RootViewController.h 文件**

```
@interface RootViewController : UITableViewController
<UITableViewDataSource, UITableViewDelegate>{
    NSMutableArray *noteList;
}
@property (nonatomic, retain)NSMutableArray *noteList;
- (IBAction)noteMove;❶
@end
```

在❶处声明了一个移动表视图单元的操作方法 noteMove。

**2. 修改 RootViewController.m 文件，实现移动操作**

选中 RootViewController.m 文件，添加如下黑体字所示的代码：

**代码 8.6 RootViewController.m 文件**

```
#import "RootViewController.h"
@implementation RootViewController
@synthesize noteList;
```

```
- (IB Action)noteMove{
    [self.tableView setEditing:
          !self.tableView.editing animated:YES];        ❶
}
- (void)viewDidLoad {
    self.title=@"Note Scan";
    NSMutableArray *array=[[NSMutableArray alloc]initWithObjects:
        @"2009-12-01",@"2009-12-02",@"2009-12-03",
        @"2009-12-04",@"2009-12-05",@"2009-12-06",nil];
    self.noteList=array;
    [array release];
    //添加移动按钮
    UIBarButtonItem *moveButton=[[UIBarButtonItem alloc]
        initWithTitle:@"Move"
        style:UIBarButtonItemStyleBordered
        target:self                                     ❷
        action:@selector(noteMove)];
    self.navigationItem.rightBarButtonItem=moveButton;
    [moveButton release];
    [super viewDidLoad];
}
- (void)didReceiveMemoryWarning
{
    [super didReceiveMemoryWarning];
}
- (void)viewDidUnload {
}
- (void)dealloc {
    [noteList release];
    [super dealloc];
}
//数据源方法
#pragma mark -
#pragma mark TableviewDataSource methods
- (NSInteger)numberOfSectionsInTableView:(UITableView *)tableView{
    return 1;
}
```

```
- (NSInteger)tableView:(UITableView *)
      tableView numberOfRowsInSection:(NSInteger)section {
      return [ self.noteList count];
}
- (UITableViewCell *)tableView:(UITableView *)
      tableView cellForRowAtIndexPath:(NSIndexPath *)indexPath {
      static NSString *CellIdentifier = @"Cell";
      UITableViewCell *cell =
            [tableView dequeueReusableCellWithIdentifier:CellIdentifier];
      if (cell == nil) {
                        cell = [[[UITableViewCell alloc]
                        initWithStyle:UITableViewCellStyleDefault
                  reuseIdentifier:CellIdentifier]
                  autorelease];
            cell.showsReorderControl=YES; ❸
      }
      NSUInteger row=[indexPath row];
      cell.textLabel.text=[noteList objectAtIndex:row];
      UIImage *image=[UIImage imageNamed:@"aiside.png"];
      cell.imageView.image=image;
      return cell;
}
//行的移动方法
- (UITableViewCellEditingStyle)tableView:
(UITableView *)tableView
editingStyleForRowAtIndexPath:
(NSIndexPath *)indexPath{                                       ❹
      return UITableViewCellEditingStyleNone;
}
- (BOOL)tableView:(UITableView *)tableView
      canMoveRowAtIndexPath:(NSIndexPath *)indexPath {
      return YES;                                               ❺
}
```

```
- (void)tableView:(UITableView *)tableView
      moveRowAtIndexPath:(NSIndexPath *)fromIndexPath
            toIndexPath:(NSIndexPath *)toIndexPath {
    NSUInteger fromRow=[fromIndexPath row];
    NSUInteger toRow =[toIndexPath row];
    id object=[[noteList objectAtIndex:fromRow] retain];        ❻
    [noteList removeObjectAtIndex:fromRow];
    [noteList insertObject:object atIndex:toRow];
    [object release];
}
@end
```

❶处添加的方法用来控制用户是否能够对表进行操作。

❷处代码为导航栏添加了一个 BarButton,其中 initWithTitle:@"Move"用来确定按钮的名称;style: UIBarButtonItemStyleBordered 确定按钮的类型;target: self 传递 self 作为目标即指向自己;action:@selector(noteMove)]用来指定触发的方法。

**Tips**

可用下面方法设置导航按钮在不同状态的名称:

```
[self. tableView setEditing:! self. tableView. editing animated:YES];
   if(self. tableView. editing)
      [self. navigationItem. rightBarButtonItem setTitle:@"Done"];
   else
      [self. navigationItem. rightBarButtonItem setTitle:@"Move"];
```

❸处的代码为行指定了标准扩展图标(图 8-15 中每行右部的图标),只有表处于编辑模式时才会显示。通过设置表视图单元的这个属性,才能启动移动控件,进入可以移动的状态。

❹处添加的方法是用来确定编辑类型。这里的返回类型是 UITableViewCellEditingStyleNone,指明为移动类型;此外还有另外两种类型: UITableViewCellEditingStyleDelete 和 UITableViewCellEditingStyleInsert,分别表示删除和插入类型。

❺处添加的方法主要是用来指定该行能够进行移动操作。

❻处添加了 tableView:moveRowAtIndexPath:fromIndexPath 方法,我们通过它来实现真正的移动。两个 NSIndexPath 实例用来存储移动行的原位置和新位置;同时声明了一个 id 类型实例对象,它用来存储将要被移动对象;最后需要将该对象从原位置上移

除，并插入到新位置上。

**3. 编译并运行**

到这里行的移动操作就完成了，编译并运行程序。当点击 Move 按钮时行的右侧会出现重新排序控件，如图 8-15 所示。可以选中任意行拖拽到表中任何位置，再次单击 Move 按钮就可以对表进行重新排序了。

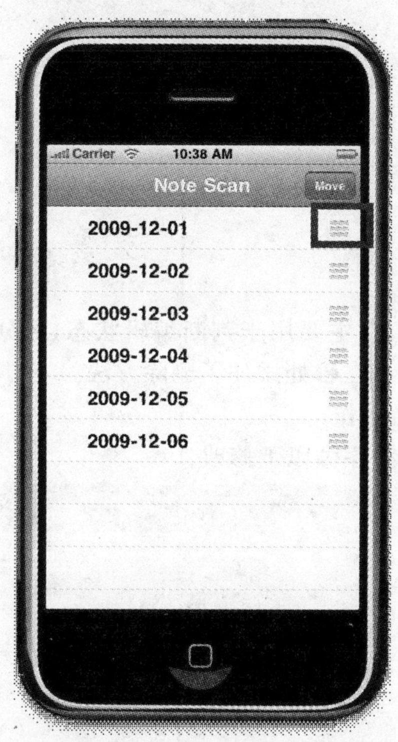


图 8-15 移动效果

图 8-16 可删除表视图(一)

### 8.3.3 删除表视图单元

接下来学习行的删除操作。这里将在原有代码基础上进行修改以实现行的删除功能，最终效果如图 8-16 所示。

要实现此功能大致需要以下几步：

- 修改 RootViewController.h 文件，声明删除方法；
- 修改 RootViewController.m 文件，实现删除方法；
- 编译并运行。

**1. 修改 RootViewController.h 文件，声明删除方法**

选中 RootViewController.h 文件，添加如下黑体字所示的代码：

**代码 8.7 RootViewController.h 文件**

```
@interface RootViewController : UITableViewController
    <UITableViewDelegate, UITableViewDataSource>{
        NSMutableArray *noteList;
        BOOL editState; ❶
}
@property (nonatomic, retain)NSMutableArray *noteList;
@property BOOL editState;
- (IBAction) noteMove;
- (IBAction) noteDelete:(id)sender; ❷
@end
```

在❶处声明了一个 BOOL 类型变量 editState，它主要用于判断编辑状态，区分同一个导航条上面的不同状态，然后决定采用何种操作。在❷处添加了删除方法。

**2. 修改 RootViewController.m 文件，实现移动操作**

接下来，选中 RootViewController.m 文件，对程序做以下修改：

**代码 8.8 RootViewController.m 文件**

```
#import "RootViewController.h"
@implementation RootViewController
@synthesize noteList;
@synthesize editState;
#pragma mark -
#pragma Table Methods
//添加行的移动与删除方法
- (IB Action)noteMove{
    editState=YES;
    [self.tableView setEditing:
    !self.tableView.editing animated:YES];
}                                                    ❶
- (IB Action)noteDelete:(id)sender{
    editState=NO;
    [self.tableView setEditing:
    !self.tableView.editing animated:YES];
}
- (void)viewDidLoad {
    self.title=@"Note";
    NSMutableArray *array=[[NSMutableArray alloc]initWithObjects:
            @"2009-12-01",@"2009-12-02",@"2009-12-03",
```

```
            @"200912-04",@"2009-12-05",@"2009-12-06",nil];
    self.noteList=array;
            [array release];
            UIBarButtonItem * moveButton=[[UIBarButtonItem alloc]
                                initWithTitle:@"Move"
                                style:UIBarButtonItemStyleBordered
                                target:self
                                action:@selector(noteMove)];
    self.navigationItem.rightBarButtonItem=moveButton;
    [moveButton release];
    UIBarButtonItem * deleteButton=[[UIBarButtonItem alloc]
                            initWithTitle:@"Delete"
                            style:UIBarButtonItemStyleBordered
                            target:self
                            action:@selector(noteDelete:)];        ❷
    self.navigationItem.leftBarButtonItem=deleteButton;
    [deleteButton release];
    editState=NO;
    [super viewDidLoad];
}
//数据源方法的实现
#pragma mark-
#pragma mark TableviewDataSource methods
- (NSInteger)numberOfSectionsInTableView:(UITableView *)tableView{
    return 1;
}
- (NSInteger)tableView:(UITableView *)
    tableView numberOfRowsInSection:(NSInteger)section {
    return [ self.noteList count];
}
- (UITableViewCell *)tableView:(UITableView *)
    tableView cellForRowAtIndexPath:(NSIndexPath *)indexPath {
    static NSString * CellIdentifier = @"Cell";
    UITableViewCell * cell =
[tableView dequeueReusableCellWithIdentifier:CellIdentifier];
    if (cell == nil) {
                    cell = [[[UITableViewCell alloc]
                    initWithStyle:UITableViewCellStyleDefault
                reuseIdentifier:CellIdentifier]
                autorelease];
```

```
            cell.showsReorderControl=YES;
        }
        NSUInteger row=[indexPath row];
        cell.textLabel.text=[noteList objectAtIndex:row];
        return cell;
}
- (void)didReceiveMemoryWarning {
      [super didReceiveMemoryWarning];
}
- (void)viewDidUnload {
}
- (void)dealloc {
      [noteList release];
      [super dealloc];
}
//委托方法的实现
#pragma mark -
#pragma mark Table Delete Methods
- (UITableViewCellEditingStyle)tableView:(UITableView *)tableView
      editingStyleForRowAtIndexPath:(NSIndexPath *)indexPath{
        if(!editState)
            return UITableViewCellEditingStyleDelete;
            else                                              ❸
            return UITableViewCellEditingStyleNone;
        }
- (BOOL)tableView:(UITableView *)tableView
canMoveRowAtIndexPath:(NSIndexPath *)indexPath
{
        if(editState)
            return YES;
        else                           ❹
            return NO;
}
- (void)tableView:(UITableView *)tableView
      moveRowAtIndexPath:(NSIndexPath *)fromIndexPath
      toIndexPath:(NSIndexPath *)toIndexPath {
      NSUInteger fromRow=[fromIndexPath row];
      NSUInteger toRow =[toIndexPath row];
      id object=[[noteList objectAtIndex:fromRow]retain];
      [noteList removeObjectAtIndex:fromRow];
```

```
        [noteList insertObject:object atIndex:toRow];
        [object release];
}
//删除方法的实现
- (void)tableView:(UITableView *)tableView
        commitEditingStyle:(UITableViewCellEditingStyle)
        editingStyle forRowAtIndexPath:
                (NSIndexPath *)indexPath {
                NSUInteger row=[indexPath row];
                [self.noteList removeObjectAtIndex:row];
                [tableView deleteRowsAtIndexPaths:
                [NSArray arrayWithObject:indexPath]
                withRowAnimation:UITableViewRowAnimationFade];          ❺
}
@end
```

图 8-17 可删除表视图(二)

在❶处设置移动方法为真,删除方法为假。

在❷处添加了删除按钮并设置它的编辑默认状态为假。

在❸处加入了判断语句,这里我们可以这样理解:如果为! editState,则进入默认的删除状态,否则进入编辑等待状态。

在❹处添加的代码用于判断是否执行移动操作,如果返回为 YES,则执行移动操作。

在❺处添加的是真正实现删除操作的方法。首先获取当前行,然后在数组中执行删除,最后是删除的动画效果。当然还有其他效果,你可以查阅帮助文档,分别试验一下效果。

**3. 编译并运行**

到这里删除操作就完成了,编译并运行程序,就会得到如图 8-16 所示的效果。当单击 Delete 按钮时进入可删除的状态,单击红色的圆形按钮,就可以删除一行了,如图 8-17 所示。

# 8.4 行的选择处理

当我们选中一行时,表视图需要确定是否选择了该行,通常需要借助委托来确定这一操作。具体来讲有两种办法可以实现,下面我们就借用在本章 8.2 节创建的简单表,来进行行的选择处理。

**1. 选中后调用方法**

在表视图单元被选中之后，需要调用委托的行选择方法。为了表明我们确实选中了一行，这里将在单击选中行后弹出一个警告，打开项目 NoteScan，然后在 NoteScanViewController.m 文件尾部的@end 之前添加如下代码：

```
- (void)tableView:(UITableView *)tableView
    didSelectRowAtIndexPath:(NSIndexPath *)indexPath{
    NSUInteger row=[indexPath row];
    NSString *rowValue=[tableData objectAtIndex:row];
    NSString *message=[[NSString alloc]
            initWithFormat:@"你选中了 %@", rowValue];
    UIAlertView *alert=[[UIAlertView alloc]
            initWithTitle:@"选中行!"
            message:message delegate:nil
            cancelButtonTitle:@"确定"
            otherButtonTitles:nil];
    [alert show];
    [message release];
    [alert release];
}
```

编译并运行程序，如图 8-18 所示，单击某一行时会弹出一个警告。

图 8-18 选中效果图

**2. 阻止行被选中的方法**

指定某一行不能被选中。接着上面的代码继续在@end之前添加如下代码：

```
- (NSIndexPath *)tableView:(UITableView *)tableView
    willSelectRowAtIndexPath:(NSIndexPath *)indexPath {
    NSUInteger row=[indexPath row];
    if(row==3)
        return nil;
     return indexPath;
}
```

通过这个方法，可以获取 indexPath 所传递的值，确定选中了哪一行。如果是选中了第四行，将它的索引设为空(nil)，那么它的返回值为 nil；如果选中其他行，将返回 indexPath 值，表示可以继续选择其他行。编译并运行，会发现无法选中第四行，但是可以选择其他行。

# 8.5 公开

所谓公开(disclosure)，就是指表视图单元右侧的细节展示按钮 或者扩展指示器 。使用公开可以使单元格进入到下一级视图。

细节展示按钮(UITableViewCellAccessoryDetailDisclosureButton)，其实是一个按钮，它能够响应用户操作，只有点击这个按钮才能够进入下一级视图。如图 8-19 所示的常见的 Wi-Fi 设置视图，就是采用这种按钮。

图 8-19 细节展示按钮

扩展指示器(UITableViewCellAccessoryDisclosureIndicator)也能够响应用户操作,与细节展示按钮不同的是,只要点击了某一行就可以进入下一级视图。如图 8-20 所示的 Settings 设置视图就采用了这种符号,点击后进入下一级子菜单选项。

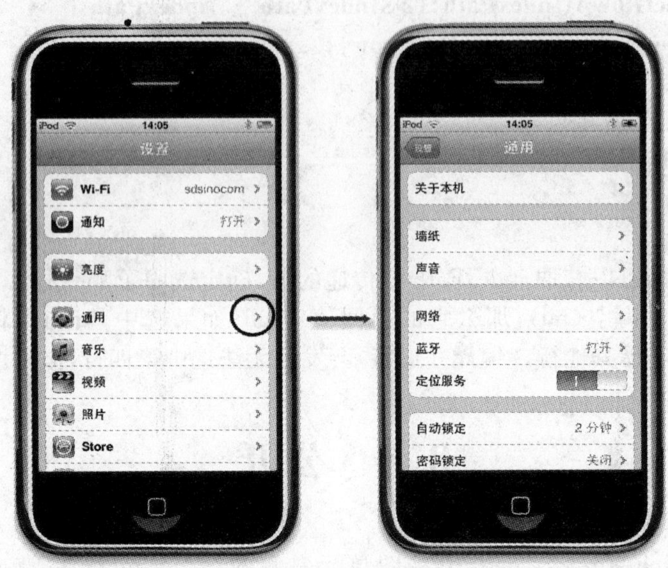


图 8-20 扩展指示器

这一节我们就主要学习一下细节展示按钮的使用方法。本节的项目将在 8.3 节创建的框架基础之上进行修改,最终效果如图 8-21 所示。

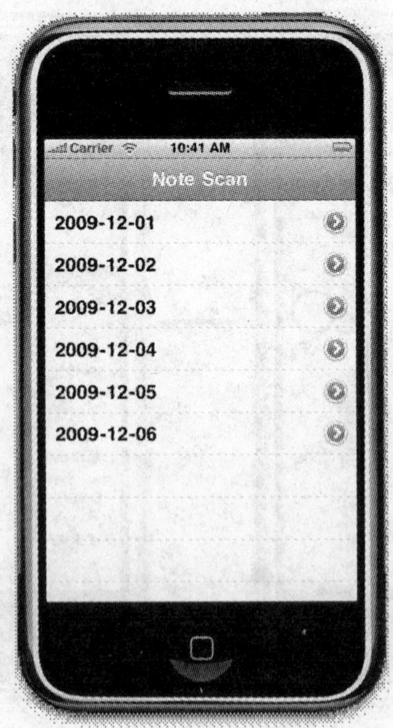


图 8-21 使用公开

本程序的步骤如下：

- 添加并编辑下级详细视图；
- 修改 RootViewController 文件，实现公开的使用；
- 编译并运行。

**1. 添加并编辑下级详细视图**

首先打开 NoteNav 项目，创建一个显示详细信息的视图，选择 Classes 文件夹，然后按下⌘+N，打开新建文件向导，在左侧的窗格里面选择 Cocoa Touches Classes，在右边的窗格里选择 UIViewController subclass，并把它命名为 NoteDetailController.m，保证选中头文件。在详细视图里面，添加一个标签，用它来显示我们选择的行。

选中 NoteDetailController.h 文件，添加如下黑体字所示的代码：

**代码 8.9 NoteDetailController.h 文件**

```
#import <UIKit/UIKit.h>
@interface NoteDetailController : UIViewController {
    IBOutlet UILabel *noteLabel; ❶
    NSString *message; ❷
}
@property (nonatomic, retain) UILabel *noteLabel;
@property (nonatomic, retain) NSString *message;
@end
```

❶处为标签声明了一个输出口，同时在❷处声明了一个字符串类型实例变量。

接下来看一看 NoteDetailController.m 文件的修改，代码如下：

**代码 8.10 NoteDetailController.m 文件**

```
#import "NoteDetailController.h"
@implementation NoteDetailController
@synthesize noteLabel;
@synthesize message;
- (void)viewWillAppear:(BOOL)animated
{
    noteLabel.text=message;                          ❶
    [super view WillAppear:animated];
}
- (void)didReceiveMemoryWarning {
    [super didReceiveMemoryWarning];
}
```

```
- (void)viewDidUnload {
}
- (void)dealloc {
    [noteLabel release];
    [message release];
    [super dealloc];
}
@end
```

在❶处使用 viewWillAppear 方法进行数据更新。因为 viewDidLoad 方法只在第一次加载视图时调用，而在这里要多次调用 NoteDetailController 视图，所以需要使用 viewWillAppear 方法。

接下来在 Resources 文件夹上单击右键，创建新的 Xib 文件，并把它命名为 NoteDetailController. xib。双击打开 NoteDetailController. xib 文件，单击 File's Owner，然后调出 Identity 窗口，把 Class 属性改为 NoteDetailController，我们还要重新连接 File's Owner 到 View 视图，因为刚才控制器类的更改打断了这个连接。

从 Library 中拖出一个 Label 到视图中，并调整它的大小，在 Identity 窗口中将对齐方式设为居中。最后一步就是连接 File's Owner 到 Label，选择 noteLabel 输出口，保存然后关闭文件。

**2. 修改 RootViewController 文件，实现公开的使用**

选中 RootViewController. h 文件，添加如下黑体字所示的代码：

**代码 8.11　RootViewController. h 文件**

```
@class NoteDetailController;
@interface RootViewController : UITableViewController
<UITableViewDataSource, UITableViewDelegate>{
    NSMutableArray * noteList;
    NoteDetailController * subController; ❶
}
@property (nonatomic, retain) NSMutableArray * noteList;
@end
```

在此文件中只是在❶处声明了一个 NoteDetailController 实例对象。接下来，选中 RootViewController. m 文件，添加如下黑体字所示的代码：

**代码 8.12　RootViewController. m 文件**

```
#import "RootViewController.h"
#import "NoteNavAppDelegate.h"
```

```
#import "NoteDetailController.h"
@implementation RootViewController
@synthesize noteList;

- (void)viewDidLoad {
    self.title=@"Note Scan";
    NSMutableArray *array=[[NSMutableArray alloc]initWithObjects:
        @"2009-12-01",@"2009-12-02",@"2009-12-03",
        @"2009-12-04",@"2009-12-05",@"2009-12-06",nil];
     self.noteList=array;
     [array release];
     [super viewDidLoad];
}
- (void)didReceiveMemoryWarning {
    [super didReceiveMemoryWarning];
}
- (void)viewDidUnload {
}

//数据源方法的实现
#pragma mark -
#pragma mark TableDataSource Methods
-(NSInteger)numberOfSectionsInTableView:
        (UITableView *)tableView
{
     return 1;
}
- (NSInteger)tableView:(UITableView *)tableView
    numberOfRowsInSection:(NSInteger)section {
        return [self.noteList count];
}
- (UITableViewCell *)tableView:(UITableView *)tableView
    cellForRowAtIndexPath:(NSIndexPath *)indexPath {
    static NSString *CellIdentifier = @"Cell";
```

```
    UITableViewCell *cell = [tableView
    dequeueReusableCellWithIdentifier:CellIdentifier];
    if (cell == nil){
        cell = [[[UITableViewCell alloc] initWithStyle:
                        UITableViewCellStyleDefault
                        reuseIdentifier:CellIdentifier] autorelease];
    }
    NSUInteger row=[indexPath row];
    NSString *rowString=[noteList objectAtIndex:row];
    cell.textLabel.text=rowString;
    [rowString release];
    [cell setAccessoryType:
        UITableViewCellAccessoryDetailDisclosureButton];❶
    return cell;
}

//委托方法的实现
#pragma mark -
#pragma mark Table Delegate Methods
- (void)tableView:(UITableView *)tableView
    didSelectRowAtIndexPath:(NSIndexPath *)indexPath{
    UIAlertView *alert=[[UIAlertView alloc]
        initWithTitle:@"查看详细信息"
        message:@"如果要查看详细信息请点击蓝色按钮"
        delegate:nil                                      ❷
        cancelButtonTitle:@"确定"
        otherButtonTitles:nil];
    [alert show];
    [alert release];
}
//点击公开按钮后触发的操作
```

```
- (void)tableView:(UITableView *)tableView
    accessoryButtonTappedForRowWithIndexPath:
    (NSIndexPath *)indexPath{
    if(subController==nil){
        subController =[[NoteDetailController alloc]
        initWithNibName:@"NoteDetailController"
        bundle:nil];
     }
     subController.title=@"详细信息";
     NSUInteger row =[indexPath row];
     NSString *selectedDate=[noteList objectAtIndex:row];          ❸
     NSSt ring *detailMessage=[[NSString alloc]initWithFormat:
          @"您选择的日期是 %@.",selectedDate];
     subController.message=detailMessage;
     subController.title=selectedDate;
     [detailMessage release];
     Note NavAppDelegate * delegate=
          [[UIApplication sharedApplication]delegate];
          [delegate.navigationController
          pushViewController:subController animated:YES];
}
- (void)dealloc {
    [subController release];
    [noteList release];
    [super dealloc];
}
@end
```

❶处添加的代码，用于为每个行后面添加一个细节展示按钮。

❷处的方法在用户选中某一行时调用，它告诉用户要单击细节展示按钮而不是点击行，并且实现了一个警告，来提示用户要进行的操作，如果直接选择了细节展示按钮，就会调用下面的方法。

❸处的方法首先检查子控制器 subController 实例变量，查看它是否为空，如果为空，就会执行初始化操作。接下来根据所选的行来为新视图设置标题，在入栈之前，为它分

配所要显示的文本，在这里设置 message 来反应单击的是哪一行的展示按钮。最后使用 UIApplication 实例获取应用程序委托，这样做是因为导航控制器是由应用程序的委托维护的，然后利用委托的导航控制器输出口将显示信息的下级视图放入导航控制器堆栈中，当点击某一行时，从栈中取出并显示。

**3. 编译并运行**

到这里对于代码的操作就完成了，现在编译并运行程序吧！单击第三行会看到图 8-22所示的效果，当点击细节展示按钮时，就会进入到下级视图，如图 8-23 所示。

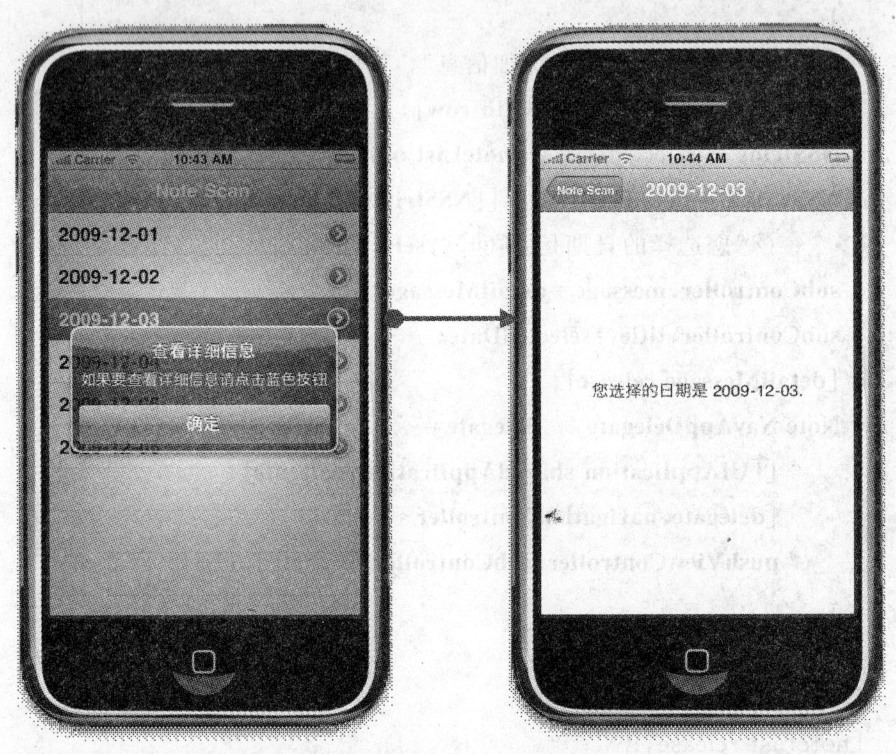


图 8-22　单击行　　　　图 8-23　下级视图

## 8.6　分组表、索引表和搜索功能的实现

前面几节我们学习了表视图的基本知识和一些简单操作，其中可以了解到表视图分为分组表和索引表两种基本样式，接下来我们就来实现简单的分组表、索引表，并添加一项常用的搜索功能。

### 8.6.1　实现分组表和索引表

在这一小节里，我们将通过构建分组表和索引表来深入了解它们的内部结构，最终要实现的效果如图 8-24 所示。

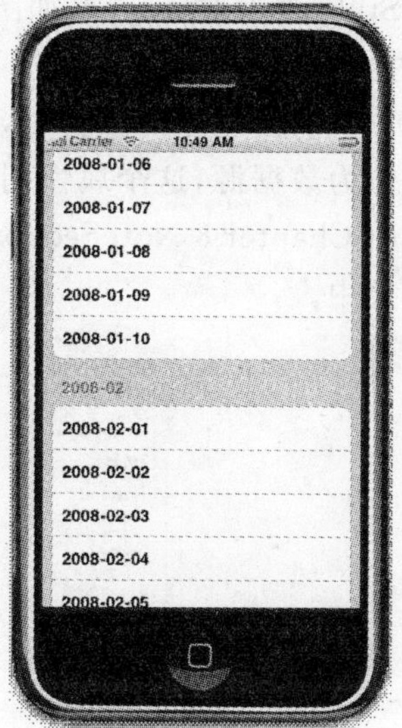


图 8-24　分组表

程序具体的步骤如下：

- 修改 NoteSectionViewController.h 文件；
- 构建分组表视图；
- 修改 NoteSectionViewController.m 文件；
- 编译并运行；
- 修改视图为索引表样式并添加索引。

#### 1. 修改 NoteSectionViewController.h 文件

首先使用 View-based Application 模板创建一个项目，并把它命名为 NoteSection。选中 NoteSectionViewController.h 文件，添加如下黑体字所示的代码：

**代码 8.13　NoteSectionViewController.h 文件**

```
#import <UIKit/UIKit.h>
@interface NoteSectionViewController : UIViewController
    <UITableViewDelegate, UITableViewDataSource> {
    NSDictionary * words;  ❶
    NSArray * keys;
}
@property (nonatomic, retain) NSDictionary * words;
@property (nonatomic, retain) NSArray * keys;
@end
```

首先添加了 UITableViewDataSource 和 UITableViewDelegate 协议，这跟我们以前所做的操作一样。接下来在❶处添加并声明了 NSDictionary 和 NSArray 实例变量，其中 NSDictionary 用于保存所有行数据，数组用于保存各个时间分区。

#### 2. 构建分组表视图

构建分组表视图时，将 Table View 拖进 View 窗口中，然后把数据源和委托连接到 File's Owner 图标上面。选中表视图，调出 Attributes 窗口，把表视图的

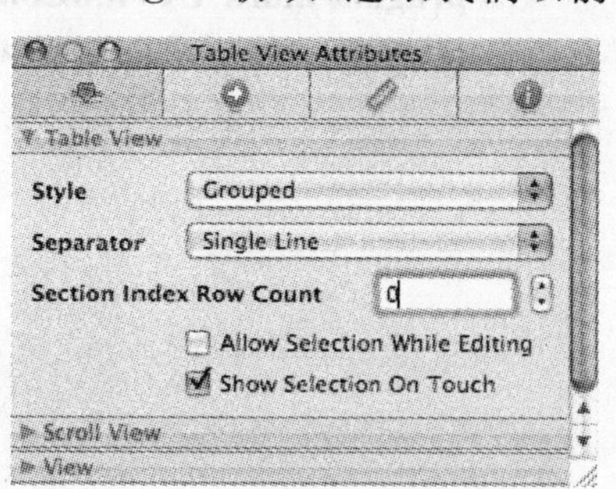


图 8-25　Style 调整

Style 改为 Grouped，如图 8-25 所示，这样做是因为在这里要创建的是一个分组表，保存并返回 Xcode。

前面创建的项目中，都是将数据直接写入程序里，而在这个项目中，将使用属性列表作为数据源，这个属性列表包含一个字典，它的结构如图 8-26 所示。首先要将随书光盘中 Chapter 8 NoteSection 文件下的 NoteSection. plist 文件添加到项目的 Resources 文件夹中。

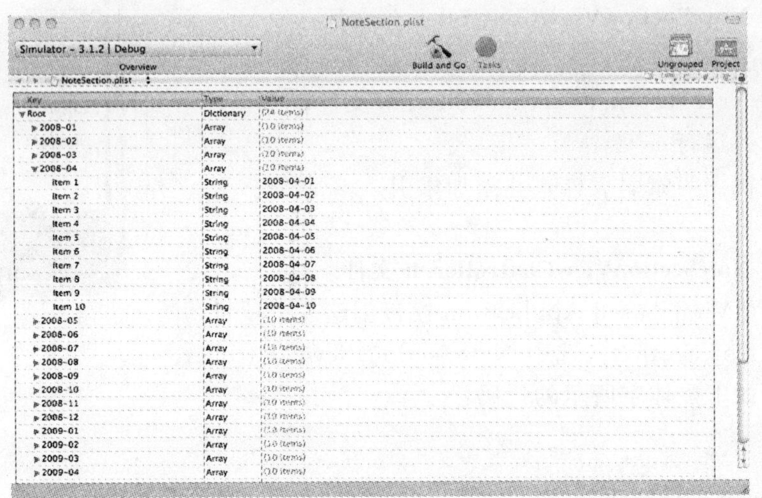

图 8-26　属性列表

**3. 修改 NoteSectionViewController. m 文件**

选中 NoteSectionViewController. m 文件，添加如下黑体字所示的代码：

**代码 8.14　NoteSectionViewController. m 文件**

```
#import "NoteSectionViewController. h"
@implementation NoteSectionViewController
@synthesize words;
@synthesize keys;
#pragma mark -
#pragma mark UIViewController Methods
//读取属性列表文件并存入字典
```

```
- (void)viewDidLoad {
    NSString * wordsPath=[[NSBundle mainBundle]
        pathForResource:@"NoteSection" ofType:@"plist"];
    NSDictionary * dictionary=[[NSDictionary alloc]
        initWithContentsOfFile:wordsPath];
    self.words=dictionary;
    [dictionary release];
    NSArray * array=[[words allKeys]
    sortedArrayUsingSelector:@selector(compare:)];
    self.keys=array;
    [super viewDidLoad];
}                                                        ❶
- (void)didReceiveMemoryWarning {
        [super didReceiveMemoryWarning];
}

- (void)viewDidUnload {
}
- (void)dealloc{
    [words release];
    [keys release];
    [super dealloc];
}
//数据源方法的实现
#pragma mark -
#pragma mark TableViewDataSource Methods
- (NSInteger)numberOfSectionsInTableView:
    (UITableView *)tableView{
        return [keys count];❷
}
- (NSInteger)tableView:(UITableView *)tableView
    numberOfRowsInSection:(NSInteger)section{
    NSString * key=[keys objectAtIndex:section];
    NSArray * wordSection=[words objectForKey:key];      ❸
    return [wordSection count];
}
```

```
-(UITableViewCell *)tableView:(UITableView *)tableView
    cellForRowAtIndexPath:(NSIndexPath *)indexPath{
    NSUInteger section=[indexPath section];
    NSUInteger row=[indexPath row];
    NSString *key=[keys objectAtIndex:section];
    NSArray *wordSection=[words objectForKey:key];
    static NSString *NoteSectionIdentifier=
        @"NoteSectionIdentifier";
    UITableViewCell *cell=[tableView
    dequeueReusableCellWithIdentifier:NoteSectionIdentifier];          ❹
    if(ce ll==nil){
        cell=[[UITableViewCell alloc]initWithStyle:
                UITableViewCellStyleDefault
                reuseIdentifier:NoteSectionIdentifier];
    }
    cell.textLabel.text=[wordSection objectAtIndex:row];
    return cell;
}
-(NSString *)tableView:(UITableView *)tableView
    titleForHeaderInSection:(NSInteger)section{
    NSString *key=[keys objectAtIndex:section];   ❺
    return key;
}
@end
```

❶处的代码，借助已经添加的属性列表文件创建了一个 NSDictionary 实例，然后获取字典中的所有键，并按照字母顺序排序，存到 NSArray 数组实例中。

❷处的方法用于指定分区的数量，这里不再是默认值 1，而是按照字典中的键数来确定分区，每一个键都拥有一个分区。

❸处的方法用来计算特定分区中的行数。首先检索对应分区的数组，然后从该数组里取出行的数量。

❹处的方法中，通过索引路径获取分区和行的值。其中分区会从字典中取出指定数组，然后使用行号来确定数组中的具体值。

❺处的方法用于为每个分区指定一个标题。

**4. 编译并运行**

到这里就完成了整个项目的操作，编译并运行，可以看到如图 8-24 所示的效果。

### 5. 修改视图为索引表样式并添加索引

接下来创建一个索引表，由于两个表有太多的相似之处，可以直接在 Interface Builder 中打开 NoteSectionViewController. xib 文件，在 Attributes 窗口中将 Style 更改为 Indexed，保存并运行，效果如图 8-27 所示，通过这样简单的操作就实现了一个索引表。

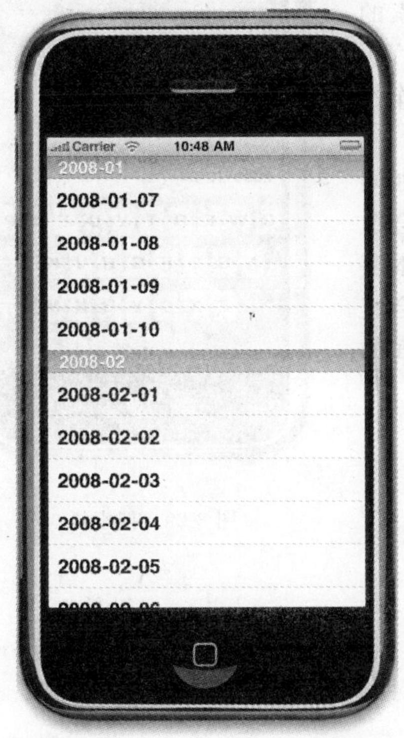


图 8-27　索引表

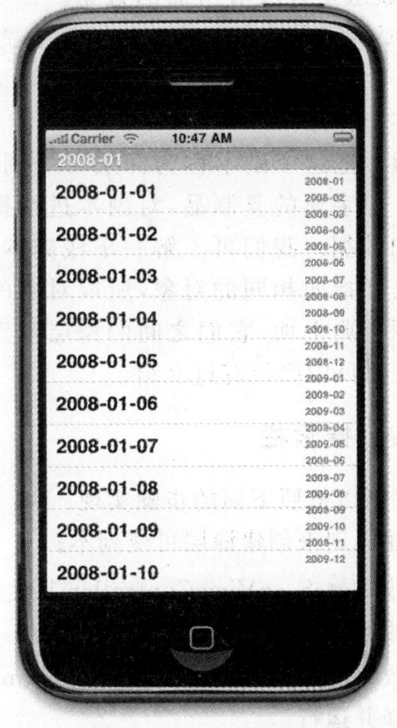


图 8-28　添加索引

如果我们所创建的行数并不是太多，所存的数据也很少，一般的索引表就可以直接存储使用，但是如果存储大量的数据，这种形式就会带来麻烦，我们需要添加一个索引以方便查询数据。只需在 NoteSectionViewController. m 文件尾部的@end 之前，添加如下代码：

```
- (NSArray *)sectionIndexTitlesForTableView:
                        (UITableView *)TableView {
    return keys;
}
```

这样就实现了一个带索引的索引表，编译并运行，会看到如图 8-28 所示的效果。这里需要声明一点，这两种表所存储的数据是相同的，只是表现形式不同而已。

## 8.6.2　搜索栏和深层可变副本

上面为了方便查询程序添加了索引功能，索引表能够方便数据查找，但是面对大量数据时，查询某个数据还是很不方便，而搜索栏却能够提供更好的用户体验。

下面我们来实现搜索功能。最终的效果如图 8-29 所示。

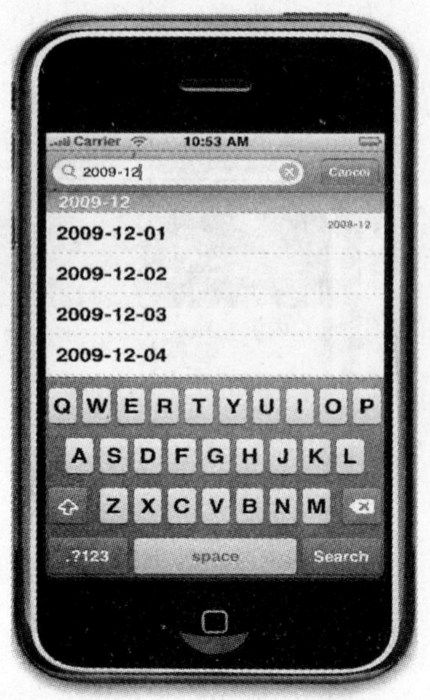


图 8-29　搜索栏

搜索栏里搜索到的数据来自于字典，字典里包含多个数组，数组里存放着所需的数据。在实现搜索的过程中，需要将符合搜索条件的数据集中显示出来，所以创建了一个字典的副本，让所有的搜索更改工作都在这个副本中进行。

我们通常把这个副本称为深层可变副本，它如同创建了一个源字典的复制品，对副本进行操作不会改变源字典的数据。我们再了解一下浅副本，浅副本字典和源字典都指向相同的对象，所以对它的操作会直接作用在源字典上面，它们之间的深层区别在前面的深拷贝与浅拷贝中已经有过介绍。

### 8.6.3　实现搜索栏

搜索栏可以依照下面的步骤实现：

- 新建类别来创建深层可变副本；
- 修改 NoteScanViewController.h 文件；
- 修改视图；
- 修改 NoteSectionViewController.m 文件实现搜索功能；
- 编译并运行。

#### 1. 新建类别来创建深层可变副本

选择 Classes 文件夹，按下⌘＋N 键，新建一个文件，选择 Other，然后选择 Empty File 创建一个空文件，命名为 DictionaryMutableDeepCopy.h；重复上述操作，再创建一个空文件，命名为 DictionaryMutableDeepCopy.m。

选中 DictionaryMutableDeepCopy.h 文件，添加如下代码：

**代码 8.15　DictionaryMutableDeepCopy.h 文件**

```
#import <Foundation/Foundation.h>
@interface NSDictionary (MutableDeepCopy)

-(NSMutableDictionary *)mutableDeepCopy; ❶
@end
```

在❶处声明了 mutableDeepCopy 方法，用来进行深层复制操作。

选中 DictionaryMutableDeepCopy.m 文件,添加如下代码:

**代码 8.16 DictionaryMutableDeepCopy.m 文件**

```
#import "DictionaryMutableDeepCopy.h"
@implementation NSDictionary (MutableDeepCopy)
//深层拷贝方法
- (NSMutableDictionary *) mutableDeepCopy
{
    NSMutableDictionary *mutableDictionary =
    [NSMutableDictionary
    dictionaryWithCapacity:[self count]];                    ❶
    NSArray *keys=[self allKeys];
    for(id key in keys)❷
      {
          id dicValue =[self valueForKey:key];
          id dicCopy =nil;
          if([dicValue respondsTo Selector:@selector
                                (mutableDeepCopy)])
              dicCopy =[dicValue mutableDeepCopy];
          else
              if([dicValue respo ndsToSelector:@selector
                                (mutableCopy)])              ❸
              dicCopy =[dicValue mutableCopy];
          if (di cCopy==nil)
              dicCopy=[ dicValue copy];
          [mutableDictionary setValue:dicCopy forKey:key];
      }
    return mutableDictionary;
}
@end
```

在❶处创建了一个新的可变字典 mutableDictionary,然后在❷处使用了一种新的语法 for(id key in keys),它的功能主要是对源字典中所有的键进行遍历,确保获得高效的循环操作,继而为每个检索到的数组创建可变副本。遍历的前提是类支持快速枚举,如 NSDictionary、NSArray 和 NSSet 都支持快速枚举。

❸处的方法首先尝试创建一个深层可变副本,如果对象没有响应 mutableDeepCopy 信息,就尝试创建可变副本;如果没有响应 mutableCopy 信息,那么它就回去创建常规副

本，这样就确保了字典中所有的对象都创建了副本。

如果一个字典内包含其他支持深层可变副本的对象，在为字典创建深层副本时也将会对字典中包含的对象创建深层副本。如果其他类引用了 DictionaryMutableDeepCopy.h 文件，就可以在任意的 NSDictionary 对象上面调用 mutableDeepCopy 方法。创建字典的深层可变副本并不是我们的目的，而只是为实现搜索栏而进行的必要步骤，接下来我们就来添加搜索栏。

**2. 修改 NoteScanViewController.h 文件**

选中 NoteScanViewController.h 文件，添加如下黑体字所示的代码：

**代码 8.17　NoteScanViewController.h 文件**

```
#import <UIKit/UIKit.h>
@interface NoteSectionViewController : UIViewController
<UITableViewDelegate, UITableViewDataSource,
UISearchBarDelegate> {❶
        IBOutlet UITableView *table;      ❷
        IBOutlet UISearchBar *search;
        NSDictionary *words;
        NSArray *keys;
        NSDictionary *allWords;
        NSMutableDictionary *words;       ❸
        NSMutableArray *keys;
}
@property (nonatomic, retain) NSDictionary *words;
@property (nonatomic, retain) NSArray *keys;
@property (nonatomic, retain) UITableView *table;
@property (nonatomic, retain) UISearchBar *search;
@property (nonatomic, retain) NSDictionary *allWords;
@property (nonatomic, retain) NSMutableDictionary *words;
@property (nonatomic, retain) NSMutableArray *keys;
- (void)resetSearch;
- (void)handleSearchForTerm:(NSString *)searchTerm;   ❹
@end
```

在❶处添加了 UISearchBarDelegate 协议，这样就让控制器类成为了搜索栏的委托。

在❷处添加了两个输出口，包括表视图的输出口、搜索栏的输出口。

在❸处更改原有的字典为可变类型用于存放与搜索标准相匹配的数据，更改数组为可变类型存放分区名称和索引值，并添加了一个附加的字典用来存放所有的数据。

在❹处声明两个方法,第一个用于取消搜索或者更改搜索的条件,第二个用于进行搜索操作。

### 3. 修改视图

双击打开 NoteSectionViewController.xib 文件,如图 8-30 所示,从 Library 中找到 Search Bar,把它拖到视图中,调整表视图和搜索栏的位置,效果如图 8-31 所示。

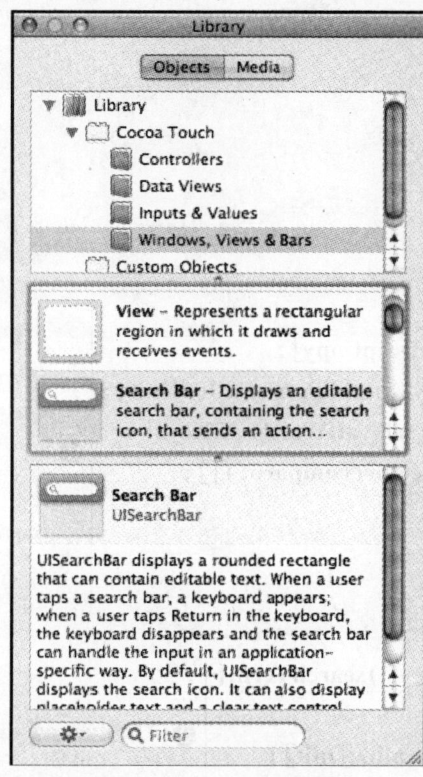


图 8-30 可删除表视图

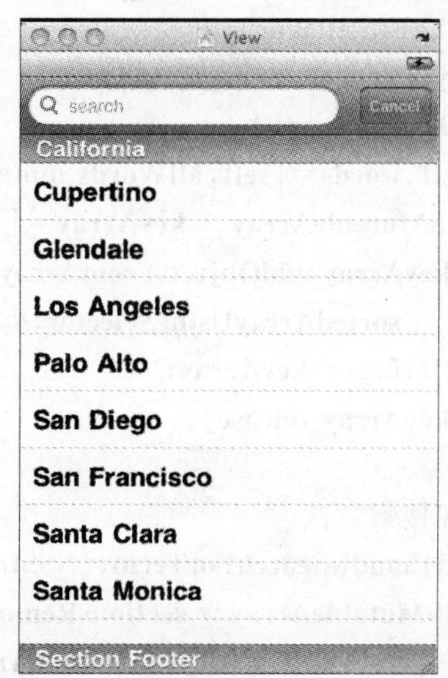


图 8-31 重新构建视图

接下来直接在 File's Owner 图标上面单击右键,拖动鼠标到表视图和搜索栏中,确定相对应的输出口。然后设置搜索栏的属性,调出 Attributes 窗口,具体属性更改如下:勾选 Shows Cancel Button 复选框,用于显示取消按钮,在 Placeholder 字段中输入 search,设置占位符为 search,如图 8-32 所示。

接下来需要为搜索栏设置委托,调出 Connections 窗口,把 delegate 连接到 File's Owner 图标,正如以前所做的一样。完成以上操作后,保存后返回 Xcode。我们继续对 NoteSectionViewController.m 进行更改。

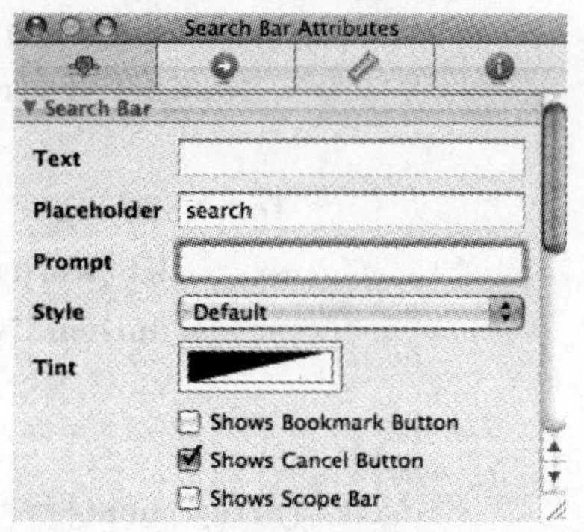


图 8-32 显示取消按钮和设置占位符

**4. 修改 NoteSectionViewController.m 文件实现搜索功能**

选中 NoteSectionViewController.m 文件，添加如下代码：

**代码 8.18 NoteScanViewController.m 文件**

```
#import "NoteSectionViewController.h"
#import "DictionaryMutableDeepCopy.h"
@implementation NoteSectionViewController
@synthesize words;
@synthesize keys;
@synthesize table;
@synthesize search;
@synthesize allWords;
//取消搜索或者改变搜索条件
- (void)resetSearch{
    self.words=[self.allWords mutableDeepCopy];
    NSMutableArray *keyArray=[[NSMutableArray alloc]init];
    [keyArray addObjectsFromArray:[[self.allWords allKeys]
        sortedArrayUsingSelector:@selector(compare:)]];          ❶
    self.keys=keyArray;
    [keyArray release];
}
//实现搜索方法
- (void)handleSearchForTerm:(NSString *)searchTerm{
    NSMutableArray *sectionsRemove=
                [[NSMutableArray alloc]init];                    ❷
    [self resetSearch];
    for(NSString *key in self.keys){
        NSMutableArray *array=[words valueForKey:key];
        NSMutableArray *toRemove=                                ❸
                [[NSMutableArray alloc]init];
        for(NSString *word in array){
            if([word rangeOfString:searchTerm
            options:NSCaseInsensitiveSearch].location            ❹
            == NSNotFound)[toRemove addObject:word];
        }
        if ([array count]==[toRemove count])
                [sectionsRemove addObject:key];                  ❺
            [array removeObjectsInArray:toRemove];
        [toRemove release];
```

```
}
        [self.keys removeObjectsInArray:sectionsRemove];
                [sectionsRemove release];                    ❻
        [table reloadData];
}
#pragma mark -
#pragma mark UIViewController Methods
- (void)viewDidLoad {
        NSString * wordsPath=[[NSBundle mainBundle]
        pathForResource:@"NoteSection" ofType:@"plist"];
        NSDictionary * dictionary=[[NSDictionary alloc]
        initWithContentsOfFile:wordsPath];
        self.words=dictionary;
        self.allWords=dictionary;
        [dictionary release];
        NSArray * array=[[wordsallKeys]sortedArrayUsingSelector:
        @selector(compare:)];
        self.keys=array;
                [self resetSearch];                          ❼
                search.autocapitalizationType=
            UITextAutocapitalizationTypeNone;
            search.autocorrectionType=UITextAutocorrectionTypeNo;
            [super viewDidLoad];
}
- (void)didReceiveMemoryWarning {
        [super didReceiveMemoryWarning];
}
- (void)viewDidUnload {
}
- (void)dealloc {
        [table release];
        [search release];
        [allWords release];
        [words release];
        [keys release];
        [super dealloc];
}
//数据源方法实现
```

```
#pragma mark -
#pragma mark TableViewDataSource Methods
- (NSInteger)numberOfSectionsInTableView:
    (UITableView *)tableView{
return [keys count];
return ([keys count]>0)?[keys count]:1;❽
}
- (NSInteger)tableView:(UITableView *)tableView
numberOfRowsInSection:(NSInteger)section {
    if([keys count]==0)
            return 0;
    NSString *key=[keys objectAtIndex:section];
    NSArray *wordSection=[words objectForKey:key];
    return [wordSection count];
}
- (UITableViewCell *)tableView:(UITableView *)tableView
      cellForRowAtIndexPath:(NSIndexPath *)indexPath{
      NSUInteger section=[indexPath section];
      NSUInteger row=[indexPath row];
      NSString *key=[keys objectAtIndex:section];
      NSArray *wordSection=[words objectForKey:key];
      static NSString *NoteSectionIdentifier=@"NoteSectionIdentifier";
      UITableViewCell *cell=[tableView
            dequeueReusableCellWithIdentifier:NoteSectionIdentifier];
      if(cell==nil){
            cell=[[UITableViewCell alloc]initWithStyle:
                        UITableViewCellStyleDefault
                        reuseIdentifier:NoteSectionIdentifier];
      }
      cell.textLabel.text=[wordSection objectAtIndex:row];
      return cell;
}
- (NSString *)tableView:(UITableView *)tableView
      titleForHeaderInSection:(NSInteger)section {
      if([keys count]==0)
            return @" ";
      NSString *key=[keys objectAtIndex:section];
      return key;
```

```
}
- (NSArray *)sectionIndexTitlesForTableView:
    (UITableView *)tableView{
    return keys;
}
//委托方法实现
 #pragma mark -
 #pragma mark TableViewDelegate Methods
 - (NSIndexPath *)tableView:(UITableView *)tableView
     willSeclectRowAtIndexPath:(NSIndexPath *)indexPath{          ❾
     [search resignFirstResponder];
     return indexPath;
 }
 #pragma mark -
 #pragma mark SearchBarDelegate Methods
 - (void)searchBarSearchButtonClicked:(UISearchBar *)searchBar{
     NSString *searchTerm=[searchBar text];
     [self handleSearchForTerm:searchTerm];
 }
 - (void)searchBar:(UISearchBar *)searchBar
     textDidChange:(NSString *)searchTerm{
     if([searchTerm length]==0)
     {
         [self resetSearch];                                          ❿
         [table reloadData];
         return;
     }
     [self handleSearchForTerm:searchTerm];
 }
 - (void)searchBarCancelButtonClicked:(UISearchBar *)searchBar{
     search.text=@"";
     [self resetSearch];
     [table reloadData];
     [searchBar resignFirstResponder];
 }
 @end
```

在❶处是取消搜索或者更改搜索条件时调用的方法。在这个方法里，首先创建了一个字典的可变副本，然后赋值给变量 words，在最后我们还对数组 keys 进行了刷新操作，这是因为我们需要把搜索时排除的分区从分区中去掉，否则屏幕就充满了标题和空白的分区。

在❷处，创建一个数组来存放我们所找到的空分区。

在❸处，快速枚举新存储的 keys 数组中的所有键。创建对应当前键的名称数组，以及一个用于存储需要从 words 数组中删除的值的数组。

在❹处是通过一个循环来实现搜索时忽略大小写，并把空分区的值放到要删除的对象数组中。

在❺处是校对将要删除的名称数组长度和名称数组长度是否相等，若相等则分区为空，把它加进键的数组中，在后面删除。

在❻处是删除空分区，释放用来存储空分区的数组，并告知表重新加载数据。这样我们就实现了搜索的方法。

在❼处我们首先把属性列表加载到 allWords 里面，然后删除了原来加载 keys 数组的代码，然后调用 resetSearch 方法完成加载并填充 words 可变字典和 keys 数组。接下来在搜索栏中进行一些限制配置，由于搜索不区分大小写，所以没有必要把用户输入的搜索内容转换为大写，当然也不需要进行自动校正。

在❽处我们对数据源方法进行一些修改。由于搜索可能会排除所有分区，但是表视图必须拥有分区，这种条件下我们必须检查删除了所有分区后的情况，如果真的全部删除，我们就为它提供一个没有行且只有一个分区的表视图。

在❾处添加的委托方法主要用于在进行搜索操作时单击任意一行都能够取消键盘，当然这一切都是通过让搜索栏放弃第一响应者的状态来实现的。

在❿处添加搜索栏的委托方法。

searchBarSearchButtonClicked:searchBar 方法：当用户单击键盘上的返回按钮或者搜索键时，会调用这个方法。它获取搜索的内容并调用前面所讲到的搜索方法，而且可以删除 words 里的空分区和不匹配内容，直接把最合适的数据呈现出来。

searchBar: searchBartextDidChange:方法：实现了 live search，也就是搜索的内容会随着输入即时地显示出来，这大大优化了用户体验，减少了我们的输入。当然获得这样的方便必然会付出一定的代价，比如可能会减弱应用程序的性能，所以我们要实现这个方法时必须考虑在设备上的稳定性。

searchBarCancelButtonClicked:方法：当用户点击搜索栏上的 Cancel 按钮时，程序会接收到通知，并把搜索的短语设置为空字符串，然后重新加载并显示所有的分类数据，最后一句用于设置搜索栏为第一响应者，取消键盘。

**5. 编译并运行**

到这里程序就完成了，保存后编译并运行，就可以看到效果了。

# 8.7 自定义表视图单元

表视图为构建应用程序界面提供了诸多方便，但是有时候还是希望能够根据需要对它进行自定义的设计，本节介绍的自定义表视图单元就能满足这种需求。

我们将对8.5节中的项目进行修改，将标题和日期显示出来并传递到详细视图里，最终的效果如图8-33所示。

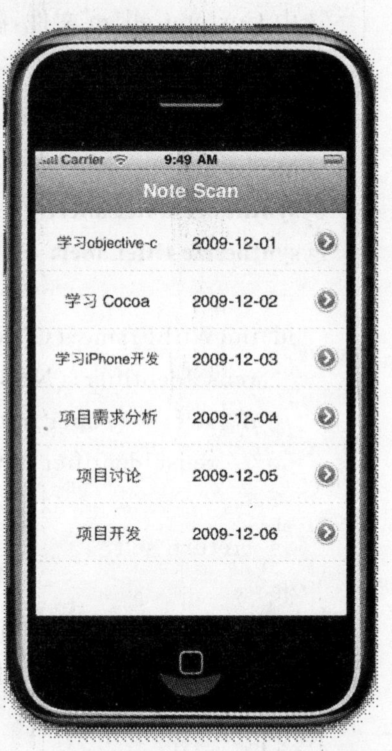

图8-33 自定义表视图

下面我们要学习的是如何利用UITableViewCell的自定义子类来构建自定义表视图单元。这里将使用Interface Builder来创建自定义表视图单元的界面，当然也可以利用代码来实现。

本程序的基本步骤如下：

- 新建UITableViewCell自定义子类；
- 利用自定义的UITableViewCell子类来修改项目；
- 编译并运行。

### 1. 新建UITableViewCell自定义子类

打开8.5节的NoteNav项目，在Xcode中右键单击Classes文件夹，选择Add→New File，在出现新建文件向导后，从左侧窗格里面选择Cocoa Touch Class，在右边窗格里点击Objective-C class，在Subclass of后选择UITableViewCell，命名为CustomCell.m，确保选中了Also Create "CustomCell.h"复选框。

创建完成后，接下来在Resources文件夹上单击右键添加文件，出现新建文件向导后，在左侧的窗格中单击User Interfaces，接下来在右侧窗格里选择Empty XIB，命名为CustomerCell.xib。到这里就创建了所有必备的文件，接下来需要创建UITableViewCell的子类。

选中CustomCell.h文件，添加如下黑体字所示的代码：

**代码8.19 CustomCell.h文件**

```
#import <UIKit/UIKit.h>
@interface CustomCell : UITableViewCell {
        IBOutlet UILabel *dateLabel;  ❶
        IBOutlet UILabel *titleLabel;
```

```
}
@property (nonatomic, retain) UILabel * dateLabel;
@property (nonatomic, retain) UILabel * titleLabel;
@end
```

在❶处声明了两个标签，用来显示日期和标题，并为它们设置输出口。

选中 CustomCell.m 文件，添加如下黑体字所示的代码：

**代码 8.20　CustomCell.m 文件**

```
#import "CustomCell.h"
@implementation CustomCell
@synthesize dateLabel;
@synthesize titleLabel;

- (id)initWithFrame:(CGRect)frame
    reuseIdentifier:(NSString *)reuseIdentifier {
    if (self = [super initWithFrame:frame
        reuseIdentifier:reuseIdentifier]) {
}

     return self;
}
- (void)setSelected:(BOOL)selected animated:(BOOL)animated {
    [super setSelected:selected animated:animated];
}
- (void)dealloc {
    [dateLabel release];
    [titleLabel release];
    [super dealloc];
}
@end
```

保存上述两个文件，然后对 CustomCell.xib 文件进行修改。双击打开 CustomCell.xib 文件，主窗口中只有 File's Owner 和 FirstResponder，我们所创建的就是一个空的 Xib 文件，如图 8-34 所示，在 Library 中找到 Table View Cell，拖进主窗口中，完成后调出 Identity 窗口，将类 UITableViewCell 改为 CustomCell。

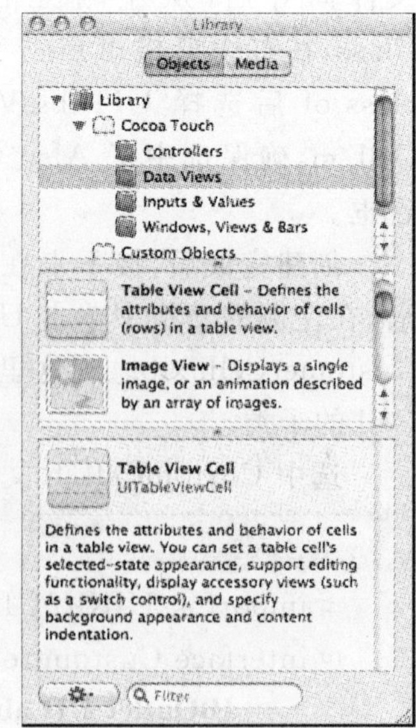


图 8-34　Table View Cell

把表视图单元的高度从 44 像素改为 60 像素，这样会更加美观，活动空间也更大。

调出 Attributes 窗口，将第一个字段 Identifier 改为 CustomCellIdentifier，这里的设置是以前常见的可重用标识符，还要确保 Accessory 弹出按钮的值为Detail Disclosure，因为这里的表格还要使用公开。

双击 CustomCell. xib 窗口中的 CustomCell，里面只有一个标记 Content View 的灰色虚线圆角矩形，如图 8-35所示。

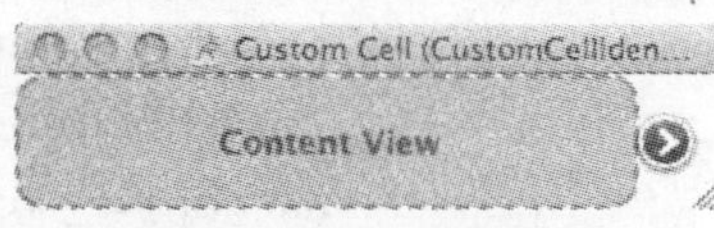


图 8-35　表视图单元

拖一个 View 到 Custom Cell 窗口中，调整 View 的大小以适应 Custom Cell 窗口。调出 Size 窗口，设置 X 为 0，Y 为 0，W 为 270，H 为 60。

拖出两个 Label 到 Custom Cell 窗口中，按照图 8-36进行布局，然后用右键把 CustomCell 图标拖动到两个标签上，分别为两个标签连接输出口，左边的标签连接的输出口为 titleLabel。右边的标签连接的输出口为 dateLabel。在这里这个表视图单元的 Xib 文件只是充当一个模板，并不用设置自己的控制器类，这也是我们没有为它连接 File's Owner 图标的原因。

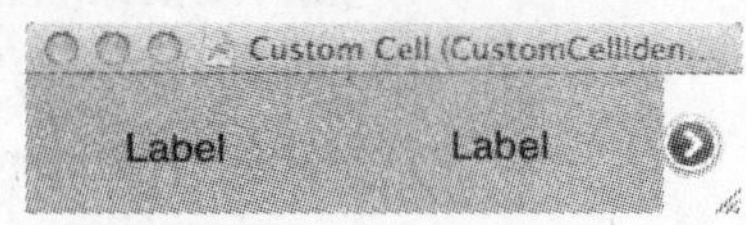


图 8-36　表视图单元布局

**2. 利用定义的 UITableViewCell 子类来修改项目**

对于 RootViewController. h 文件，我们没有进行修改，直接来看更改较多的 RootViewController. m 文件。

选中 RootViewController. m 文件，添加如下黑体字所示的代码：

**代码 8.21　RootViewController. m 文件**

```
#import "RootViewController.h"
#import "NoteNavAppDelegate.h"
#import "NoteDetailController.h"
#import "CustomCell.h"
@implementation RootViewController
@synthesize noteList;

//构建表中数据
- (void)viewDidLoad {
    self.title=@"Note Scan";
```

```
NSDictionary *row1=[[NSDictionary alloc]
    initWithObjectsAndKeys:
        @"2009-12-01",@"Date",
        @"学习 objective-c",@"Title",nil];
    NSDictionary *row2=[[NSDictionary alloc]
    initWithObjectsAndKeys:
        @"2009-12-02",@"Date",
        @"学习 Cocoa",@"Title",nil];
    NSDictionary *row3=[[NSDictionary alloc]
    initWithObjectsAndKeys:
        @"2009-12-03",@"Date",
        @"学习 iPhone 开发",@"Title",nil];
    NSDictionary *row4=[[NSDictionary alloc]
    initWithObjectsAndKeys:
        @"2009-12-04",@"Date",
        @"项目需求分析",@"Title",nil];
    NSDictionary *row5=[[NSDictionary alloc]
        initWithObjectsAndKeys:
        @"2009-12-05",@"Date",
        @"项目讨论",@"Title",nil];
    NSDictionary *row6=[[NSDictionary alloc]
        initWithObjectsAndKeys:
        @"2009-12-06",@"Date",
        @"项目开发",@"Title",nil];
    NSArray *array=[[NSArray alloc]initWithObjects:
        row1,row2,row3,row4,row5,row6,nil];
    self.noteList=array;
    [row1 release];
    [row2 release];
    [row3 release];
    [row4 release];
    [row5 release];
    [row6 release];
    [array release];                                    ❶
[super viewDidLoad];
}
```

```
- (void)didReceiveMemoryWarning {
    [super didReceiveMemoryWarning];
}
- (void)viewDidUnload {
}
//数据源方法实现
#pragma mark Table view methods
- (NSInteger)numberOfSectionsInTableView:
        (UITableView *)tableView {
    return 1;
}
- (NSInteger)tableView:(UITableView *)tableView
    numberOfRowsInSection:(NSInteger)section {
    return [self.noteList count];
}

//用表视图单元构建表视图
- (UITableViewCell *)tableView:(UITableView *)tableView
    cellForRowAtIndexPath:(NSIndexPath *)indexPath{
     static NSString *CustomCellIdentifier =
            @"CustomCellIdentifier";
     CustomCell *cell=(CustomCell *)[tableView
          dequeueReusableCellWithIdentifier:(
               CustomCellIdentifier];
     if (cell == nil){
          NSArray *nib=[[NSBundle mainBundle]loadNibNamed:
               @"CustomCell" owner:self options:nil];
          cell=[nib objectAtIndex:0];
     }
     NSUInteger row=[indexPath row];
     NSDictionary *rowdata=[self.noteList objectAtIndex:row];
     cell.dateLabel.text=[rowdata objectForKey:@"Date"];
     cell.titleLabel.text=[rowdata objectForKey:@"Title" ];
     [cell setAccessoryType:
          UITableViewCellAccessoryDetailDisclosureButton];
     return cell;
}
```

❷

```
//委托方法实现
#pragma mark -
#pragma mark Table Delegate Methods
-(void)tableView:(UITableView *)tableView
didSelectRowAtIndexPath:(NSIndexPath *)indexPath{
    UIAlertView *alert=[[UIAlertView alloc]
                    initWithTitle:@"查看详细信息"
                    message:@"如果要查看详细信息请点击蓝色按钮"
                    delegate:nil
                    cancelButtonTitle:@"确定"
                    otherButtonTitles:nil];
    [alert show];
    [alert release];
}
//点击公开按钮后进入下级视图
-(void)tableView:(UITableView *)tableView
    accessoryButtonTappedForRowWithIndexPath:
    (NSIndexPath *)indexPath {
        if (subController==nil) {
            subController =[[NoteDetailController alloc]
            initWithNibName:@"NoteDetailController"
                        bundle:nil];
        }
        subController.title = @"详细信息";
        NSUInteger row = [indexPath row];
        NSDictionary *selectedDate = [noteList objectAtIndex:row];
        NSString *selectStr = [selectedDate objectForKey:@"Date"];
        NSString *titleStr = [selectedDate objectForKey:@"Title"];
        NSStrin g *detailMessage = [[NSString alloc] initWithFormat:
                →(@"详细内容:%@---->%@",selectStr,titleStr];
        subController.message=detailMessage;
        [detailMessage release];
        NSString *selectedDate=[noteList objectAtIndex:row];
        NSString *detailMessage=[[NSString alloc]initWithFormat:
            @"您选择的日期是 %@.",selectedDate];
```

❸

```
        subController.message=detailMessage;
        subController.title=selectedDate;
        [detailMessage release];
        NoteNavAppDelegate *delegate=
            [[UIApplication sharedApplication]delegate];
        [delegate.navigationController pushViewController:subController
        animated:YES];
}
- (void)dealloc {
        [subController release];
        [noteList release];
        [super dealloc];
}
- (CGFloat)tableView:(UITableView *)tableView
heig htForRowAtIndexPath:(NSIndexPath *)indexPath{      ❹
    return 60;
}
@end
```

在❶处的 ViewDidLoad 方法中，创建了一系列的字典，每个字典中都包含了某一行的日期和标题信息，然后把所有的字典都存入数组中，这些数据都将成为表中要显示的数据。

在❷处的 tableView:cellForRowAtIndexPath:方法中，由于我们在 Xib 文件里设计构建了表视图单元，所以可以直接从 Xib 文件中加载，后面根据键值从字典里获取相应的值赋予各自标签。

在❸处对 tableView:accessoryButtonTappedForRowWithIndexPath:方法进行了修改，用 indexPath 参数确定表请求的行，并为该行获取正确的字典，然后根据键—值对应读取时间、标题。

在❹处添加方法用来设置行的高度，因为我们对表视图单元默认的高度值进行了修改，所以需要添加这个方法来通知表视图，否则表视图不会留出足够的显示空间。

**3. 编译并运行**

到这里自定义表视图单元的项目就完成了，编译并运行，单击一行时，就会看到如图 8-37 所示的效果；单击对应的蓝色按钮会进入下一级详细信息视图，如图 8-38 所示。

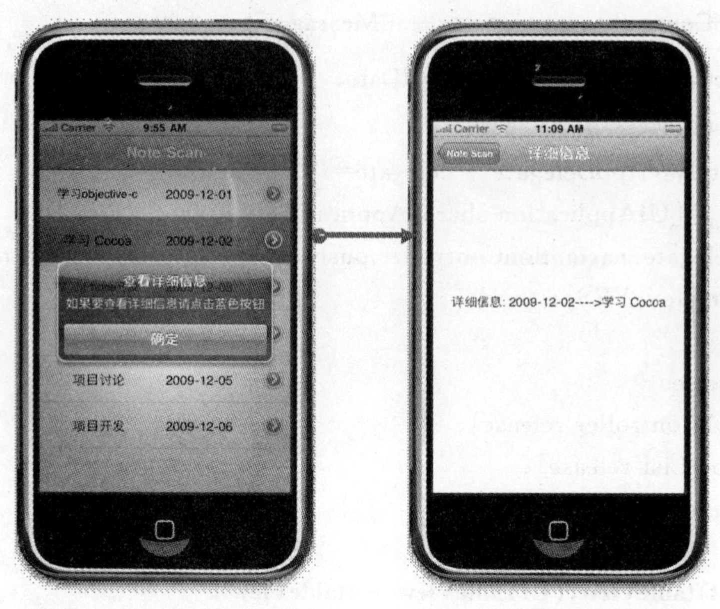


图 8-37 单击一行　　　　图 8-38 详细信息视图

# 8.8 可编辑的详细窗格

在这一节里我们将利用前面所学的自定义表视图单元的知识来创建可编辑的详细窗格，这也是表视图应用中很常见很重要的一部分。在这里仍然采用基于导航的应用程序模板来创建新项目，并命名为NoteRecord，在开始程序操作之前，先来看一下最终要实现的效果，如图 8-39 所示。

图 8-39 可编辑详细窗格

在这个项目中，首先用自定义表视图单元来创建表格，第一行用来编辑标题，第二行进入子视图后可以用选取器来选取时间，第三行进入子视图后可以用校验来标明选中何种类别，第四行进入子视图后可以利用 UITextView 来编辑详细内容。

## 8.8.1 编辑自定义表视图单元

首先来创建自定义的表视图单元，这跟前面所学习的内容一样，具体步骤如下：

· 新建文件，修改 CustomCell. h 文件；

- 修改 CustomCell. m 文件。

### 1. 新建文件，修改 CustomCell. h 文件

在 Xcode 中，打开 Classes 和 Resources 文件夹，在 Classes 文件夹上面单击右键 Add→New File，然后选择 Objective-C class，同时将 Subclass of 改为 UITableViewCell，点击 Next，命名为 CustomCell. m。然后在 Resources 文件夹上面单击右键创建一个 Empty XIB 文件，命名为 CustomCell. xib，这里我们不再介绍具体的创建步骤，如果有疑问可以查阅上一节的内容。

选中 CustomCell. h 文件，添加如下黑体字所示的代码：

**代码 8.22 CustomCell. h 文件**

```
#import <UIKit/UIKit.h>
@interface CustomCell : UITableViewCell {
IBOutlet UILabel *noteLabel;            ]❶
IBOutlet UITextField *noteTextField;    ]
}
@property (nonatomic ,retain)UILabel *noteLabel;
@property (nonatomic ,retain)UITextField *noteTextField;
@end
```

❶处为标签和文本框声明输出口。从开始的效果图里可以看出，表的每一行都显示时间和标题，其中标题为可以编辑的内容，所以我们创建一个标签来显示时间，创建一个文本框供用户输入标题内容。在 Xib 文件中的操作也跟前面没有多大区别，这里的表视图单元高度采用默认值，没有过多的修改，具体的内容在上一节里已经介绍，这里不再赘述。

### 2. 修改 CustomCell. m 文件

接下来选中 CustomCell. m 文件，添加如下黑体字所示的代码：

**代码 8.23 CustomCell. m 文件**

```
#import "CustomCell.h"
@implementation CustomCell
@synthesize noteLabel;
@synthesize noteTextField;
-(id)initWithFrame:(CGRect)frame
```

```
        reuseIdentifier:(NSString *)reuseIdentifier {
    if (self = [super initWithFrame:
        frame reuseIdentifier:reuseIdentifier]) {
    }
    return self;
}
- (void)setSelected:(BOOL)selected animated:(BOOL)animated {
        [super setSelected:selected animated:animated];
}
- (void)dealloc {
        [noteLabel release];
        [noteTextField release];
        [super dealloc];
}
@end
```

这里没有新的操作，只是添加了几行最基本的代码。接下来的一小节，我们将要编辑设定时间视图。

## 8.8.2　编辑设定时间视图

标题项并没有下级子视图，所以我们接下来编写设定时间的子视图，具体步骤如下：

- 修改 SetDateViewController. h 文件；
- 修改 SetDateViewController. xib 文件，构建视图；
- 修改 SetDateViewController. m 文件。

图 8-40　选取时间

### 1. 修改 SetDateViewController. h 文件

在 Classes 文件夹上单击右键新建文件，选择 UIViewController subclass 选项，并直接勾选 With XIB for user interface 选项创建 Xib 文件，单击 Next 将文件命名为SetDateViewController. m。在这里我们将要创建如图 8-40 所示的子视图。

选中 SetDateViewController. h 文件，添加如下黑体字所示的代码：

**代码 8.24 SetDateViewController.h 文件**

```
#import <UIKit/UIKit.h>
@interface SetDateViewController : UIViewController {
    IBOutlet UILabel *setDateLabel;
    IBOutlet UIDatePicker *datePicker;      ❶
    NSString *time;
}
@property (nonatomic,retain) UILabel * setDateLabel;
@property (nonatomic,retain) UIDatePicker *datePicker;
@property (nonatomic,retain) NSString *time;
- (IBAction)setDate:(id)sender; ❷
@end
```

在❶处声明一个标签用来在子视图中显示所选时间,还声明了一个时间选取器(Date Picker)对象,最后声明的字符串用来存储时间,在❷处声明的方法主要用于设定时间。

**2. 修改 SetDateViewController.xib 文件,构建视图**

双击 SetDateViewController.xib 文件,构建如图8-41所示的视图。首先从 Library 中分别拖出两个 Label 和一个 DatePicker 添加到 View 中,按照图中所示安排各个控件的位置,并在第一个 Label 中输入"您选择的时间是:"。选中 Date Picker 更改它的 Mode 属性为 Date 型,分别连接 Label、DatePicker 到 File's Owner,为它们分别设置输出口,并将 setDate 方法连接到 Date Picker 上面(单击右键 Date Picker 拖动到 File's Owner 上面,选择 setDate 方法)。这样就完成了对 SetDateViewController.xib 文件的修改,保存文件。

图 8-41 时间视图

**3. 修改 SetDateViewController.m 文件**

接下来看一下 SetDateViewController.m 文件中的修改:

**代码 8.25 SetDateViewController.m 文件**

```
#import "SetDateViewController.h"
@implementation SetDateViewController
@synthesize setDateLabel;
@synthesize datePicker;
```

```
@synthesize time;
- (IBAction)setDate:(id)sender{
    NSString *str = [NSString stringWithFormat:@"%@",
    [datePicker date]];
    [setDateLabel setText:[str substringToIndex:10]];
    time = setDateLabel.text;
}                                                              ❶
- (id)initWith NibName:(NSString *)nibNameOrNil
          bundle:(NSBundle *)nibBundleOrNil {
    if (self = [super initWithNibName:
        nibNameOrNil bundle:nibBundleOrNil]) {
        self.title = @"时间设置";                              ❷
    }
    return self;
}
- (void)viewDidLoad {
    self.setDateTextField.text =@"";
    NSDate *now = [[NSDate alloc]init];
    [datePicker setDate:now animated:YES];                     ❸
    [now release];
    [super viewDidLoad];
}
- (void)didReceiveMemoryWarning {
    [super didReceiveMemoryWarning];
}
- (void)viewDidUnload {
}
- (void)dealloc {
    [setDateLabel release];
    [datePicker release];
    [time release];
    [super dealloc];
}
@end
```

首先在❶处添加 setDate 方法。通过时间选取器来获取时间，并把它存入 time 字符串中，这里只截取了获取到的完整时间的前10位。在❷处添加的方法是用来设置导航栏上面对应视图的标题。

在 viewDidLoad 方法中，先利用❸处的代码设定所显示时间为当前时间选取器里面的默认时间。

### 8.8.3 编辑设定类型视图

构建类别选择视图的方法与上一小节设定时间视图的方法一致，把它命名为 SetTypeViewController. m，最终效果如图 8-42 所示。

图 8-42 设定类别视图

程序的大致步骤如下：

- 修改 SetTypeViewController. h 文件；
- 构建视图，修改 SetTypeViewController. m 文件。

**1. 修改 SetTypeViewController. h 文件**

首先选中 SetTypeViewController. h 文件，添加如下黑体字所示的代码：

**代码 8.26 SetTypeViewController. h 文件**

```
#import <UIKit/UIKit.h>
@interface SetTypeViewController : UIViewController {
    IBOutlet UITableView *setType;      ┐
    NSArray *setList;                   │ ❶
    NSIndexPath *lastIndexPath;         │
    NSString *types;                    ┘
}
@property (nonatomic, retain) UITableView *setType;
@property (nonatomic, retain) NSArray *setList;
@property (nonatomic, retain) NSIndexPath *lastIndexPath;
@property (nonatomic, retain) NSString *types;
@end
```

由于最终要显示的类别选择视图是一个表视图，所以在❶处的代码中，首先声明一个 UITableView 类型的输出口，接下来声明数组 setList 来存储所选的类型，然后声明 NSIndexPath 类型变量来标记当前选中项，最后设置了一个 types 字符串来存储所选类型，用于传递到主视图。

**2. 构建视图，修改 SetTypeViewController. m 文件**

接下来我们对 SetTypeViewController. xib 文件进行设置。双击打开文件，从 Library 窗口中拖出一个 Table View 放到主视图中，并将 Style 选项更改为 Grouped，连接数据源和委托，然后再单击右键把 File's Owner 拖到表视图中，为表视图确定 setType 输出口，保存文件。

选中 SetTypeViewController.m 文件，添加如下黑体字所示的代码：

**代码 8.27　SetTypeViewController.m 文件**

```
#import "SetTypeViewController.h"
@implementation SetTypeViewController
@synthesize setType;
@synthesize setList;
@synthesize lastIndexPath;
@synthesize types;
-(id)initWithNibName:(NSString *)nibNameOrNil
            bundle:(NSBundle *)nibBundleOrNil {
    if (self=[super initWithNibName:
            nibNameOrNil bundle:nibBundleOrNil]) {
        self.title=@"设定类别";
    }
    return self;
}
-(void)didReceiveMemoryWarning {
    [super didReceiveMemoryWarning];
}
-(void)viewDidUnload {
}
-(void)dealloc {
    [setType release];
    [setList release];
    [lastIndexPath release];
    [types release];
    [super dealloc];
}

//数据源方法的实现
#pragma mark-
#pragma mark setTypeTable dataSource Methods
-(NSInteger)numberOfSectionsInTableView:(UITableView *)tableView{
```

```
    return 1;
}
-(NSInteger)tableView:(UITableView *)tableView
    numberOfRowsInSection:(NSInteger)section {
        return 4;
}
-(UITableViewCell *)tableView:(UITableView *)tableView
    cellForRowAtIndexPath:(NSIndexPath *)indexPath {
    static NSString *setTypeCellIdentifier = @"setTypeCellIdentifier";
    UITableViewCell *cell=[tableView
            dequeueReusableCellWithIdentifier
            :setTypeCellIdentifier];
    if(cell == nil) {
        cell = [[UITableViewCell alloc] initWithStyle:
                UITableViewCellStyleDefault
                reuseIdentifier:setTypeCellIdentifier];
    }
    if(indexPath.row == 0) {
        cell.textLabel.text = @"运动";
    }
    if(indexPath.row == 1) {
        cell.textLabel.text = @"娱乐";
    }
    if(indexPath.row == 2) {
        cell.textLabel.text = @"工作";
    }
    if(indexPath.row == 3) {
        cell.textLabel.text = @"学习";
    }
        return cell;
    }                                                          ❶
//委托方法的实现
#pragma mark-
#pragma mark setTypeTable Delegate Methods
-(void)tableView:(UITableView *)tableView didSelectRowAtIndexPath:
```

```
(NSIndexPath *)indexPath {
    int newRow = [indexPath row];
    int oldRow = [lastIndexPath row];
    if (newRow != oldRow||(lastIndexPath==nil)) {
        UITableViewCell *newCell =
        [tableView cellForRowAtIndexPath:indexPath];
        newCell.accessoryType=
            UITableViewCellAccessoryCheckmark;                    ❷
        UITableViewCell *oldCell =
            [tableView cellForRowAtIndexPath:lastIndexPath];
        oldCell.accessoryType = UITableViewCellAccessoryNone;
        lastIndexPath = indexPath;
        self.types = newCell.textLabel.text;
        [tableView deselectRowAtIndexPath:
            indexPath animated:YES];
    }
}
@end
```

在❶处直接为每一行赋值。❷处首先获取当前新行和上一次选中的行，然后比较两行的索引值，如果不相等，则获取当前选中的单元，制定一个校验标记作为它的扩展图标；如果相等则获取上一次选中的单元，把它的扩展标记设置为无。这里值得注意的是，当第一次进入这个视图点击第一行时（lastIndexPath＝＝nil），也应该为它设定扩展标记，否则就无法直接选中这一行。我们需要在设定之后存储刚才在 lastIndexPath 中选中的索引路径，以便在下一次选中行时使用它。最后取消刚才选中的行。

图 8-43　设定内容视图

### 8.8.4　编辑详细内容视图

再创建一个控制器，并把它命名为 SetContentViewController.m，我们要在这个视图里添加一个 Text View，用于编辑详细内容，最终效果如图 8-43 所示。

本程序步骤如下：

- 修改 SetContentViewController. h 文件；
- 编辑 SetContentViewController. xib 文件，构建视图；
- 修改 SetContentViewController. m 文件。

**1. 修改 SetContentViewController. h 文件**

选中 SetContentViewController. h，添加如下黑体字所示的代码：

**代码 8.28　SetContentViewController. h 文件**

```
#import <UIKit/UIKit.h>
@interface SetContentViewController : UIViewController {
    IBOutlet UITextView *textView; ❶
}
@property (nonatomic, retain) UITextView *textView;
@end
```

在❶处声明了一个 UITextView 类型的输出口，它提供用户输入的区域。

**2. 编辑 SetDateViewController. xib 文件，构建视图**

双击打开 SetContentViewController. xib 文件对它进行编辑。从 Library 中拖出一个 Text View 放到视图中，调出 Attributes 窗口，在这里设置它占满全部窗口，然后把随书光盘中 Chapter 8 NoteRecord 文件夹下的 background. png 图片导入项目中，接下来添加一个 Image View，并把它的尺寸设置为占满全部窗口，选中 Image View，在 Attributes 窗口中，设置 Image 为 background. png；然后，选择工具栏中的 Layout 选项，选择 Sent To Back，将 Image View 设置为背景显示。连接 File's Owner 到 Text View，选中输出口 textView，构建效果如图 8-44 所示，保存文件。

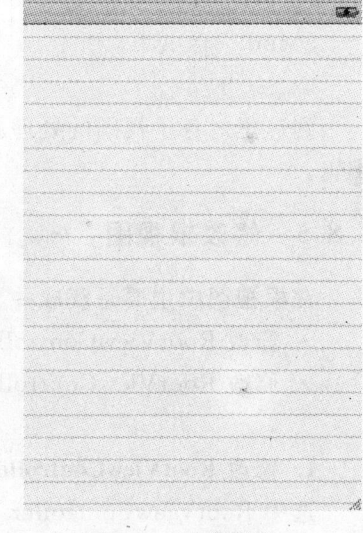

图 8-44　内容视图

**3. 修改 SetDateViewController. m 文件**

选中 SetContentViewController. m 文件，添加如下黑体字所示的代码：

**代码 8.29　SetContentViewController. m 文件**

```
#import "SetContentViewController.h"
@implementation SetContentViewController
@synthesize textView;
```

```
-(id)initWithNibName:(NSString *)nibNameOrNil
     bundle:(NSBundle *)nibBundleOrNil {
     if (self = [super initWithNibName:nibNameOrNil
      bundle:nibBundleOrNil]) {
          self.title=@"请输入内容";
}
     return self;
}

-(void)didReceiveMemoryWarning {
     [super didReceiveMemoryWarning];
}
-(void)viewDidUnload {
}
-(void)dealloc {
     [textView release];
     [super dealloc];
}
@end
```

到这里对各子视图的处理就完成了。稍微休息一下，接下来对最重要的根视图进行编辑。

### 8.8.5　修改根视图

在根视图中主要是获取并显示各个子视图的数据，具体步骤如下：

- 修改 RootViewController.h 文件；
- 修改 RootViewController.m 文件。

**1. 修改 RootViewController.h 文件**

选中 RootViewController.h 文件，添加如下黑体字所示的代码：

**代码 8.30　RootViewController.h 文件**

```
#import "SetDateViewController.h"
#import "SetTypeViewController.h"      ❶
#import "SetContentViewController.h"

@interface RootViewController : UITableViewController
```

```
    <UITableViewDelegate,UITableViewDataSource>{ ❷
    SetDateViewController * setDateController;
    SetTypeViewController * setTypeController;
    SetContentViewController * setContentController;
    NSString * myTitle; ❸
}
@property (nonatomic, retain)
    SetDateViewController * setDatecontroller;
@property (nonatomic, retain)
    SetTypeViewController * setTypeController;
@property (nonatomic, retain)
    SetContentViewController * setContentController;
@property (nonatomic, retain) NSString * myTitle;
-(void)myChange:(id)sender; ❹
@end
```

首先在❶处把各子视图的头文件导入，然后在❷处根据表视图的要求让类遵守两个协议，声明各子视图控制器类。

注意在❸处声明了一个字符串 myTitle，它将用来存储添加的标题，具体使用将在 RootViewController. m 文件中用到时讲解。在❹处添加了一个 myChange 方法，它将用来响应 Text Field 值的变化并将标题存储到 myTitle 中。这样做的原因是标题行的特殊性，它没有下级子视图，所以在操作时需要特别注意。

在开始编辑 RootViewController. m 文件之前，首先应该把模板里的表改为分组表，双击打开 RootViewController. xib 文件，确保选中表视图，调出 Attributes 窗口，在 Style 选项中，将表的类型改为 Grouped。

**2. 修改 RootViewController. m 文件**

选中 RootViewController. m 文件，添加如下黑体字所示的代码：

**代码 8.31 RootViewController. m 文件**

```
#import "RootViewController.h"
#import "CustomCell.h" ❶
@implementation RootViewController
@synthesize setDateController;
@synthesize setTypeController;
@synthesize setContentController;
@synthesize myTitle;
```

```
-(void)viewDidLoad {
    self.title=@"信息记录";❷
    [super viewDidLoad];
}
-(void)viewWillAppear:(BOOL)animated {❸
    [self.tableView reloadData];
}
-(void)didReceiveMemoryWarning {
    [super didReceiveMemoryWarning];
}
-(void)viewDidUnload {
}
-(void)myChange:(id)sender{
    UITextField *myTextField=(UITextField *)sender;      ❹
    self.myTitle=myTextField.text;
}

//数据源方法实现
#pragma mark-
#pragma mark Table Data Source Methods
-(NSInteger)numberOfSectionsInTableView:(UITableView *)tableView{
    return 1;
}
-(NSInteger)tableView:(UITableView *)tableView
        numberOfRowsInSection:(NSInteger)section {
    return 4;
}
//构建根视图里的表
-(UITableViewCell *)tableView:(UITableView *)tableView
        cellForRowAtIndexPath:(NSIndexPath *)indexPath{
    static NSString *recordNoteViewControllerCell =
        @"recordNoteViewControllerCell";
    CustomCell *cell = (CustomCell *)[tableView
    dequeueReusableCellWithIdentifier:
        recordNoteViewControllerCell];
    if (cell == nil){
```

```
        NSArray *nib = [[NSBundle mainBundle]
            loadNibNamed:@"CustomCell"
                owner:self
                options:nil];
        cell = [nib objectAtIndex:0];
    }

    if([indexPath row] == 0) {
        [cell.contentView addSubview:cell.noteLabel];
        cell.noteLabel.text = @"标题:";
        [cell.contentView addSubview:cell.noteTextField];
        cell.noteTextField.placeholder=@"请输入标题";      ❺
        [cell.noteTextField addTarget:self
         action:@selector(myChange:)
        forControlEvents:UIControlEventEditingChanged];
        cell.noteTextField.text=myTitle;
    }
    if(indexPath.row==1) {
        [cell.contentView addSubview:cell.noteLabel];
        cell.noteLabel.text = @"时间:";
        [cell.noteTextField setEnabled:NO];
        [cell.contentView addSubview:cell.noteTextField];   ❻
        cell.noteTextField.placeholder=@"请选择时间";
        cell.noteTextField.text = setDatecontroller.time;
        cell.accessoryType
            =UITableViewCellAccessoryDisclosureIndicator;
    }
    if(indexPath.row==2) {
        [cell.contentView addSubview:cell.noteLabel];
        cell.noteLabel.text = @"类别:";
        [cell.noteTextField setEnabled:NO];
        [cell.contentView addSubview:cell.noteTextField];
        cell.noteTextField.placeholder=@"请选择类别";
        cell.noteTextField.text =setTypeController.types;
        cell.accessoryType
            =UITableViewCellAccessoryDisclosureIndicator;
```

```
        }
        if([indexPath row]==3) {
            [cell.contentView addSubview:cell.noteLabel];
            cell.noteLabel.text = @"内容:";
            [cell.noteTextField setEnabled:NO];
            [cell.contentView addSubview:cell.noteTextField];
            cell.noteTextField.placeholder=@"请输入内容";
            cell.noteTextField.text =
            setContentController.textView.text;
            cell.accessoryType
                =UITableViewCellAccessoryDisclosureIndicator;
        }
        return cell;
}
//委托方法实现
#pragma mark-
#pragma mark Table View Delegate Methods
-(void)tableView:(UITableView *)tableView
    didSelectRowAtIndexPath:(NSIndexPath *)indexPath{
    if(indexPath.row==1){
        if (self.setDatecontroller ==nil) {
        SetDateViewController * nextController =
        [[[SetDateViewController alloc]
            initWithNibName:@"SetDateViewController"
                bundle:nil] autorelease];
        self.setDateController = nextController;
    }
    [self.navigationController pushViewController:
        setDatecontroller animated:YES];
    }
    if (indexPath.row==2){
        if (setTypeController == nil) {
        SetTypeViewController * nextController=
        [[[SetTypeViewController alloc]
            initWithNibName:@"SetTypeViewController"
                bundle:nil] autorelease];
```

```
        self.setTypeController = nextController;
    }
    [self.navigationController pushViewController:
        setTypeController animated:YES];
    }
    if (indexPath.row==3){
        if (setContentController == nil) {
            SetContentViewController * nextController =
            [[[SetContentViewController alloc]
                initWithNibName:@"SetContentViewController"
                    bundle:nil]autorelease];
            self.setContentController = nextController;
        }
        [self.navigationController pushViewController:
            setContentController animated:YES];
    }
}
-(void)dealloc {
    [setContentController release];
    [setTypeController release];
    [setDatecontroller release];
    [myTitle release];
    [super dealloc];
}
@end
```

虽然在这个文件中添加了很多的代码，但是对于绝大多数的方法我们都已经很熟悉，而且代码的结构也很清晰，我们来逐步分析一下。

在❶处导入了 CustomCell.h 文件，由于根视图的表通过构建自定义表视图单元实现，所以要先导入这个头文件。

在❷处的 viewDidLoad 方法中为根视图导航栏设置了标题。

在❸处使用方法重新加载了表视图的数据，因为根视图从下级视图中获取数据后，程序要重新加载表视图，才能把更新的数据显示在表中。

通过调用❹处的方法来获取 Label 内的值，并把它存入前面声明的 myTitle 中，然后在绘制行时把值赋予 noteTextField。

由于前面把表视图更改为分组表了，所以在数据源方法中，首先把表设置为一组（默认)、四行。在方法 tableView:cellForRowAtIndexPath 中我们使用表视图单元来配置表,这些内容在自定义表视图单元中已经讲过,这里不再赘述。需要注意的是,在最后分别为四个行赋值时,标题和其他三项并不相同。

再来看❺处的绘制标题栏语句。首先判断所要赋值的行,在这里为第一行,然后在所选行的 Label 上面赋值为“标题”,设置 Text Field 的默认值为"请输入标题",然后将 Label 和 Text Field 添加到 contentView 中。为了给 Text Field 赋值,使用了下面的语句:

```
[cell.noteTextField addTarget:self
        action:@selector(myChange:)
      forControlEvents:UIControlEventEditingChanged];
      cell.noteTextField.text=myTitle;
```

它的作用是当检测到 Text Field 内的值发生变化时,就调用前面介绍的 myChange:方法将值传递给 myTitle,最后再把值传递回 noteTextField,这样就实现了标题行的值的保存,不会出现单击其他行进入下级视图再返回时,标题无法保存的状况。

接下来绘制其他行时使用的方法与❻处绘制时间行的方法一样,通过[**cell.noteTextField setEnabled:NO**]设置 Text Field 为不可编辑状态,因为它只是用来显示子视图的内容。最后在表行上添加一个扩展指示器,它将带领我们进入下级详细视图。

在表视图的委托方法 tableView:didSelectRowAtIndexPath 中,分别为每一行单独设置单击选中后进入下级视图的操作,与以往不同的是这里采用逐行处理的方式。

**3. 编译并运行**

到这里程序就完成了,编译并运行程序,效果如图 8-45 所示。

图 8-45 最终效果

# 8.9 表视图的美化

我们已经对表视图的构建和基本操作方法都有了深入的了解,在这一节中,我们将利用简单的委托方法来美化表格。

## 8.9.1 在行左侧添加图像

想要向每一行添加一个图像,可以借助 UITableViewCell 子类来实现,如果只是让图像位于每一行的左侧,则可以借助表视图单元默认的 image 属性来实现，这一节里我们就来实现在行的左侧添加图片。

打开在 8.2 节里创建的 NoteScan 项目，将在这个项目的基础上进行修改。找到随书光盘中的 aiside. png 图像文件，把它添加到项目中的 Resources 文件夹里，当然也可以添加自己喜欢的其他图标。

在行的左侧添加图片的代码非常简单，在 NoteScanViewController. m 文件里添加如下黑体字所示的代码：

```
-(UITableViewCell *)tableView:(UITableView *)tableView
    cellForRowAtIndexPath:(NSIndexPath *)indexPath {
    ......
    NSUInteger row=[indexPath row];
    cell.textLabel.text=[tableData objcetAtIndex :row];
    UIImage *image=[UIImage imageNamed:@"aiside.png"];
    cell.imageView.image=image;
    return cell;
}
```

这样就把表视图单元的 image 属性设置为 aiside. png 了。编译并运行程序，可以看到如图 8-46 所示的效果。还可以为每一行设置不同的图像，当然要费些功夫，不过可以自己试试看！

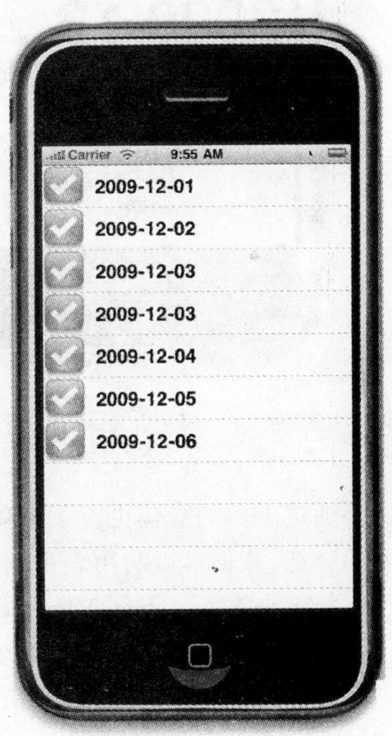


图 8-46 添加图标

## 8.9.2 利用委托配置表视图

在实现了一个简单的表时，我们可以清楚地看到，程序中主要实现了 UITableViewDataSource 方法，因为较为简单的表视图只是利用数据源提供绘制表时所需的数据，利用委托来完成外观配置和处理某些用户交互。接下来就学习几个简单的配置选项。

**1. 设置字体大小和行高**

有很多时候我们希望更改表视图中文字的大小，以符合特定应用程序的需要。要做到这一点，只需添加如下黑体字所示的代码：

```
-(UITableViewCell *)tableView:(UITableView
    *)tableView
    cellForRowAtIndexPath:(NSIndexPath *)
    indexPath {
    ......
    NSUInteger row = [indexPath row];
```

```
    cell.textLabel.text = [tableData objcetAtIndex :row];
    cell.textLabel.font = [UIFont boldSystemFontOfSize:60];
    return cell;
}
```

在这里设置字体大小为 60 像素，编译并运行程序，可以看到表中的数字已经变大，如图 8-47 所示。

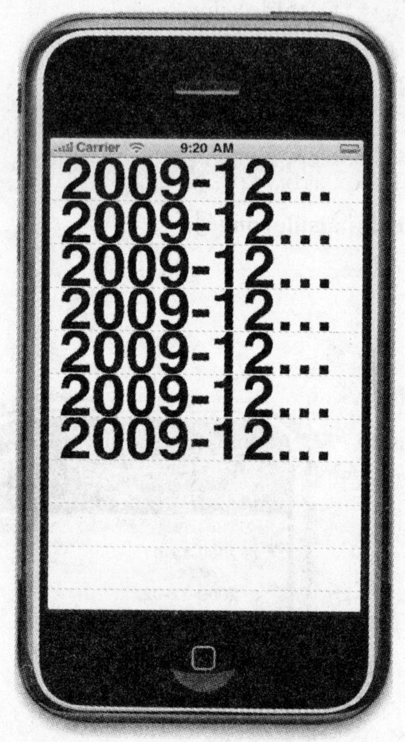


图 8-47　设置字体大小

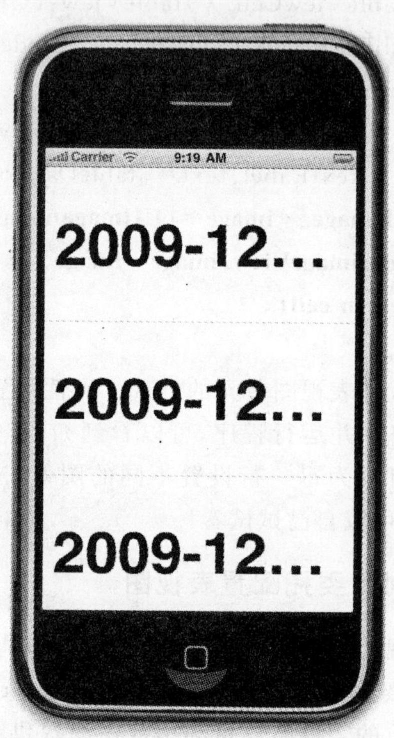


图 8-48　设置行高

通过上面的操作更改了字体的大小，但是文本和图像占满了整个表格，这样的大小并不合适，我们可以利用表视图委托协议中的方法来设置合适的行高。

接下来在 NoteScanViewController.m 文件里添加代码，具体代码如下：

```
#pragma mark -
#pragma mark TableViewDelegate Methods
-(CGFloat)tableView:(UITableView *)tableView
    heightForRowAtIndexPath:(NSIndexPath *)indexPath {
    return 160;
}
```

通过上面这一段代码，我们把表的行高设置为 160 像素。编译并运行程序，可以看到如图 8-48 所示的效果。

**2. 设置行缩进级别**

我们通常所见的表行显示数据时，都是有规则地排列的，但是有时候也有一些表需要处理缩进级别以显示不同的层次，就如同文件夹系统一样，它需要用缩进来表示不同的文件夹层次。下面我们就利用委托来实现行的缩进，具体代码如下：

```
-(NSInteger)tableView:(UITableView *)tableView
    indentationLevelForRowAtIndexPath:
    (NSIndexPath *)indexPath{
        NSUInteger row=[indexPath row];
        return row;
}
```

需要注意的是，在这里将每一行的缩进级别设置为它们各自的行号，也就是第 0 行缩进级别为 0，第 1 行的缩进级别为 1，缩进级别逐级递增，行也就会相应地逐级右移。编译并运行程序，可以看到如图 8-49 所示的效果。

其实在常见的应用程序中，表视图一般需要通过委托来进行美化，这里所讲的内容只是基本的表视图美化效果，如果想要做出华丽的效果，需要我们精心设计。

图 8-49 设置缩进

## 8.10 小结

在本章中，我们学习了如何构建表视图以及它的一些操作，其中需要注意的是：数据源和协议方法的使用、合理搭配使用表视图的各种操作、自定义表视图单元，它们是本章的重点。

在本章学习的这些内容，是本书的核心知识之一，也是以后开发复杂的 iPhone 应用程序的基础。

# 第 9 章　数据持久性存储

**本章内容**

- ■ 持久性保存的概念
- ■ 使用属性列表文件保存数据
- ■ 使用归档保存数据
- ■ 使用嵌入式数据库保存数据
- ■ 使用 Core Data 保存数据

在 iPhone 应用程序开发中，数据持久性保存是很重要的内容。iPhone 开发中我们可以采取多种机制来持久保存数据，本章将重点介绍其中四种，分别是属性列表、对象归档、iPhone 的嵌入式关系数据库(SQLite3)和 Core Data，当然这也只是几种常用的数据保存机制，其他像传统的 C 语言 I/O 调用等就不再介绍，因为它需要很大的代码量，使用不便。

## 9.1　应用程序沙盒

我们采用的这几种数据持久性存储机制，使每个应用程序都拥有一个 Documents 文件夹，主要用来存储应用程序的数据。要想了解数据存储的问题，首先应当知道它的存储位置。在 iPhone 中应用程序数据存储在当前用户文件目录的如下路径中：

**Library/Application Support/iPhoneSimulator/User/Applications**

按照这个路径打开，就会发现一系列的文件夹，如图 9-1 所示，每一个文件夹都包含一个应用程序和其对应的文件夹。我们通常所做的项目，编译运行后都会在此创建这样一个文件夹，用于保存项目的数据。

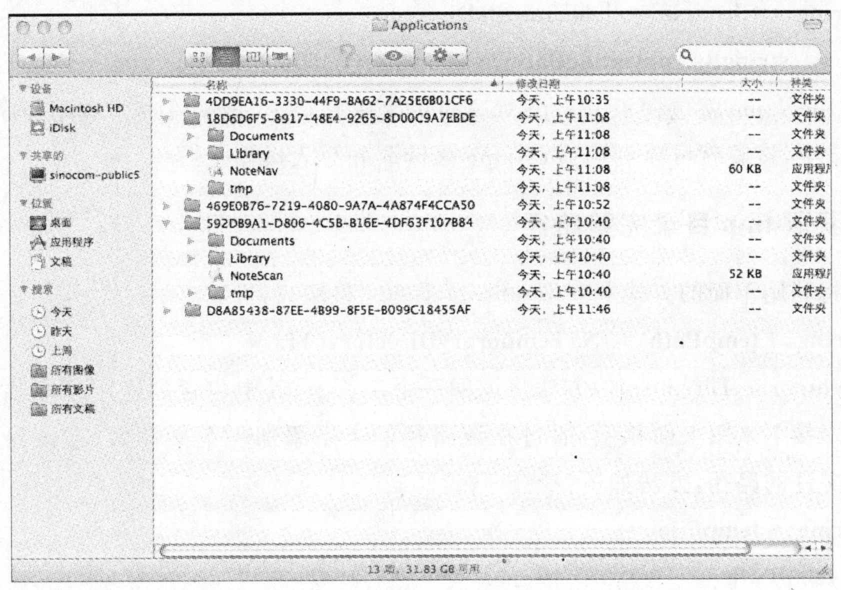

图 9-1 文件系统

按上面路径打开其中任何一个 iPhone 应用程序文件夹，你会发现其中又有三个文件夹：Documents、Library 和 tmp。我们前面提过 Documents 目录用于存储应用程序数据（iPhone 自带应用程序例外，如 Settings），tmp 目录用于存储应用程序的临时文件。如果要在程序中实现持久性存储，必须要检索并获取存储目录的完整路径，然后才能够进行数据的读取和写入。所以接下来我们来学习一下如何获取 Documents 目录和 tmp 目录的完整路径。

## 9.1.1 获取 Documents 目录完整路径

在 Xcode 中自动生成的应用程序文件夹是一个随机的名称，但可以利用一个简单的方法来获取 Documents 目录的完整路径，代码如下：

```
NSArray * paths = NSSearchPathForDirectoriesInDomains
          (NSDocumentDirectory, NSUserDomainMask, YES);
NSString * documentsDirectory = [paths objectAtIndex: 0];
```

我们首先使用一个名为 NSSearchPathForDirectoriesInDomains 的 C 函数来查找目录，NSDocumentDirectory 表明正在查找 Documents 目录的路径；常量 NSUserDomainMask 是一个枚举值，获取的目录数为 1，用于将搜索范围限制在沙盒范围之内。把得到的值付给一个数组，接着取出数组的第一个元素，即是 Documents 文件夹的路径。

获取路径之后，我们要创建一个新的文件，用于写入和读出数据，新的文件名称按如下方式定义：

```
NSString * filename = [documentsDirectory
        stringByAppendingPathComponent:@"file.plist"];
```

这样 fileName 就包含了 file. plist 文件的完整路径了，该文件存于应用程序的 Documents 目录中，接着就可以利用 fileName 来读取和写入操作了。

### 9.1.2　获取 tmp 目录完整路径

我们将利用下面的方式来获取 tmp 目录的完整路径：

```
NSString * tempPath = NSTemporaryDirectory();
```

NSTemporaryDirectory()方法将返回一个字符串，其中包含到应用程序临时目录的完整路径。接下来跟上面相似，通过在该路径的末尾添加一个文件名，在该目录中创建一个到该文件的路径，方法如下：

```
NSString * tempFile =
    [tempPath
stringByAppendingPathComPonent:@"tempFile.plist"];
```

## 9.2　文件保存策略

在本章中要学习的四种文件持久保存机制，采用的文件保存方法都不一样。对于属性列表和归档要考虑将数据存储在一个文件还是多个文件中；如果采用嵌入式数据库(SQLite3)存储，只需创建一个数据库文件，进行数据存储和检索即可；而本章采用的 Core Data 是一个管理对象生命周期、对象图和持久性的框架，它对于数据持久性存储也是通过实体完成的。下面我们就来比较一下，使用单个文件和使用多个文件持久保存数据的不同之处。

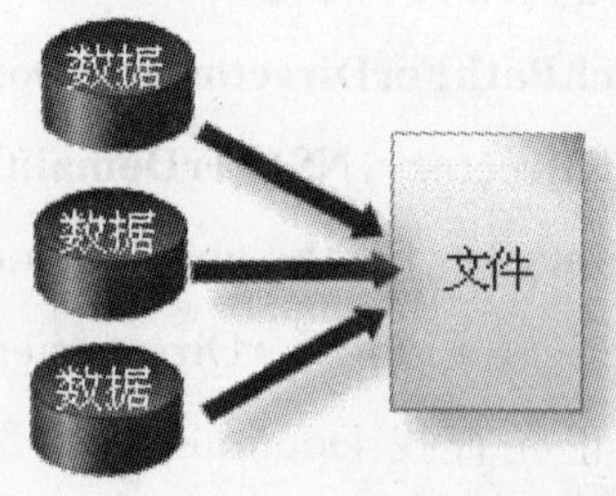


图 9-2　单文件持久性存储

单个文件持久性存储是比较常用的一种方式，如图 9-2 所示。首先创建一个根对象，一般为数组或者字典类型，当然也可以基于自定义类，接下来就用应用程序的数据来填充根对象。保存数据时，用代码把根对象里的全部内容重新写入单个文件，应用程序再重新启动时会将该文件的全部内容读入内存，在退出时注销全部内容，这样就实现了持久性存储。

单文件持久性存储缺点很明显，就是必须将全部的应用程序数据存储到内存中，而且必须把所有数据全部写入文件系统，不管我们更改了多少数据，都必须进行这些操作，所以使用此方法时，应确保应用程序管理的数据不超过几兆字节。

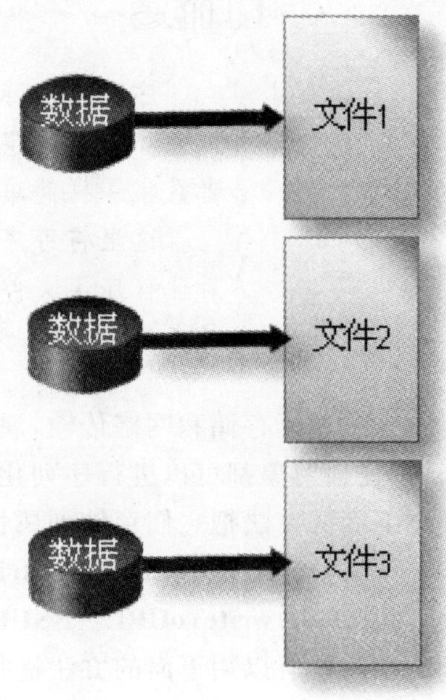

图 9-3　多文件持久性存储

多个文件持久性存储是一个复杂方法，它将数据保存到多个文件中，如图 9-3 所示。使用这种方式存储的优势是，它允许应用程序只加载用户请求的数据，而这些数据都保存在不同的文件中，更改时只需保存更改的文件即可，它还允许开发人员在收到内存不足的通知时释放内存。

多文件持久性存储的缺点是使应用程序变得非常复杂，对于我们所学习的部分，建议采用更为简单的单个文件持久性存储。

# 9.3　使用属性列表保存应用程序数据

从本节开始，我们把数据持久性存储到应用程序中。首先学习使用属性列表持久保存数据的方法，在这里将使用上一章 8.8 节可编辑详细内容窗格的项目来实现持久性存储。

## 9.3.1　属性列表序列化

在前面的一些项目中，我们已经使用过属性列表，属性列表非常简单实用，只要字典或数组中仅包含特定可序列化的对象，就可以将字典或数组实例写入属性列表，从属性列表里创建它们。

**Tips**

序列化实际上就是把对象的状态信息转换为可以传输和存储的字节类型序列，让我们能像存储文本或者数字一样简单地存储对象。

这里将可序列化的对象写到属性列表文件中，然后让程序去文件中读取序列化的对象并还原为原来的状态。

为了便于存储和网络传输，属性列表对象在写入过程中已经被转换为字节流。一般来讲，任何对象都可以进行序列化，但是只能把某些对象的实例放到集合类中，然后可以使用下面的方法把它们存储到属性列表中：

```
-(BOOL)writeToFile:(NSString *)aPath atomically:(BOOL)flag;
-(BOOL)writeToURL:(NSURL *)aURL atomically:(BOOL)flag;
```

相应地可以用下面的方法进行读取：

```
-(id)initWithContentsOfFile:(NSString *)aPath;
-(id)initWithContentsOfURL:(NSURL *)aURL;
```

在前面章节中学习了 Objective-C 类，例如：NSArray、NSMutableArray、NSDictionary、NSMutaleDictionary、NSString、NSMutableString、NSData、NSMutableData、NSDate、NSNumber，这些类都可以按上面的方法进行序列化。

使用属性列表也存在一些问题，我们无法将自定义的对象序列化到属性列表中，而且不能使用如 NSURL、NSImage、NSColor 等这些没有在之前的可序列化对象列表中指定的类，这些限制导致我们无法轻松地创建派生和计算的属性，当然除去这点，属性列表在应用程序中还是非常有用的。

### 9.3.2　属性列表在应用程序中的使用

在这一小节中，我们在 8.8 节实现的项目基础上，利用属性列表实现持久保存应用程序数据，也就是将更改的信息保存下来，在程序重新启动时数据已经更新。首先将原项目备份，然后看看本节要实现的效果，如图 9-4 所示。

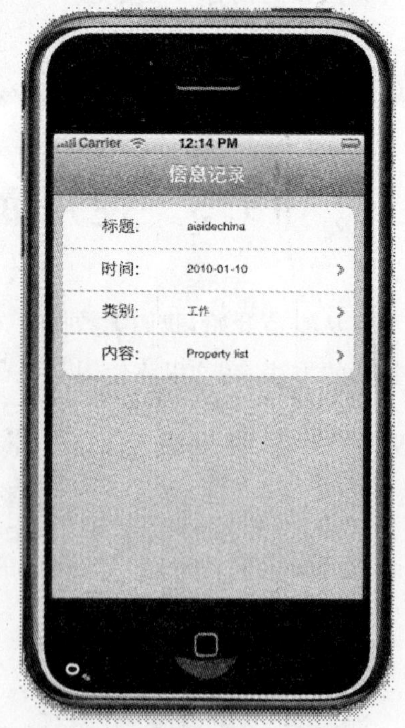


图 9-4　属性列表存储效果

了解了最终要实现的效果，接下来看看如何实现它。我们把程序分为以下几个步骤：

- 修改 RootViewController.h 文件，定义存储的

必要方法；

- 修改 RootViewController.m 文件，利用属性列表实现数据存储；
- 编译并运行。

**1. 修改 RootViewController.h 文件，定义存储的必要方法**

打开 NoteRecord 项目，我们对于大部分文件都不会改动，只是对根控制器文件进行修改。选中 RootViewController.h 文件，进行如下修改：

**代码 9.1 RootViewController.h 文件**

```
#import "SetDateViewController.h"
#import "SetTypeViewController.h"
#import "SetContentViewController.h"
#define kFilename @"data.plist" ❶
@interface RootViewController : UITableViewController
    <UITableViewDelegate, UITableViewDataSource>{
    SetDateViewController *setDatecontroller;
    SetTypeViewController *setTypeController;
    SetContentViewController *setContentController;
    NSString *myTitle;
    NSArray *array1; ❷
}
@property (nonatomic, retain)
        SetDateViewController *setDatecontroller;
@property (nonatomic, retain)
        SetTypeViewController *setTypeController;
@property (nonatomic, retain)
        SetContentViewController *setContentController;
@property (nonatomic, retain) NSString *myTitle;
@property (nonatomic, retain) NSArray *array1;
-(NSString *)dataFilePath; ❸
@end
```

首先在❶处用一个常量 kFilename 定义了文件名，接下来在❷处声明了一个数组用来存储从下级子视图中获取的数据，最后在❸处定义的方法用于返回数据文件的完整路径名。

**2. 修改 RootViewController. m 文件，利用属性列表实现数据存储**

由于 RootViewController. xib 文件没有任何改动，所以直接修改 RootViewController. m 文件实现使用属性列表持久存储应用程序数据。

选中 RootViewController. m 文件，修改如下：

**代码 9.2 RootViewController. m 文件**

```
#import "RootViewController.h"
#import "CustomCell.h"
@implementation RootViewController
@synthesize setDatecontroller;
@synthesize setTypeController;
@synthesize setContentController;
@synthesize myTitle;
@synthesize array1;
//获取存储目录完整路径名
-(NSString *)dataFilePath {
    NSArray *paths=NSSearchPathForDirectoriesInDomains
        (NSDocumentDirectory,NSUserDomainMask,YES);
    NSString *documentsDirectory=[paths objectAtIndex:0];      ❶
    return [documentsDirectory
        stringByAppendingPathComponent :kFilename];
}
//退出程序时保存数据
```

```
-(void)applicationWillTerminate:(NSNotification *)notification {
    NSString *str1 = [[(CustomCell *)
            [self.tableView cellForRowAtIndexPath:
    [NSIndexPath indexPathForRow:0 inSection:0]]
            noteTextField] text];
    NSString *str2 = [[(CustomCell *)
            [self.tableView cellForRowAtIndexPath:
    [NSIndexPath indexPathForRow:1 inSection:0]]
            noteTextField] text];
    NSString *str3 = [[(CustomCell *)
            [self.tableView cellForRowAtIndexPath:
    [NSIndexPath indexPathForRow:2 inSection:0]]
            noteTextField] text];                              ❷
    NSString *str4 = [[(CustomCell *)
            [self.tableView cellForRowAtIndexPath:
    [NSIndexPath indexPathForRow:3 inSection:0]]
            noteTextField] text];
    NSMutableArray *array=[[NSMutableArray alloc]init];
    [array addObject:str1];
    [array addObject:str2];
    [array addObject:str3];
    [array addObject:str4];
    [array writeToFile:[self dataFilePath]atomically:YES];
    [array release];
}

//重新启动后从属性列表文件中读取数据来初始化
#pragma mark -
-(void)viewDidLoad {
    self.title=@"信息记录";
    NSString *filePath=[self dataFilePath];
    if([[NSFileManager defaultManager]
        fileExistsAtPath:filePath]){
        NSArray *array=[[NSArray alloc]    ❸
            initWithContentsOfFile:filePath];
        self.array1 = array;
        [array release];
```

```
    }
    self.myTitle=[array1 objectAtIndex:0]; ❹
    UIApplication *app=[UIApplication sharedApplication];
    [[NSNotificationCenter defaultCenter]
            addObserver:self
            selector:@selector(applicationWillTerminate:)          ❺
            name:UIApplicationWillTerminateNotification
            object:app];
    [super viewDidLoad];
}
-(void)viewWillAppear:(BOOL)animated {
    [self.tableView reloadData];
}
//数据源方法的实现
#pragma mark-
#pragma mark Table Data Source Methods
-(NSInteger)numberOfSectionsInTableView:(UITableView *)tableView
{
    return 1;
}
-(NSInteger)tableView:(UITableView *)tableView
     numberOfRowsInSection:(NSInteger)section {
     return 4;
}
-(UITableViewCell *)tableView:(UITableView *)tableView
    cellForRowAtIndexPath:(NSIndexPath *)indexPath {
    Static NSString *recordNoteViewControllerCell
                = @"recordNoteViewControllerCell";
    CustomCell *cell = (CustomCell *)[tableView
    dequeueReusableCellWithIdentifier:recordNoteViewControllerCell];
    if (cell == nil) {
        NSArray *nib = [[NSBundle mainBundle]
                        loadNibNamed:@"CustomCell"
                            owner:self
                            options:nil];
        cell = [nib objectAtIndex:0];
    }
```

```
        if([indexPath row] == 0) {
            [cell.contentView addSubview:cell.noteLabel];
            cell.noteLabel.text = @"标题:";
            [cell.contentView addSubview:cell.noteTextField];
            cell.noteTextField.placeholder=@"请输入标题";
            [cell.noteTextField addTarget:self
                      action:@selector(myChange:)
             forControlEvents:UIControlEventEditingChanged];
            cell.noteTextField.text=myTitle;
    }
    //绘制时间行
    if(indexPath.row==1) {
        [cell.contentView addSubview:cell.noteLabel];
        cell.noteLabel.text = @"时间:";
        [cell.noteTextField setEnabled:NO];
        [cell.contentView addSubview:cell.noteTextField];
        cell.noteTextField.placeholder=@"请选择时间";
        cell.noteTextField.text = setDatecontroller.time;
        if(setDatecontroller != nil)
            cell.noteTextField.text = setDatecontroller .time;
        else                                                        ❻
            cell.noteTextField.text = [array1 objectAtIndex:1];
        cell.accessoryType=
            UITableViewCellAccessoryDisclosureIndicator;
    }
    //绘制类别行
    if (indexPath.row==2) {
        [cell.contentView addSubview:cell.noteLabel];
        cell.noteLabel.text = @"类别:";
        [cell.noteTextField setEnabled:NO];
        [cell.contentView addSubview:cell.noteTextField];
        cell.noteTextField.placeholder=@"请选择类别";
        cell.noteTextField.text =setTypeController.types;
        if(setTypeController != nil)
            cell.noteTextField.text = setTypeController.types;
        else
            cell.noteTextField.text =[array1 objectAtIndex:2];
```

```
            cell.accessoryType=
                    UITableViewCellAccessoryDisclosureIndicator;
        }
        //绘制详细内容行
        if([indexPath row]==3) {
            [cell.contentView addSubview:cell.noteLabel];
            cell.noteLabel.text = @"内容:";
            [cell.noteTextField setEnabled:NO];
            [cell.contentView addSubview:cell.noteTextField];
            cell.noteTextField.placeholder = @"请输入内容"
            cell.noteTextField.text = setContentController.textView.text;
            if(setContentController !=nil)
                cell.noteTextField.text=
                    setContentController.textView.text;
                else
                    cell.noteTextField.text = [array1 objectAtIndex:3];
                cell.accessoryType
                    =UITableViewCellAccessoryDisclosureIndicator;
        }
        return cell;
    }
    #pragma mark-
    #pragma mark Table View Delegate Methods
    ……
    -(void)dealloc {
        [array1 release];
        [setContentController release];
        [setTypeController release];
        [setDatecontroller release];
        [myTitle release];
        [super dealloc];
    }
    @end
```

首先结合程序来分析一下利用属性列表实现数据持久性存储的原理。如果要实现数据持久性存储,必须首先获取存储目录完整路径名,在❶处添加的方法就是通过查找文档目录并附加 kFilename 来实现的。

❷处添加的方法 applicationWillTerminate:(NSNotification *)notification 就用来实现保存功能。此方法不需要在头文件中声明,因为这是 API 提供的方法,当程序退出时会自动查找这个方法并执行。在这个方法中,使用指向 NSNotification 的指针作为参数,实际上这个方法就是一个通知,而且所有的通知都使用一个 NSNotification 实例作为参数。这一段代码实现真正的写入操作,将可变数组里的内容写入属性列表文件,从而使用属性列表将数据保存下来。

在 viewDidLoad 方法中,首先通过❸处的代码检查数据文件是否存在,如果不存在,我们就不再尝试加载它;如果存在,就用该文件的内容实例化数组 array1,也就是把上面保存在数组里的对象复制到四个文本字段中去。值得注意的是,在这里我们实例化了数组 array1,使用数组 array1 所获取的值来构建后面的表视图,所以在 table:cellForRowAtIndexPath方法中进行如下操作:

如果表行没有下级视图(标题行),则在❹处指定 myTitle 的值为 array1 中存储的标题值,之所以把它放到 ViewDidLoad 方法中是因为要在每次加载视图时都把 array1 中的值赋给 myTitle,这样标题行中的值总是 array1 中存储的值,也就实现了持久性存储。

如果表行能够进入下级视图(时间行、类型行、详细内容行),则在❻ 处进行判断,如果控制器的内容没有改变,没有进入下级视图,则直接将数组 array1 里的值赋给 noteTextField;如果发生改变(setDateController =! nil),进入下级子视图,则将子视图中改变的值赋给 noteTextField。

再次回到 viewDidLoad 方法中,来看❺处的代码。从属性列表中获得数据之后,就获得了应用程序实例的引用,我们使用这个引用来订阅通知,并告知通知中心来调用该方法。这里使用默认的 NSNotification 实例和 addobserver:selector:name:object:方法来实现这些功能。

在这个方法中,observer 设为 self,这就意味着 RootViewController 是我们要通知的对象;传递 selector 给刚才编写的 applicationWillTerminate 方法,告诉通知中心在发布通知时调用这个方法;参数 name 是想要接收的通知的名称;最后的 object 是从中获取通知的对象,如果它的值为 nil,则通知我们发布(UIApplicationWillTerminateNotification)的时间和方法。

**3. 编译并运行**

到这里操作就完成了,编译并运行,将表中的各个选项填写完整,然后点击模拟器下方的 Home 按钮,当再次打开程序时就会看到数据已经被保存下来。

## 9.4 使用归档持久保存应用程序数据

所谓归档其实就是另一种形式的序列化,当然它的应用范围比起属性列表更加广泛,基本上常见的支持数据存储的 Foundation 类和 Cocoa Touch 类都支持归档。我们可以将复杂的对象写入归档文件,然后再读取出来,这样就简单地实现了数据持久性存储。

当然要实现归档，必须符合 NSCoding 协议才能实现编码，如果要实现对象的复制，还要符合 NSCopying 协议，这两个协议在归档中发挥着巨大的作用，接下来我们就了解一下这两个协议和其中的一些重要方法。

### 9.4.1　NSCoding 协议和 NSCopying 协议

要实现对象的归档，符合 NSCoding 协议是必备的前提，因为它提供了归档必备的两个方法，一个是用来将对象编码到归档中，另一个是借助对归档的解码来创建一个新对象。NSCopying 协议的复制方法也经常使用，下面我们来详细了解一下这些方法。

**1. 将对象编码到归档中**

编码方法和解码方法都传递一个 NSCoder 实例，通常采用键－值编码来对对象和标量(如 int 和 float)进行编码和解码。编码方法大致可以写成如下方式：

```
-(void)encodeWithCoder:(NSCoder *)encoder{
    [encoder  encodeObject:object1 forKey:kObject1Key];
    [encoder  encodeObject:object2  forKey:kObject2Key];
    [encoder  encodeInt:oneInt  forKey:kOneIntKey];
    [encoder  encodeFloat:oneFloat  forKey:kOneFloatKey];
}
```

**2. 将归档解码创建新文件**

在完成编码之后，我们还需要初始化 NSCoder 中的对象，以还原以前归档的对象，这个方法分为两种方式。如果父类不符合 NSCoding，就只对子类解码，方法如下：

```
-(id)initWithCoder:(NSCoder *)decoder{
    if(self =[super init]){
        self.object1 = [decoder decoderObjectForKey:kObject1Key];
        self.object2 = [decoder decoderObjectForKey:kObject2Key];
        sSelf.oneInt = [decoder decoderIntForKey:kOneIntKey];
        self.oneFloat=[decoder decoderFloatForKey:kOneFloatKey];
    }
        return self;
}
```

在这里使用[super init]实现了对象实例的初始化，没有调用父类的 initWithCoder，因为 NSObject 没有实现它。但是有时候会碰到某些类的父类也符合 NSCoding 协议，在实现 NSCoding 时，采用的方法如下：

```
-(id)initWithCoder:(NSCoder *)decoder{
    if(self =[super initWithCoder:decoder]){
        self. object1 = [decoder decoderObjectForKey:kObject1Key];
        self. object2= [decoder decoderObjectForKey:kObject2Key];
        self.oneInt = [decoder decoderIntForKey:kOneIntKey];
```

```
        self.oneFloat= [decoder decoderFloatForKey:kOneFloatKey];
    }
        return self;
}
```

可以看出两种方法的不同之处就在于父类是否符合 NSCoding 协议。到这里我们就可以使用这两种方法来对所有对象的属性进行编码和解码了,然后可以对对象进行归档,进而实现归档文件的写入和读取。

**3. NSCopying 协议的复制方法**

前面介绍过,NSCopying 协议在归档中是非常重要的协议,它通常与 NSCoding 协议一起出现,这是因为 NSCopying 协议有一个 copyingwithzone:方法,该方法可以用于复制对象,方法的代码如下:

```
-(id) copyWithZone:(NSZone *)zone {
    MyClass *copy= [[[self calss] allocWithZone: zone] init];
    copy.object1= [self.object1 copy];
    copy.object2= [self.object2 copy];
    copy.oneInt =self.oneInt;
    copy.oneFloat =self.oneFloat;
    return copy;
}
```

这里创建了一个同类的新实例对象,并把该实例对象的属性都设置为与该对象相同的值,这就实现了复制,但是我们并没有直接释放新创建的对象,而是调用复制方法将其释放。

### 9.4.2 归档的实现与取消

前面我们学习了实现归档要符合的两个协议即 NSCoding 和 NSCopying,接下来学习本节的重点:归档的实现和取消。

归档的实现其实相对容易,我们把代码分成几部分进行解释。

**1. 实现归档**

```
NSMutableData *theData = [[NSMutableData alloc] init];
NSKeyedArchiver *archiver = [[NSKeyedArchiver alloc]
                          initForWritingWithMutableData:theData];
```

首先我们创建了一个 NSMutableData 实例,用于包含编码的数据,接下来创建 NSKeyedArchiver 实例来将对象归档到 NSMutableData 实例中。

下一步,我们要使用键-值编码来对需要包含在归档文件中的对象进行归档。举个例子:

```
[archiver encodeObject: myobject forKey:kDataKey];
```

最后,完成对象的编码之后,需要告诉归档程序已经完成,把 NSMutableData 实例写入文件系统，然后对对象进行内存清理。代码如下：

```
[archiver  finishEncoding];
BOOL success = [data writeToFile:@"/path/to/archive"
        atomically:YES];
[archiver release];
[data release];
```

这样我们就完成了对数据对象的归档。接下来要从归档重新变回对象，也就是对数据对象取消归档。

**2. 取消归档**

```
NSData *theData = [NSData alloc] initWithContentsOfFile:path];
NSKeyedUnarchiver *unarchiver = [[NSKeyedUnarchiver alloc]
            initForReadingWithData:theData];
```

完成归档的取消后，就需要使用前面归档时使用过的键值(kDataKey)来从程序中读取对象：

```
self.object =[unarchiver decodeObjectForKey :kDataKey];
```

最后一步就是告诉归档程序完成了读取操作，清理内存即可。

```
[unarchiver finishDecoding];
[unarchiver release];
[data release];
```

到现在为止，我们对于归档的方法有了大体的认识，下面我们就讲解如何在应用程序中用归档来实现数据持久存储。

### 9.4.3 归档在应用程序中的使用

下面，我们将在 9.3.2 小节的程序基础之上进行修改，实现使用归档来持久保存应用程序数据，最终效果与 9.3.2 小节相同，如图 9-5 所示。现在就开始分步实现这个项目：

- 自定义一个 NoteMsg 类作为数据模型；
- 修改 RootViewController.h 文件；
- 修改 RootViewController.m 文件实现持久性存储。

图 9-5 使用归档存储效果

**1. 自定义一个 NoteMsg 类作为数据模型**

首先创建一个新文件作为数据模型，用来放置源程序字典中的数据。打开 NoteRecord 项目，然后右键单击 Classes 文件夹，在新建文件向导中选择

NSObject subclass，命名为 NoteMsg. m，确保同时创建头文件。

选中 NoteMsg. h 文件，添加如下黑体字所示的代码：

**代码 9.3 NoteMsg. h 文件**

```
#define kStrTitleKey @"StrTitle"
#define kStrDateKey @"StrDate"
#define kStrTypesKey @"StrTypes"
#define kStrContentKey @"StrContent"
#import <Foundation/Foundation.h>
@interface NoteMsg : NSObject
    <NSCoding, NSCopying> {❶
NSString * strTitle;
NSString * strDate;        ❷
NSString * strTypes;
NSString * strContent;
}
@property (nonatomic, retain) NSString * strTitle;
@property (nonatomic, retain) NSString * strDate;
@property (nonatomic, retain) NSString * strTypes;
@property (nonatomic, retain) NSString * strContent;
@end
```

在❶处需要让这个类遵循 NSCoding 协议和 NSCopying 协议，这是实现归档的前提。在❷处表示上面创建的一个简单类用来充当数据模型，这个类具有四个字符串属性，分别用来存储标题、时间、类型和详细内容。

选中 NoteMsg. m 文件，添加如下黑体字所示的代码：

**代码 9.4 NoteMsg. m 文件**

```
#import "NoteMsg.h"
@implementation NoteMsg
@synthesize strTitle;
@synthesize strDate;
@synthesize strTypes;
@synthesize strContent;
```

```
#pragma mark NSCoding
-(void)encodeWithCoder:(NSCoder *)encoder {
    [encoder encodeObject:strTitle forKey:kStrTitleKey];
    [encoder encodeObject:strDate forKey:kStrDateKey];          ❶
    [encoder encodeObject:strTypes forKey:kStrTypesKey];
    [encoder encodeObject:strContent
            forKey:kStrContentKey];
}
-(id)initWithCoder:(NSCoder *)decoder {
    if (self=[super init]) {
        self.strTitle=[decoder
            decodeObjectForKey:kStrTitleKey];
        self.strDate=[decoder
            decodeObjectForKey:kStrDateKey];
        self.strTypes=[decoder                                  ❷
            decodeObjectForKey:kStrTypesKey];
        self.strContent=[decoder
            decodeObjectForKey:kStrContentKey];
    }
    return self;
}
#pragma mark-
#pragma mark NSCopying
-(id)copyWithZone:(NSZone *)zone {
    NoteMsg *copy=[[[self class] allocWithZone:zone] init];
    strTitle= [self.strTitle copy];
    strDate= [self.strDate copy];                               ❸
    strTypes= [self.strTypes copy];
    strContent= [self.strContent copy];
    return copy;
}
@end
```

这里添加的方法，我们前面已经学习过，首先使用❶处方法对四个字符串进行编码，然后使用❷处方法进行解码，它们共同属于 NSCoding 协议。最后使用了❸处属于 NSCopying 协议的方法，把四个字符串复制到刚刚新建的 NoteMsg 对象中，这样就创建了可以归档的数据对象，接下要做的就是使用这些对象来实现持久保存程序数据。

### 2. 修改 RootViewController.h 文件

选中 RootViewController.h 文件，进行如下修改：

**代码 9.5 RootViewController.h 文件**

```
#import "SetDateViewController.h"
#import "SetTypeViewController.h"
#import "SetContentViewController.h"
#define kFilename @"data.plist"
#define kFilename @"archive"   ❶
#define kDataKey @"Data"   ❷
@class NoteMsg;❸
@interface RootViewController : UITableViewController
    <UITableViewDelegate, UITableViewDataSource>{
    SetDateViewController *setDatecontroller;
    SetTypeViewController *setTypeController;
    SetContentViewController *setContentController;
    NSArray *array1;
    NSString *myTitle;
    NoteMsg *noteMsg;
}
@property (nonatomic, retain)
        SetDateViewController *setDatecontroller;
@property (nonatomic, retain)
        SetTypeViewController *setTypeController;
@property (nonatomic, retain)
        SetContentViewController *setContentController;
@property (nonatomic, retain) NSArray *array1;
@property (nonatomic, retain) NSString *myTitle;
@property (nonatomic, retain) NoteMsg *noteMsg;
-(void)myChange:(id)sender;
-(NSString *)dataFilePath;
-(void)applicationWillTerminate:(NSNotification *)notification;
@end
```

在这里我们采用归档来实现应用程序数据的持久性存储，所以为了防止混乱，就必须要先删除原有的属性列表文件名，然后在❶处定义我们自己的新文件名，还要在❷处定义一个常量用来作为编码和解码的键值。我们在❸处定义了 NoteMsg 类，这样可以方便地调用对象，不必每次都初始化了。完成这些之后，我们就开始实现应用程序数据的持久性储存。

**3. 修改 RootViewController.m 文件实现持久性存储**

选中 RootViewController.m 文件，进行如下修改：

**代码 9.6　RootViewController.m 文件**

```
#import "RootViewController.h"
#import "CustomCell.h"
#import "NoteMsg.h" ❶
@implementation RootViewController
@synthesize setDatecontroller;
@synthesize setTypeController;
@synthesize setContentController;
@synthesize array1;
@synthesize myTitle;
@synthesize noteMsg;

//获取存储目录完整路径名
-(NSString *)dataFilePath {
    NSArray *paths=NSSearchPathForDirectoriesInDomains
    (NSDocumentDirectory, NSUserDomainMask, YES);
    NSString *documentsDirectory=[paths objectAtIndex:0];
    return [documentsDirectory
        stringByAppendingPathComponent:kFilename];
}

//退出程序时保存数据
-(void)applicationWillTerminate:(NSNotification *)notification {
    NSString *str1= [[(CustomCell *)
        [self.tableView cellForRowAtIndexPath:
    [NSIndexPath indexPathForRow:0 inSection:0]]
         noteTextField] text];
    NSString *str2 = [[(CustomCell *)
        [self.tableView cellForRowAtIndexPath:
    [NSIndexPath indexPathForRow:1 inSection:0]]
        noteTextField] text];
    NSString *str3 = [[(CustomCell *)
```

```
        [self.tableView cellForRowAtIndexPath:
    [NSIndexPath indexPathForRow:2 inSection:0]]
        noteTextField] text];
    NSString * str4 = [[(CustomCell * )
        [self.tableView cellForRowAtIndexPath:
    [NSIndexPath indexPathForRow:3 inSection:0]]
        noteTextField] text];

//将数据写入自定义类 NoteMsg 的对象中,实现归档
NoteMsg * tempNoteMsg = [[NoteMsg alloc] init];
tempNoteMsg.strTitle=str1;
tempNoteMsg.strDate=str2;
tempNoteMsg.strTypes=str3;
tempNoteMsg.strContent=str4;
NSMutableData * data=[[NSMutableData alloc]init];
NSKeyedArchiver * archiver=[[NSKeyedArchiver alloc]
    initForWritingWithMutableData:data];                    ❷
[archiver encodeObject:tempNoteMsg forKey:kDataKey];
[archiver finishEncoding];
[data writeToFile:[self dataFilePath] atomically:YES];
[archiver release];
[data release];
[tempNoteMsg release];

NSMutableArray * array=[[NSMutableArray alloc]init];
[array addObject:str1];
[array addObject:str2];
[array addObject:str3];
[array addObject:str4];
[array writeToFile:[self dataFilePath]atomically:YES];
[array release];
}

//取消归档,还原数据
#pragma mark-
-(void)viewDidLoad {
    self.title=@"信息记录";
```

```
    NSString *filePath=[self dataFilePath];
    if([[NSFileManager defaultManager]fileExistsAtPath:filePath]) {

            NSData *data=[[NSMutableData alloc]
                    initWithContentsOfFile:[self dataFilePath]];
            NSKeyedUnarchiver *unarchiver=
                    [[NSKeyedUnarchiver alloc]
                    initForReadingWithData:data];                    ❸
        self.noteMsg=[unarchiver decodeObjectForKey:kDataKey];
            [unarchiver finishDecoding];
            [unarchiver release];
            [data release];
            NSArray *array=[[NSArray alloc]
                initWithContentsOfFile:filePath];
            self.array1 = array;
            [array release];
    }
    self.myTitle=[array1 objectAtIndex:0];
    self.myTitle=noteMsg.strTitle; ❹
    UIApplication *app=[UIApplication sharedApplication];
    [[NSNotificationCenter defaultCenter]addObserver:self
                selector:@selector(applicationWillTerminate:)
                name:UIApplicationWillTerminateNotification
                object:app];
    [super viewDidLoad];
}
-(void)viewWillAppear:(BOOL)animated {
    [self.tableView reloadData];
}
-(void)didReceiveMemoryWarning {
    [super didReceiveMemoryWarning];
}
-(void)viewDidUnload {
}
-(void)myChange:(id)sender {
    UITextField *myTextField=(UITextField *)sender;
```

```
    self.myTitle=myTextField.text;
}
//数据源方法的实现
#pragma mark-
#pragma mark Table Data Source Methods
-(NSInteger)numberOfSectionsInTableView:(UITableView *)tableView {
    return 1;
}
-(NSInteger)tableView:(UITableView *)
    tableView numberOfRowsInSection:(NSInteger)section{
    return 4;
}
//绘制表视图
-(UITableViewCell *)tableView:(UITableView *)tableView
    cellForRowAtIndexPath:(NSIndexPath *)indexPath{
    static NSString *recordNoteViewControllerCell =
        @"recordNoteViewControllerCell";
    CustomCell *cell = (CustomCell *)[tableView
        dequeueReusableCellWithIdentifier:
        recordNoteViewControllerCell];
    if (cell == nil){
            NSArray *nib = [[NSBundle mainBundle]
                loadNibNamed:@"CustomCell"
                         owner:self
                options:nil];
        cell = [nib objectAtIndex:0];
    }
    if([indexPath row] == 0) {
        [cell.contentView addSubview:cell.noteLabel];
        cell.noteLabel.text = @"标题:";
        [cell.contentView addSubview:cell.noteTextField];
        cell.noteTextField.placeholder=@"请输入标题";
        [cell.noteTextField addTarget:self action:@selector(myChange:)
        forControlEvents:UIControlEventEditingChanged];
        cell.noteTextField.text=myTitle;
    }
    if(indexPath.row==1) {
```

```
        [cell.contentView addSubview:cell.noteLabel];
        cell.noteLabel.text = @"时间:";
        [cell.noteTextField setEnabled:NO];
        [cell.contentView addSubview:cell.noteTextField];
        cell.noteTextField.placeholder=@"请选择时间";
        if(setDatecontroller != nil)
        cell.noteTextField.text = setDatecontroller.time;
        else
            cell.noteTextField.text = [array1 objectAtIndex:1];
            cell.noteTextField.text = noteMsg.strDate; ❺
        cell.accessoryType
            =UITableViewCellAccessoryDisclosureIndicator;
    }
    if(indexPath.row==2) {
        [cell.contentView addSubview:cell.noteLabel];
        cell.noteLabel.text = @"类别:";
        [cell.noteTextField setEnabled:NO];
        [cell.contentView addSubview:cell.noteTextField];
        cell.noteTextField.placeholder=@"请选择类别";
        if(setTypeController != nil)
            cell.noteTextField.text = setTypeController.types;
        else
            cell.noteTextField.text =[array1 objectAtIndex:2];
            cell.noteTextField.text = noteMsg.strTypes;
        cell.accessoryType
            =UITableViewCellAccessoryDisclosureIndicator;
    }
    if([indexPath row]==3) {
        [cell.contentView addSubview:cell.noteLabel];
        cell.noteLabel.text = @"内容:";
        [cell.noteTextField setEnabled:NO];
        [cell.contentView addSubview:cell.noteTextField];
        cell.noteTextField.placeholder=@"请输入内容";
        if(setContentController !=nil)
            cell.noteTextField.text=
                  setContentController.textView.text;
        else
```

```
            cell.noteTextField.text = [array1 objectAtIndex:3];
            cell.noteTextField.text = noteMsg.strContent;
        cell.accessoryType
            =UITableViewCellAccessoryDisclosureIndicator;
    }
    return cell;
}
#pragma mark-
#pragma mark Table View Delegate Methods

……

-(void)dealloc {
    [array1 release];
    [setContentController release];
    [setTypeController release];
    [setDatecontroller release];
    [recordList release];
    [myTitle release];
    [noteMsg release];
    [super dealloc];
}
@end
```

在这一部分代码中,修改并不是特别多,主要集中在两个方面,一方面是删除原来利用属性列表持久保存应用程序数据的代码,另一方面是添加实现归档和取消归档读取数据的代码。

首先,在❶处导入 NoteMsg.h 文件,在 applicationWillTerminate 方法中,删除将数据写入属性列表文件的代码;接下来在❷处定义一个 NoteMsg 类型的新对象 tempNoteMsg,将获取的各个属性值存入新建对象的相应字符串中。对存入数据的归档操作也在这个方法中实现。

实现了对获取数据的归档之后,我们如果想在程序重新启动时,再次读取这些数据,就必须还原归档的数据,进行取消归档的操作,具体到程序中,就是指使用与归档操作相同的键值来从解压程序中读取数据。

在 viewDidLoad 方法中,我们也是先删除掉有关属性列表的代码,接下来在❸处添加取消归档的方法,注意由于我们在文件开始就声明了 NoteMsg 类,所以这里可以直接引用它的对象。

最后一步就是在 table:cellForRowAtIndexPath 方法中，使用❺处的代码把刚刚获取的值存入相应的字符串中，注意对于标题行的操作仍然是在 ViewDidLoad 方法中的❹处将获取的值赋予 myTitle。这样，我们就实现了利用归档持久保存应用程序数据。

**4. 编译并运行程序查看效果**

运行程序，效果与 9.3.2 小节的最终效果相同。在 9.3 和 9.4 两节里我们实现的是相同的功能，就是把数据持久保存下来，以便在重新启动程序时能够再次读取数据。

Tips

归档与属性列表比较：使用归档保存持久性数据，可以在归档集合类的同时，归档其中包含的所有对象，只要对象符合 NSCoding 协议即可，而且无论我们添加多少对象，把它们写入文件系统的方式都完全相同，反观属性列表的使用，工作量会随着添加对象而不断增加。

# 9.5 使用 SQLite3 持久保存应用程序数据

学习过属性列表和归档之后，我们对于数据的持久性存储已经有了一定的了解。提到数据存储，一般我们首先想到的是数据库。所以这一节里，我们学习一下 iPhone 的嵌入式数据库 SQLite3，并利用它来实现数据的持久性存储。

## 9.5.1 SQLite3 简介

首先来了解一下 SQL 语言。SQL 是一种数据库查询语言，用于存取数据以及查询、更新和管理关系数据库系统，因为强大的查询功能和简单的语法，已经成为主流数据库的标准语言。这里我们要学习的 SQLite3 是一种嵌入式的数据库，无需服务器支持，它将 SQL 语句嵌入到一般通用编程语言程序中去，SQL 语句负责对数据库中数据的提取及操作，它所提取的数据将逐行提交给程序，程序中其他语句负责数据的处理。

SQLite3 的程序接口是基于 C 语言的，它在存储和检索大量数据上非常有效，而且能够聚合复杂的数据，更快地处理数据获取结果。与使用对象处理数据相比，最大的优点就是不必把所有对象都加载到内存中，而只是提取符合特定条件的对象。

在本节的学习中，我们将接触到以下内容：设置 SQLite3、创建存放数据的表、保存数据和从数据库中检索数据。

## 9.5.2 基本数据库操作

**1. 数据库的打开和关闭方法**

如果在指定的位置上存在数据库，就可以直接打开；如果不存在，我们应当创建一个新的数据库。我们可以使用下面的命令来实现打开数据库的操作，代码如下：

```
sqlite3 *database;
int result=sqlite3_open("/path/to/database/file", &database);
```

在这里如果 result 等于 SQLITE_OK，就表示数据库已经成功打开。请注意，括号内数据库的路径必须作为 C 语言字符串传递，因为我们说过 SQLite3 的程序接口是基于 C 语言的，所以不会识别 Objective-C 中的 NSString。可以采用下面的方法从 NSString 实例生成 C 语言字符串：

```
char * stringPath=[pathString UTF8String];
```

于是上面的 result 语句就变为：

```
int result=sqlite3_open([pathString UTF8String], &database);
```

当我们打开 SQLite3 数据库并执行完所有的操作之后，可以调用下面的语句来关闭数据库：

```
sqlite3_close(database);
```

**2. 在数据库中新建表**

了解了数据库的开启和关闭，接下来就看一看数据是如何在数据库中存储的。其实很简单，数据被存放在数据库中的表里，可以使用下面的语句来新建一个表，并且把这个新建的表通过 sqlite3_exec 函数传递到已经打开的数据库中。代码如下：

```
char *errorMsg;
const char *creatSQL="CREATE TABLE IF NOT EXISTS PEOPLE
                ↳(ID INTEGER PRIMARY KEY AUTOINCREMENT,
                ↳ FIELD_DATA TEXT) ";
int result = sqlite3_exec(database, createSQL, NULL, NULL, %errorMsg);
```

其中需要注意的是 FIELD_DATA TEXT 是我们创建的文本字段。执行该操作之前，需要检查 result 的值是否为 SQLite3_OK，如果命令没有成功运行，对所发生问题的描述将包含在 errorMsg 中。我们在这里讲解的函数 sqlite3_exec 的使用，是针对 SQLite3 运行任何不返回数据的命令如查询、插入和删除操作的。

我们使用 SQLite3 的主要目的是对于数据的操作，其中最主要的还是写入、检索、读取操作，前面所学习的是数据的写入操作，接下来我们要做的就是学习一下如何从 SQLite3 中检索数据。

**3. 在数据库中检索数据**

在开始检索语句之前，首先生成一个 NSString 类型的查询语句，代码如下：

```
NSString *query =@"SELECT ID, FIELD_DATA
                    ↳ FROM FIELDS ORDER BY ROW";
sqlite3_stmt *statement;
int result = sqlite3_prepare_v2 (database, [query UTF8String],
                              ↳(-1, &statement, nil);
```

如果代码中的 result 值等于 SQLITE_OK，就表示数据库语句已经准备好了，接下来我们就可以开始单步调试结果集，并从中获取数据了。例如下面这段代码：

```
while (sqlite3_step(statement) == SLITE_ROW) {
    char * titleChar = (char * )sqlite3_column_text(statement, 1);
    NSString * title = [[NSString alloc] initWithUTF8String:titleChar];
    [title release];
}
sqlite3_finalize(statement);
```

上面的代码就是进行单步调试结果集，并且从中检索 int 和 NSString 类型的值，因为从数据库里获取的字符串是 C 类型的，所以必须要转换为 NSString 类型。

### 9.5.3 在项目中使用 SQLite3 的开发流程

通过前面的学习，我们对于 SQLite3 已经有了一定的了解，下面，我们就通过新建一个项目实现在应用程序中使用 SQLite3 来持久保存应用程序数据。首先来了解一下项目中各部分的功能及组合关系，如图 9-6 所示。

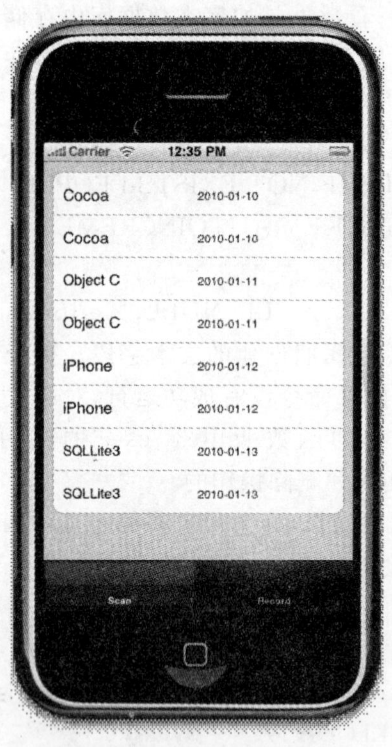


图 9-6 Scan 标签视图

图 9-7 Record 标签视图

可以看出该项目的基本框架是由工具栏控制的两个标签视图来组成的。第一个视图是浏览标签视图，用来显示持久存储下来的数据；切换到第二个标签视图，我们会看到非常熟悉的内容，如图 9-7 所示，用它来写入数据。

我们先在第二个 Record 标签视图之内填充数据，然后通过 SQLite3 持久保存下来，切换标签后在第一个 Scan 标签视图内显示出来。了解了项目的大体框架和基本功能之后，我们通过图 9-8 来了解整个项目的流程，我们将基于这个流程，分步创建项目来实现使用 SQLite3 持久存储应用程序数据。

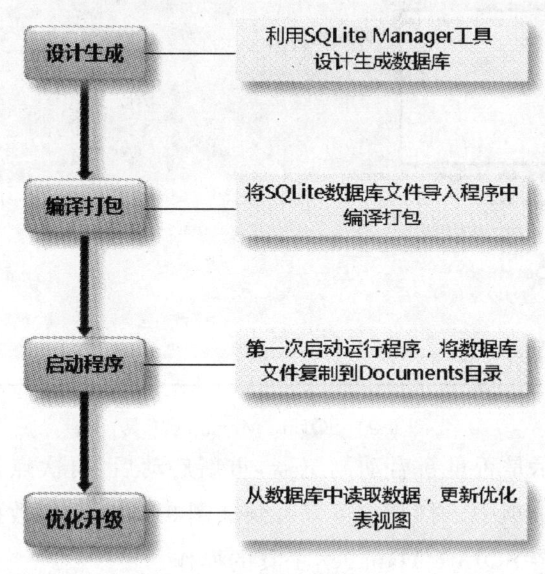


图 9-8 开发流程

## 9.5.4 设计生成数据库

我们可在程序中使用代码来创建 SQLite3 数据库，但在真正的项目开发中更为常用的是使用数据库工具创建 SQLite3 数据库。在这里我们使用 FireFox SQLite Manager 工具实现，它是 FireFox 浏览器提供的非常方便的数据库管理工具，如果你还没有 FireFox 浏览器，建议下载安装，然后按照下面的步骤实现开发流程的第一步：设计生成数据库。

**1. 下载安装 SQLite Manager 工具**

首先，打开火狐浏览器，在工具下面选择附加组件，在搜索框中输入 SQLite，选择并安装 SQLite Manager，如图 9-9 所示。

图 9-9　SQLite Manager 工具

安装完成后会提示是否重新启动浏览器，重新启动后，再次点击工具会看到 SQLite Manager 已经出现在选项中。打开它可以看到如图 9-10 所示的数据库工具界面，到这里我们就完成了下载安装 SQLite Manager 工具的操作。

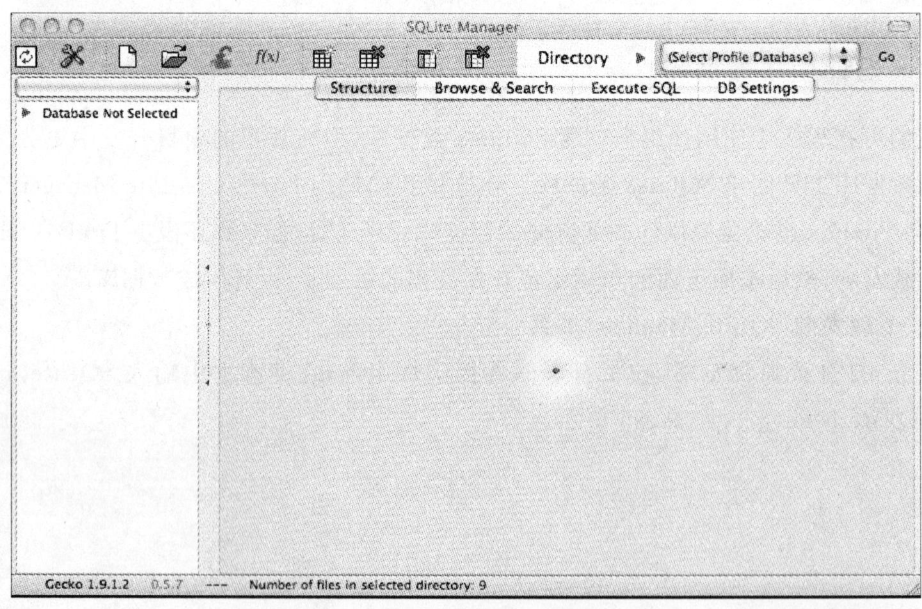


图 9-10　SQLite Manager 工具界面

**2. 创建 SQLite3 数据库**

首先，在 SQLite Manager 中，选择空白新建图标，新建一个数据库，会出现如图 9-11 所示的对话框，输入定义的数据库名 NoteSQL. db。

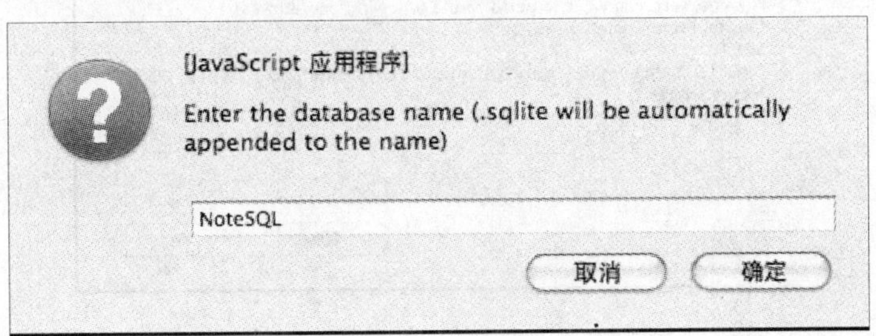


图 9-11　命名数据库

然后，点击 Create Table 图标，创建一个新的表并命名为 NoteSQL。编辑表中的内容，如图 9-12 所示。

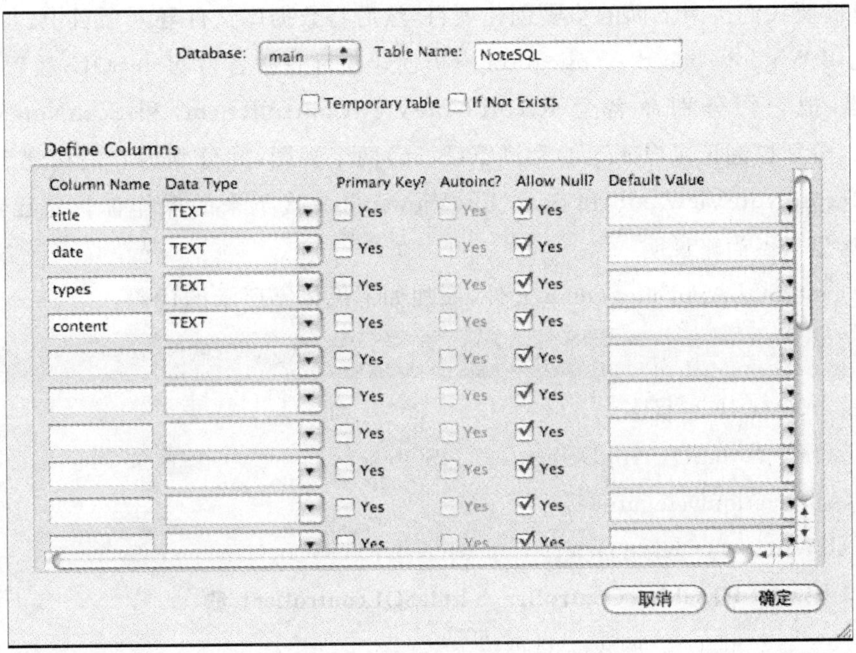

图 9-12　在数据库内新建表

完成上述操作后，单击确定，弹出如图 9-13 所示的对话框，单击 YES 就完成了数据库的创建工作。

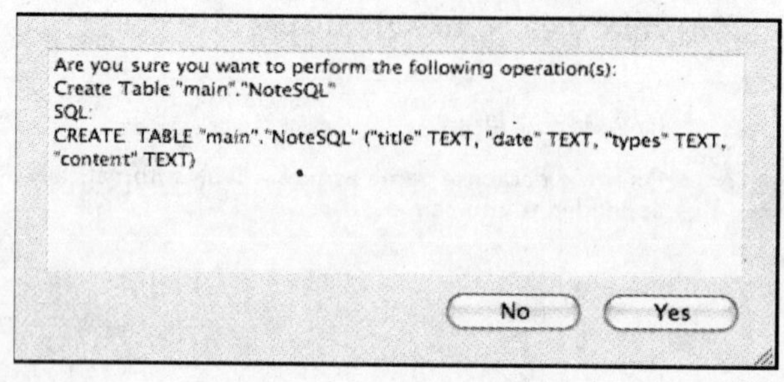

图 9-13　完成新建表

### 9.5.5　创建项目并把数据库文件导入项目中

完成了数据库的创建之后，接下来进行流程的第二步：将数据库导入项目中，编译打包。这里需要我们按照下面的步骤创建项目，然后将数据库文件导入项目中。

首先用 Window-based Application 模板创建项目并命名为 NoteSQL；然后添加两个子控制器，把它们分别命名为 RecordNoteViewController.m，和 ScanNoteViewController.m，分别控制记录(Record)和浏览(Scan)两个视图，并分别为它们构建 Xib 文件，命名为 RecordNoteView.xib 和 ScanNoteView.xib。接下来的操作你肯定已经非常熟悉，我们这里只做简略说明。

选中 NoteSQLAppDelegate.h 文件，添加如下黑体字所示的代码：

**代码 9.7　NoteSQLAppDelegate.h 文件**

```
#import <UIKit/UIKit.h>
@interface NoteSQLAppDelegate : NSObject
<UIApplicationDelegate>{
    UIWindow * window;
    IBOutlet UITabBarController * noteSQLcontroller; ❶
}
@property (nonatomic, retain) IBOutlet UIWindow * window;
@property(nonatomic, retain)
    UITabBarController * noteSQLcontroller;
@end
```

首先在❶处添加了一个 TabBarController，并为它设置输出口。

接下来双击打开 MainWindow. xib 文件，从 Library 中拖动一个 Tab Bar Controller 到主窗口，然后进行设置，选中相应的标签，选择各自的控制器类和 Xib 文件，让每个标签能够控制相应的视图，同时把各个标签的名字相应地改成 Scan 和 Record。如图 9-14 所示。

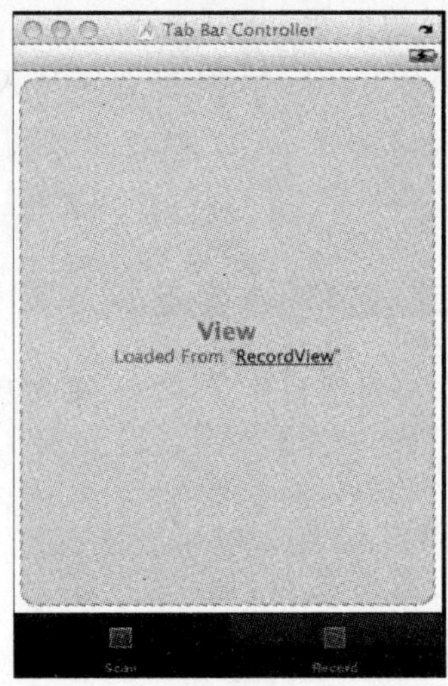


图 9-14 构建 TabBar 视图

接下来向 NoteSQLAppDelegate. m 添加如下黑体字所示的代码：

**代码 9.8 NoteSQLAppDelegate. m 文件**

```
#import "NoteSQLAppDelegate.h"
@implementation NoteSQLAppDelegate
@synthesize window;
@synthesize noteSQLcontroller;
-(void)applicationDidFinishLaunching:(UIApplication *)application {
    [window addSubview:noteSQLcontroller.view]; ❶
    [window makeKeyAndVisible];
}
-(void)dealloc {
    [noteSQLcontroller release];
    [window release];
    [super dealloc];
}
@end
```

至此，在❶处向窗口中添加 noteSQLcontroller 视图的操作就完成了。接下来实现 Record 标签视图来向数据库中写入数据，在这里与前面的项目不同的是，添加了一个按钮(Bar ButtonItem)，命名为“添加”，在数据添加完后，点击这个按钮实现数据的写入。到这里基本上构建起了这个项目的框架。

接下来将前面创建的 NoteSQL. db 数据库文件导入项目中，注意确保选中 Copy items into destination group's folder(if need)，这样做是为了保证数据库文件被复制到项目中。

### 9.5.6　用数据库写入和读取应用程序数据

上面，数据库文件已经被导入到项目中了，接下来按以下步骤使用数据库实现应用程序数据的写入和读取：

- 修改 NoteSQLAppDelegate 文件，添加数据库复制方法；
- 在 Record 标签视图里将数据写入 SQLite3；
- 在 Scan 标签视图里从 SQLite3 中读取数据；
- 编译并运行程序。

**1. 修改 NoteSQLAppDelegate 文件，添加数据库复制方法**

在 NoteSQLAppDelegate 文件中，我们将添加几个方法来实现编译程序时将数据库从项目中复制到可写的 Documents 目录下。

我们继续向 NoteSQLAppDelegate. h 文件中添加如下黑体字所示的代码：

**代码 9.9　NoteSQLAppDelegate. h 文件**

```
#import <UIKit/UIKit.h>
#import <sqlite3.h>❶
@interface NoteSQLAppDelegate :
    NSObject <UIApplicationDelegate> {
    UIWindow *window;
    IBOutlet UITabBarController *noteSQLcontroller;
    sqlite3 *database; ❷
}
@property (nonatomic, retain) IBOutlet UIWindow *window;
@property (nonatomic, retain)
    UITabBarController *noteSQLcontroller;
-(void)createDatabaseIfNeeded:(NSString *)filename; ❸
```

```
-(void)openDB; ❹
@end
```

在完成项目框架构建之后，我们需要添加数据库复制方法。在❶处导入数据库头文件，每当我们要使用数据库时都需要导入这个头文件。接下来在❷处定义了一个实例变量，用它指向应用程序的数据库。在❸处添加的方法用于在需要时复制数据库，在❹处的代码是实现打开数据库的方法。

接下来选中 NoteSQLAppDelegate.m 文件，添加如下黑体字所示的代码：

**代码 9.10 NoteSQLAppDelegate.m 文件**

```
#import "NoteSQLAppDelegate.h"
@implementation NoteSQLAppDelegate
@synthesize window;
@synthesize noteSQLcontroller;
-(void)applicationDidFinishLaunching:(UIApplication *)application {
    [window addSubview:noteSQLcontroller.view];
    [self createDatabaseIfNeeded:@"NoteSQL.db"];
    [self openDB];                                    ❶
    [window makeKeyAndVisible];
}

//数据库的复制操作
```

```
-(void)createDatabaseIfNeeded:(NSString *)filename{
    NSAutoreleasePool *pool=[[NSAutoreleasePool alloc] init];
    BOOL success;
    NSFileManager *fileManager =
    [NSFileManager defaultManager];
    NSError *error;
    NSArray *paths = NSSearchPathForDirectoriesInDomains
        (NSDocumentDirectory, NSUserDomainMask, YES);
    NSString *documentsDirectory = [paths objectAtIndex:0];
    NSString *writableDBPath = [documentsDirectory
    stringByAppendingPathComponent:filename];
    success = [fileManager fileExistsAtPath:writableDBPath];
    if (success)
        return;
    NSString *defaultDBPath =
        [[[NSBundle mainBundle] resourcePath]
    stringByAppendingPathComponent:filename];
    success = [fileManager copyItemAtPath:defaultDBPath
            toPath:writableDBPath error:&error];
    if (!success) {
        NSAssert1(0, @"Failed to create writable
            database file with message '%@'.",
        [error localizedDescription]);
    }
    [pool release];
}                                                        ❷
//打开数据库
-(void)openDB {
    NSArray *paths = NSSearchPathForDirectoriesInDomains
    (NSDocumentDirectory, NSUserDomainMask, YES);
        NSString *documenthPath = [[paths objectAtIndex:0]
    stringByAppendingPathComponent:@"NoteSQL.db"];
    int returnValue = sqlite3_open
        ([documenthPath UTF8String], &database);
    if (returnValue != SQLITE_OK) {
        sqlite3_close(database);
        NSAssert1(0, @"Failed to open database
        with message '%s'.", sqlite3_errmsg(database));
    }                                                    ❸
```

```
}
//关闭数据库
-(void) applicationWillTerminate: (UIApplication * )application{
    if(sqlite3_close(database) != SQLITE_OK)
        NSAssert1(0,@"Error while closing the
            connction to the db. %s",sqlite3_errmsg(database));      ❹
}
-(void)dealloc {
    [noteSQLcontroller release];
    [window release];
    [super dealloc];
}
@end
```

在这里我们着重学习一下两个新加入的方法，在❶处的代码是对于新添加方法的调用，在❷处添加的方法是实现数据库的复制操作。在❸处的方法是用来打开数据库，前面数据库的基本操作中已经介绍过这个方法，最后通过❹处的方法来关闭数据库，如果失败则返回错误信息。

**2. 在 Record 标签视图里将数据写入 SQLite3**

对于 Record 标签视图的操作，我们从 RecordNoteViewController 开始代码的添加，当然在这之前，必须要先导入可编辑的表视图单元里面的部分文件。除去 NoteRecordAppDelegate. h 文件和 NoteRecordAppDelegate. m 文件外，其他构建视图的文件都需要导入，然后稍加修改就可以构建完整的 Record 标签视图了，当然最主要的还是有关 SQL 数据库的数据写入操作。

选中 RecordNoteViewController. h 文件，添加如下黑体字所示的代码：

**代码 9.11　RecordNoteViewController. h 文件**

```
#import <UIKit/UIKit.h>
#import "RootViewController.h" ❶
@interface RecordNoteViewController : UIViewController {
    UINavigationController *recordController; ❷
    RootViewController *rootViewController; ❸
}
@property (nonatomic, retain)
    UINavigationController *recordController;
@property (nonatomic, retain)
    RootViewController *rootViewController;
```

```
@end
```

在❶处导入 RootViewController.h 文件，并在❷处对导航控制器类进行了声明。接下来在❸处声明了一个根控制器，用于管理导航控制器下面的各个子视图，这与我们前面所做项目进行的操作相同。接下来选中 RootViewController.m 文件，添加如下黑体字所示的代码：

**代码 9.12　RecordNoteViewController.m 文件**

```
#import "RecordNoteViewController.h"
@implementation RecordNoteViewController
@synthesize recordController;
@synthesize rootViewController;
-(void)viewDidLoad {
    RootViewController * tempView =
     [[RootViewController alloc]
    initWithNibName:@"RootViewController"bundle:nil];        ❶
    self.rootViewController = tempView;
    [tempView release];

    UINavigationController * nav =
        [[UINavigationController alloc]
        initWithRootViewController:rootViewController];      ❷
    self.recordController = nav;
    [nav release];
    [self.view addSubview:self.recordController.view];
    [super viewDidLoad];
}
-(void)didReceiveMemoryWarning {
        [super didReceiveMemoryWarning];
}
-(void)viewDidUnload {
}
-(void)dealloc {
        [rootViewController release];
        [recordController release];
        [super dealloc];
}
```

```
end
```

这里在❶和❷处分别对根控制器、导航控制器进行了初始化，完成这些操作后保存文件。

在这里说明一下，开始代码添加操作之前需要添加 SQLite 类库。右键单击 Frameworks，然后按照下面的路径添加：

**Developer/Platforms/iPhoneSimulator. platform/Developer/SDKs/iPhoneSimulator3. 1. 2. sdk/usr/lib**

当然这里最后面的 SDK 版本要根据你自己使用的版本而定。最后选择 libsqlite3. dylib 文件，注意一定要选择正确，只有选择 libsqlite3. dylib 才能使我们的 SQLite3 类库始终指向最新版本的 SQLite3，还需要将 Reference Type 改为 Relative to Current SDK。下面就可以开始向 RootViewController. h 文件添加代码了。

选中 RootViewController. h 文件，添加如下黑体字所示的代码：

**代码 9. 13 RootViewController. h 文件**

```
#import <UIKit/UIKit.h>
#import "SetDateViewController.h"
#import "SetTypeViewController.h"
#import "SetContentViewController.h"
#import <sqlite3.h> ❶
@interface RootViewController : UIViewController
    <UITableViewDelegate, UITableViewDataSource,
        UINavigationControllerDelegate>{
    IBOutlet UITableView *rootNoteView; ❷
    SetDateViewController *setDatecontroller;
    SetTypeViewController *setTypeController;
    SetContentViewController *setContentController;
    NSString *myTitle;
    sqlite3 *database; ❸
}
@property (nonatomic, retain) UITableView *rootNoteView;
@property (nonatomic, retain) NSString *myTitle;
@property (nonatomic, retain)
     SetDateViewController *setDatecontroller;
@property (nonatomic, retain)
     SetTypeViewController *setTypeController;
@property (nonatomic, retain)
```

```
    SetContentViewController * setContentController;
-(void)myChange:(id)sender;
-(NSString * )dataFilePath; ❹
-(void)addRecord; ❺
@end
```

首先在❶处导入了 sqlite3 类的头文件，而且在❷处声明了一个表，接下来在❸处我们声明了一个 sqlite3 型的实例变量 database，它将指向应用程序的数据库。下面添加的两个方法，我们并不陌生，❹用于获取数据文件的完整路径名，❺用于完成编辑后添加记录。

选中 RootViewController.m 文件，添加如下黑体字所示的代码：

**代码 9.14　RootViewController.m 文件**

```
#import "RootViewController.h"
#import "CustomCell.h"
@implementation RootViewController
@synthesize rootNoteView;
@synthesize setDatecontroller;
@synthesize setTypeController;
@synthesize setContentController;

//返回数据文件的完整路径名
-(NSString * )dataFilePath {
    NSArray * paths = NSSearchPathForDirectoriesInDomains
(NSDocumentDirectory, NSUserDomainMask, YES);
    NSString * documentsDirectory = [paths objectAtIndex:0]; ❶
return[documentsDirectory
    stringByAppendingPathComponent:@"NoteSQL.db"];
}

//添加记录，写入数据库
```

```
-(void)addRecord {
    NSString * titleStr = [[(CustomCell * )
        [self.rootNoteView cellForRowAtIndexPath:
        [NSIndexPath indexPathForRow:0 inSection:0]]
            noteTextField] text];
    NSString * dateStr = [[(CustomCell * )
        [self.rootNoteView cellForRowAtIndexPath:
        [NSIndexPath indexPathForRow:1 inSection:0]]
            noteTextField] text];
    NSString * typesStr= [[(CustomCell * )
        [self.rootNoteView cellForRowAtIndexPath:
        [NSIndexPath indexPathForRow:2 inSection:0]]
noteTextField] text];
    NSString * contentStr = [[(CustomCell * )
        [self.rootNoteView cellForRowAtIndexPath:
        [NSIndexPath indexPathForRow:3 inSection:0]]
            noteTextField] text];
    if (([titleStr length]<=0) || ([dateStr length]<=0) ||
        ([typesStr length]<=0) || ([contentStr length]<=0)){
        UIAlertView * alert = [[UIAlertView alloc]
initWithTitle:@"警告"
                message:@"请填写完整!"
                delegate:self
                cancelButtonTitle:@"确定"
                otherButtonTitles:nil];
        [alert show];
        [alert release];
    }                                                    ❷

//打开数据库
else{
    if (sqlite3_open([[self dataFilePath]
            UTF8String], &database) != SQLITE_OK){
        sqlite3_close(database);
        NSAssert(0, @"Failed to open database");       ❸
    }
//插入操作
```

```
    NSString * insertSQL =
        [[NSString alloc] initWithFormat:
            @"INSERT INTO NoteSQL
                (title,date,types,content)
                VALUES('%@','%@','%@','%@');"
                ,titleStr,dateStr,typesStr,contentStr];          ❹
    char * errorMsg2;
    if (sqlite3_exec (database, [insertSQL UTF8String],
            NULL, NULL, &errorMsg2) != SQLITE_OK) {
            NSAssert1(0, @"Error updating tables:
            %s", errorMsg2);
            sqlite3_free(errorMsg2);
    }
    //添加信息成功提示
    else {
            UIAlertView * alert =[[UIAlertView alloc]
                        initWithTitle:@"信息"
                        message:@"添加成功!"
                        delegate:self
                        cancelButtonTitle:@"确定"
                        otherButtonTitles:nil];                 ❺
            [alert show];
            [alert release];
        }
        [insertSQL release];
        sqlite3_close(database);
    }
}

//设置导航栏标题
-(id)initWithNibName:(NSString *)nibNameOrNil
bundle:(NSBundle *)nibBundleOrNil {
    if (self = [super initWithNibName:
nibNameOrNil bundle:nibBundleOrNil]) {
        self.title=@"添加记录";                                  ❻
    }
    return self;
}
-(void)viewWillAppear:(BOOL)animated {
    [self.tableView reloadData]
```

```
    [self.rootNoteView reloadData]; ❼
    [super viewWillAppear:YES];
}
//导航控制器控制的两个视图传值,根视图重新加载数据
-(void)navigationController:(UINavigationController *)
    navigationController
     willShowViewController:(UIViewController *)
    viewController animated:(BOOL)animated {                    ❽
    [self.rootNoteView reloadData];
}
//"添加"按钮
-(void)viewDidLoad {
    self.navigationController.delegate =self;
    UIBarButtonItem *add = [[UIBarButtonItem alloc]
                        initWithTitle:@"添加"
                        style:UIBarButtonItemStyleBordered
                        target:self                              ❾
                        action:@selector(addRecord)];
    self.navigationItem.rightBarButtonItem = add;
    [add release];
    self.myTitle=nil;
    [super viewDidLoad];
}
-(void)didReceiveMemoryWarning {
    [super didReceiveMemoryWarning];
}
-(void)viewDidUnload {
}
-(void)myChange:(id)sender{
    UITextField *myTextField=(UITextField *)sender;
    self.myTitle=myTextField.text;
}
#pragma mark-
#pragma mark Table Data Source Methods
if([indexPath row] == 0) {
    [cell.contentView addSubview:cell.noteLabel];
    cell.noteLabel.text = @"标题:";
```

```
    [cell.contentView addSubview:cell.noteTextField];
    cell.noteTextField.placeholder=@"请输入标题";
    [cell.noteTextField addTarget:self
    action:@selector(myChange:)
    forControlEvents:UIControlEventEditingChanged];
    cell.noteTextField.text=myTitle;
    if(myTitle!=nil)
        cell.noteTextField.text=myTitle;      ⑩
    else
        cell.noteTextField.text=nil;
}
#pragma mark Table View Delegate Methods
……
-(void)dealloc {
        [setContentController release];
        [setTypeController release];
        [setDatecontroller release];
        [rootNoteView release];
        [super dealloc];
}
@end
```

在❶处的方法我们已经用过多次，用来返回数据文件的完整路径名，只是注意一点，在最后附加到目录后面的文件名是所创建的数据库名NoteSQL.db。

-(void)addRecord{}这个方法比较复杂，下面我们来剖析这段代码。

首先在❷处创建了四个字符串实例，来存储从表视图单元中获取的字符串，然后进行如下判断：如果有任何一项内容没有填写，即某一个字符串长度不大于0，就弹出警告，来提醒我们填写完整。记录填写完整，获取之后，我们就需要把这些数据写入数据库。

在❸处打开数据库。接下来需要把数据写入这个表中，通过❹处代码创建数据库插入语句。接下来立即执行❺处创建的插入操作，把数据写入数据库中。值得注意的是，在插入数据时，必须按照创建表时的顺序，因为数据库中数据是按照顺序存储的。这样我们就将数据写入了数据库，这是实现数据持久存储的基础。

接下来通过❻处的代码设置导航栏的标题为“添加记录”，在❼处将Record视图中的数据在Scan视图中显示。需要注意的是，在❽处添加的代码是实现在Scan视图中导航控制器控制的两个视图间传值，将各子视图的值传递给Record视图的表并显示出来。

在❾处给导航栏右侧添加了一个“添加”按钮，注意我们添加了一行代码self.myTitle=nil;是为了第一次启动项目在❿处绘制表视图标题行时，不会出现数据库读取错误。

完成了写入操作之后，就应该将这些存储下来的数据读出，这需要在 Scan 标签视图中进行操作。在 Record 标签视图中实现数据写入，然后在 Scan 标签视图中实现数据读取，这符合我们最初对于项目的规划。

**3. 在 Scan 标签视图里从 SQLite3 中读取数据**

在对 Scan 标签进行操作之前，我们首先要把数据规范地存入一个类以便于管理，这个类就是所需的数据模型，这与我们前面在归档里使用的方式类似。创建一个 NSObject subclass，命名为 recordInfo.m。

选中 recordInfo.h 文件，添加如下代码：

**代码 9.15 recordInfo.h 文件**

```
#import <Foundation/Foundation.h>
@interface recordInfo : NSObject {
    NSInteger indexID;
    NSString * title;
    NSString * date;        ❶
    NSString * types;
    NSString * content;
}
@property NSInteger indexID;
@property (nonatomic, retain) NSString * title;
@property (nonatomic, retain) NSString * date;
@property (nonatomic, retain) NSString * types;
@property (nonatomic, retain) NSString * content;
@end
```

在❶处分别定义了四个字符串，来存储标题、日期、类型、详细内容，接下来向 recordInfo.m 添加代码如下：

**代码 9.16 recordInfo.m 文件**

```
#import "recordInfo.h"
@implementation recordInfo
@synthesize indexID;
@synthesize title;
@synthesize date;
@synthesize types;
@synthesize content;
@end
```

这个文件内容非常简单,我们不再解释,完成这些后就应该开始操作 Scan 标签视图。

选中 ScanNoteViewController.h 文件,添加如下黑体字所示的代码:

**代码 9.17 ScanNoteViewController.h 文件**

```
#import <UIKit/UIKit.h>
#import <sqlite3.h> ❶
@interface ScanNoteViewController : UIViewController
    <UITableViewDelegate, UITableViewDataSource>{
    IBOutlet UITableView *mainTable;
    NSMutableArray *dateShow;                     ❷
    sqlite3 *database;
    }
@property (nonatomic, retain) UITableView *mainTable;
@property (nonatomic, retain) NSMutableArray *dateShow;
-(NSString *)dataFilePath; ❸
-(NSMutableArray *)getArray; ❹
@end
```

我们在这里需要创建一个基本的表视图,然后通过这个表视图来显示我们在 Record 标签视图里添加的时间和标题。具体来看我们添加的代码:在❶处导入了 sqlite3 类;在❷处添加了一个表视图,声明了一个可变数组,声明了一个 sqlite3 的常量,并且添加了两个方法;❸处方法是用于返回数据文件的完整路径名的方法;在❹处添加的方法是从数据库中获取存储数据的方法。

选中 ScanNoteViewController.m 文件,添加如下黑体字所示的代码:

**代码 9.18 ScanNoteViewController.m 文件**

```
#import "ScanNoteViewController.h"
#import "CustomCell.h"
#import "recordInfo.h"
@implementation ScanNoteViewController
@synthesize mainTable;
@synthesize dateShow;
//返回数据文件的完整路径名
```

```
-(NSString * )dataFilePath {
    NSArray * paths = NSSearchPathForDirectoriesInDomains
          (NSDocumentDirectory, NSUserDomainMask, YES);
    NSString * documentsDirectory = [paths objectAtIndex:0];          ❶
    return [documentsDirectory
        stringByAppendingPathComponent:@"NoteSQL.db"];
}
//打开数据库并从中读取数据
-(NSMutableArray * )getArray {
    NSMutableArray * mutArray = [[NSMutableArray alloc] init];
    if (sqlite3_open([[self dataFilePath] UTF8String],
                  &database) != SQLITE_OK) {                             ❷
        sqlite3_close(database);
        NSAssert(0, @"Failed to open database");
    }
    NSString * query = @"SELECT * FROM
                  ↳ NoteSQL order by date"; ❸
    sqlite3_stmt * statement;
    if(sqlite3_prepare_v2(database, [query UTF8String],
                  ↳-1, &statement, nil) == SQLITE_OK) {
        while (sqlite3_step(statement) == SQLITE_ROW)
        {
            char * titleChar = (char * )sqlite3_column_text
                (statement, 1);
            char * dateChar = (char * )sqlite3_column_text                 ❹
                (statement, 2);
            char * typesChar = (char * )
                sqlite3_column_text(statement, 3);
            char * contentChar = (char * )
                sqlite3_column_text(statement, 4);
        //C 字符串转换成 NSString
        NSString * title = [[NSString alloc]
            initWithUTF8String:titleChar];          ❺
        NSString * date = [[NSString alloc]
            initWithUTF8String:dateChar];
        NSString * types = [[NSString alloc]
            initWithUTF8String:typesChar];
```

```
                NSString * content = [[NSString alloc]
                  initWithUTF8String:contentChar];
                recordInfo * record = [[recordInfo alloc] init];
                record.title = [title copy];
                record.types = [types copy];
                record.date = [date copy];                          ❻
                record.content = [content copy];
                [mutArray addObject:record];
            }
            sqlite3_finalize(statement);
        }
        return mutArray;
}
-(id)initWithNibName:(NSString *)nibNameOrNil
        bundle:(NSBundle *)nibBundleOrNil {
    if (self = [super initWithNibName:
        nibNameOrNil bundle:nibBundleOrNil]) {
        self.title = @"清单列表";
    }
    return self;
}
-(void)viewDidLoad
{
    self.navigationItem.rightBarButtonItem = self.editButtonItem;
    [super viewDidLoad];
}
//更新表内数据
-(void)viewWillAppear:(BOOL)animated{ ❼
    self.dateShow = [self getArray];
}
-(void)didReceiveMemoryWarning {
        [super didReceiveMemoryWarning];
}
-(void)viewDidUnload {
}
-(void)dealloc {
    [dateShow release];
```

```
    [mainTable release];
    [super dealloc];
}
//数据源方法的实现
#pragma mark-
#pragma mark Table Data Source Methods
-(NSInteger)numberOfSectionsInTableView:(UITableView *)tableView
{
    return 1;
}
-(NSInteger)tableView:(UITableView *)tableView
    numberOfRowsInSection:(NSInteger)section {
        return [self.dateShow count];
}
-(UITableViewCell *)tableView:(UITableView *)tableView
    cellForRowAtIndexPath:(NSIndexPath *)indexPath {
    static NSString *scanNoteViewControllerCell =
        @"scanNoteViewControllerCell";
    CustomCell *cell = (CustomCell *)[tableView
    dequeueReusableCellWithIdentifier:scanNoteViewControllerCell];
    if (cell == nil){
        NSArray *nib=[[NSBundle mainBundle]
                    loadNibNamed:@"CustomCell"
                    owner:self
                    options:nil];
        cell=[nib objectAtIndex:0];
    }
    [cell.contentView addSubview:cell.noteLabel];
    cell.noteLabel.text=[[dateShow
     objectAtIndex:indexPath.row]title];
    [cell.contentView addSubview:cell.noteTextField];
        recordInfo *tempRecord =
        [self.dateShow objectAtIndex:indexPath.row];
    cell.noteTextField.text=tempRecord.date;
    [cell.noteTextField setEnabled:NO];
    return cell;
}
```

@end

其实在这个文件里，我们用了大量的代码来创建表视图，而本节的重点是对于SQLite3 的操作，所以在这里我们并不会特别介绍表视图的创建，而只介绍 SQLite3 的操作。

首先，在❶处我们添加第一个方法，用来返回数据文件的完整路径名。

接下来在❷处打开数据库，如果开启失败，就会触发断言，设置断言对于开发人员调试程序是非常有帮助的。接下来我们就开始单步调试每个返回的行，然后从数据库中检索数据。在❸处主要是用于准备数据库检索的语句，然后利用❹处的代码将获取的字段数据都作为C语言字符串检索出来。检索出来的C语言字符串，必须通过❺处的代码转换成 NSString 类型。❻处将已经转换成 NSString 类型的字符串复制到 recordInfo 类型实例中。从数据库中读取数据后在表中重新显示数据时，需要用❼处的方法来显示getArray里读取的数据。

**4. 编译并运行程序**

编译并运行程序，点击右边的 Record 标签视图，切换到添加记录界面，在里面修改数据后点击添加按钮，出现如图 9-15 所示的效果，如果数据填充不完整则会出现如图9-16所示的效果。

图 9-15　数据添加成功

图 9-16　数据填写不完整

添加完数据之后，切换到 Scan 标签，我们就能够看到如图 9-17 所示的效果，当然具体内容你可以自己随意编写。

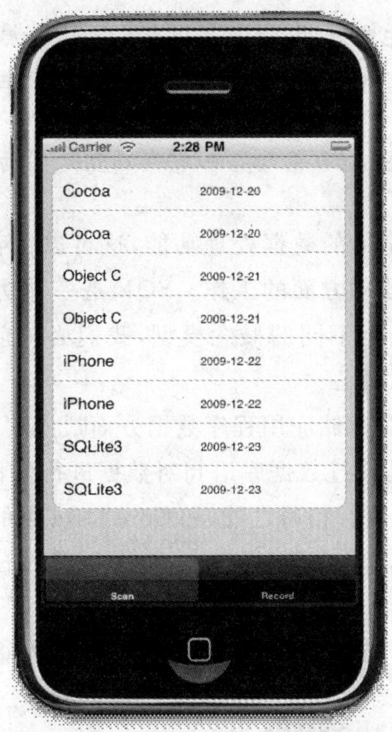


图 9-17 在数据库内新建表

## 9.6 使用 Core Data 持久保存应用程序数据

在 iPhoneSDK 升级到 3.0 版本之后，增加了一些新的特性，而 Core Data 就是其中特别重要的一种，与属性列表和归档相比，Core Data 的功能更加强大，比起 SQLite3 又更直观容易。它给我们提供了一种更加直观的设计数据模型的方式，同时也是一种对应用程序数据进行持久性存储的新方式。

在本节中，我们将主要学习如何利用 Core Data 来实现应用程序中数据的持久性存储。

### 9.6.1 Core Data 简介

Core Data 是一个用来管理对象生命周期、对象图和持久化的框架，它拥有的特性包括：undo 和 redo 的管理、值的自动正当性检查、保证对象关系的完整性、将对象的状态改变通知其他对象以及与 Cocoa 绑定的集成等。

Tips

**对象生命周期：**对象从诞生（由 alloc 或者 new 方法实现）、生存（进行操作）、交友（参数）到生命结束（被释放）的整个过程。

**对象图：**对象的集合和它们相互之间的关系。

Core Data 拥有数据库全部的数据处理功能，这可以让我们更加专注于对象及程序本身的逻辑，而省去了数据处理方面的工作。SQL 是一种功能强大的数据库语言，但是它跟我们所接触的对象是完全不同的概念模型，我们必须前后不停地去照顾数据的处理，而 Core Data 却让人非常省心。

SQLite3 和 Core Data 在存储应用程序数据方面最主要的不同之处是它们进行操作的目标是不同的，SQLite3 是通过数据库语句对数据进行操作，而 Core Data 是通过实体对对象进行操作。当然两者也有相通之处，如 Core Data 中的 Entity 和 Attribute 的概念与 SQLite3 中的表和列的概念十分相似，Core Data 在 iPhone 中默认的最终存储文件是 SQLite3 文件。

下面我们就借助图 9-18 来了解一下利用 Core Data 存储应用程序数据的原理。

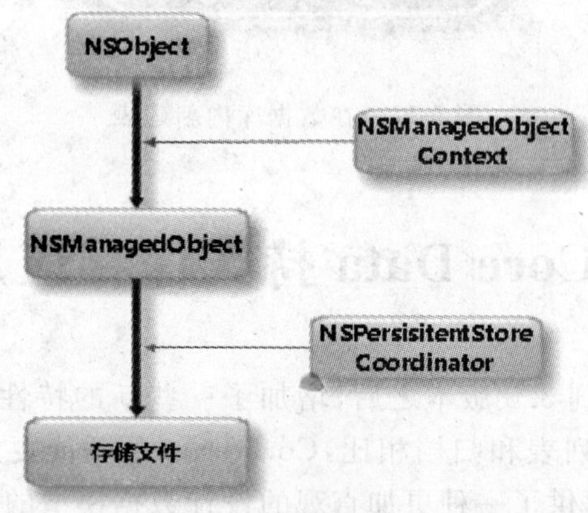


图 9-18　Core Data 存储原理

概括地讲，Core Data 就是应用程序数据存储时的“中间转换器”。你或许不太喜欢使用 SQLite3 来存储数据，因为它们的数据库操作语言比较繁琐，那么我们可以借助 Core Data 把数据都转换成可操作的对象，在数据存储、写入时直接操作对象，这样更加简便。

如图 9-18 所示，如果要用 Core Data 把对象存入存储文件中，就必须借助 NSManagedContext 把对象（object）变为可操作的对象（managed object），这个过程就如同实例化一个类，而这个所谓的“类”被称为实体，通常我们利用 NSManagedContext 来对对象进行删除、添加、移动等操作。接下来在把可操作的对象(managed object)存入数据文件的过程中,需要借助持久性协调器(NSPersistentStoreCoordinator)来实现存储的连接协调工作如确定存储的位置等,这样就完成了利用Core Data来存储应用程序数据。

接下来了解一下利用Core Data存储应用程序数据过程中几个需要重点关注的概念。

**1. 实体(Entity)**

实体是这个过程中最基本的元素,它通过NSEntityDescription类告诉我们该利用Core Data存储哪些数据。我们可以创建包含多个属性的实体,并把它映射到一个或多个类上,注意这些类都必须继承自NSManagedObject,正是这个类来使用实体管理各种数据。

可以看到每一个实体都有很多的属性,这些属性包含实体的基本属性(Attribute),还有它与其他实体的关系(Relationship)以及像实体名称和实体表示类这样的元数据(Fetched Property)。

**2. NSManagedObject**

它是一个实体或数据文件中的一条记录,对应着NSEntityDescription,提供所需元数据,它并非实体而是根据实体创建的对象。在Core Data中必须把对象(object)转换成可操作对象(managed object),才能够操作对象。

Tips

managed object 和实体共同指向数据模型对象,实体是对于对象的描述,而 managed object 是借助实体创建的一个实例,所以在模型编辑器中创建实体,而在代码中创建 managed object,区别就像是类和类的实例对象。

**3. NSManagedObjectModel**

它是程序中所用到实体的集合,包含一个或多个NSEntityDescription对象,其中每个NSEntityDescription对象内都有一个表示实体属性的NSPropertyDescription实例(实体属性描述)。每个描述负责指定实体的属性(NSAttributeDescription)、与其他实体的关系(NSRelationshipDescription)、需要从其他对象中获得的属性(NSFetchedPropertyDescription),如图9-19所示。

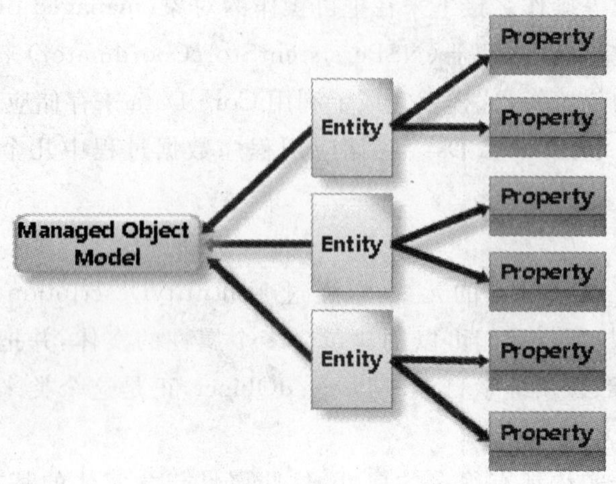


图 9-19　Managed Object Model

创建模型的方法如下：

**managedObjectModel = [[NSManagedObjectModel mergedModelFromBundles:nil]retain];**

这个方法通过合并在应用程序的资源文件夹中找到的所有模型来创建 NSManagedObjectModel实例。

**4. NSManagedObjectContext**

它是程序中的"对象空间"，通过它来进行 Managed Object 的获得、修改、保存、undo、redo 等比较高级的操作。这也是 Core Data 最为关键的地方，如图 9-20 所示。

Core Data 中获取 NSManagedObjectContext 对象是通过分配并初始化 NSManagedObjectContext实例实现，代码如下：

**managedObjectContext = [[NSManagedObjectContext alloc] init];**
**[managedObjectContext setPersistentStoreCoordinator: coordinator];**

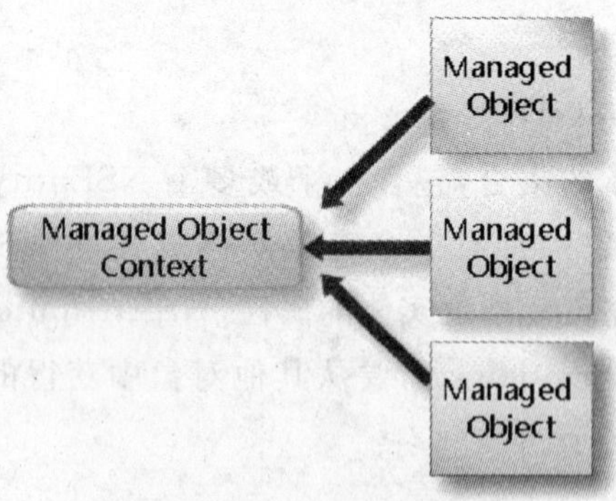


图 9-20　NSManagedObjectContext

**5. NSPersistentStoreCoordinator**

管理比较低级的数据存储文件的读、写、添加、删除操作，通常用来建立 NSManagedObject和存储文件之间的联系。

利用下面的代码，从应用程序 Documents 目录中加载数据库，这个数据库就用来存储可操作对象(managed object)：

```
NSURL * storeUrl = [NSURL fileURLWithPath:
                    [[self applicationDocumentsDirectory]
                    stringByAppendingPathComponent:
                    @"NoteCoreData.sqlite"]];
```

获取到 URL 之后，就需要创建一个 NSPersistentStoreCoordinator 的实例：

```
persistentStoreCoordinator = [[NSPersistentStoreCoordinator alloc]
        initWithManagedObjectModel: [self managedObjectModel]];
```

上述代码表示的是使用 managedObjectModel 方法的返回值来初始化 NSPersistentStoreCoordinator 对象。managedObjectModel 用来保存数据，而 persistentStoreCoordinator 用来管理数据。

上面介绍的三个类共同构成了 Core Data 的 API，也称 Core Data 栈，只有先了解了如何创建 Core Data 栈，才能够对 NSManagedObjectContext 进行操作，并将它存入存储文件。

**6. NSFetchedResultsController**

NSFetchedResultsController 是使用 Core Data 进行数据写入和读取时，位于中间的一个调配器。它通过使用获取请求(NSFetchRequest)和 NSManagedObjectContext 来获取数据，其中获取请求又包含 NSSortDescriptor 和 NSPredicate 对象。图 9-21 就表示了它们之间的关系。

图 9-21　NSFetchedResultsController

## 9.6.2　Core Data 在应用程序中的使用

下面我们将使用 Core Data 来实现应用程序数据的持久性存储。效果与前面使用 SQLite3 存储的效果一致，在 Record 标签视图下进行的操作如图 9-22 所示，在这里将利用 Core Data 进行保存，在 Scan 标签里面显示出来的效果如图 9-23 所示。

图 9-22 Record 标签视图

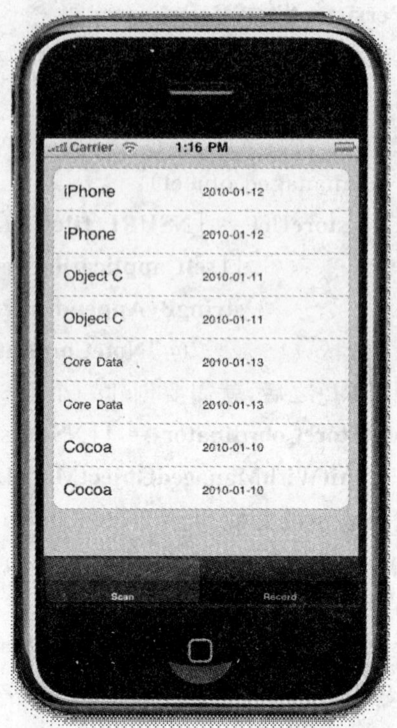


图 9-23 Scan 标签视图

其实这个项目里的大部分内容跟上一节一样，我们都已经实现，这里的焦点就是如何利用 Core Data 实现数据的持久性存储。项目实现的具体步骤如下：

- 构建带有 TabBarController 的项目框架；
- 利用 Core Data 写入数据；
- 利用 Core Data 读取数据；
- 编译并运行程序。

**1. 构建带有 TabBarController 的项目框架**

使用 Window-based-Application 模板新建项目，并命名为 NoteCoreData。这里注意，一定要勾选 Use Core Data for storage 选项，如图 9-24 所示，因为这是本节项目的基础。然后将上一节项目中的 Scan 文件夹和 Record 文件夹全部导入到本项目中，再将项目中的 MainWindow. xib 文件删除，导入上一节项目中的 MainWindow. xib 文件，这样就省去了工具栏的设置。

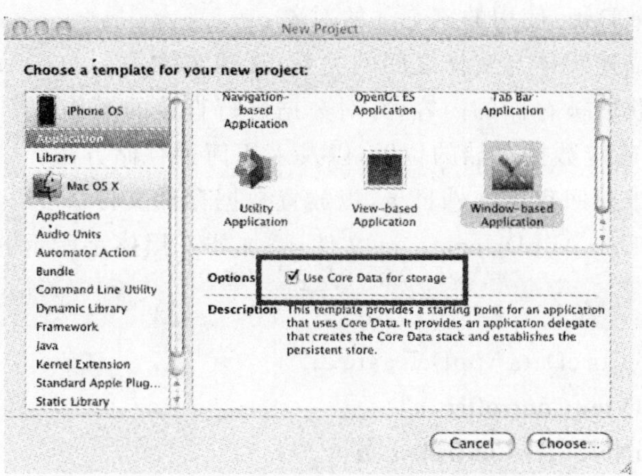


图 9-24 新建项目

单击展开 Classes 文件夹，选中 NoteCoreDataAppDelegate.h 文件，添加如下黑体字所示的代码：

**代码 9.19 NoteCoreDataAppDelegate.h 文件**

```
@interface NoteCoreDataAppDelegate : NSObject
    <UIApplicationDelegate> {
    NSManagedObjectModel * managedObjectModel;
    NSManagedObjectContext * managedObjectContext;
    NSPersistentStoreCoordinator * persistentStoreCoordinator;
    IBOutlet UITabBarController * noteCoreDataController;
    UIWindow * window;
}
@property (nonatomic, retain, readonly)
   NSManagedObjectModel * managedObjectModel; ❶
@property (nonatomic, retain, readonly)
   NSManagedObjectContext * managedObjectContext; ❷
@property (nonatomic, retain, readonly)
   NSPersistentStoreCoordinator * persistentStoreCoordinator; ❸
@property (nonatomic, retain)
       UITabBarController * noteCoreDataController;
@property (nonatomic, retain) IBOutlet UIWindow * window;
-(NSString * )applicationDocumentsDirectory; ❹
@end
```

在这段代码里我们要关注的，并不是添加的 TabBarController，而是原有的陌生代码，它们声明了 Core Data 使用的三个类的对象。

❶处的类用于描述实体与实体之间的关系，管理实体。

❷处的类用于获取被存储的内容，并对数据进行直接的操作。

❸ 处的类用于确定数据存储的位置，以及采用何种存储方法。

❹处的方法用于返回程序文档目录，以放置数据存储文件。

选中 NoteCoreDataAppDelegate. m 文件，添加如下黑体字所示的代码：

**代码 9.20 NoteCoreDataAppDelegate. m 文件**

```
#import "NoteCoreDataAppDelegate.h"
#import "RootViewController.h"
#import "ScanNoteViewController.h"
#import "RecordNoteViewController.h"
@implementation NoteCoreDataAppDelegate
@synthesize window;
@synthesize noteCoreDataController;

//创建 NSManagedObjectContext 对象，声明视图控制器对象
#pragma mark-
#pragma mark Application lifecycle
-(void)applicationDidFinishLaunching:
    (UIApplication *)application {
    NSManagedObjectContext *context=
        [self managedObjectContext];
    NSArray *array=[noteCoreDataController
        viewControllers];
    ScanNoteViewController *scan=                                  ❶
        (ScanNoteViewController *)[array objectAtIndex:0];
    scan.managedObjectContext=context;
    [scan release];
    [window addSubview:[noteCoreDataController view]];
    [window makeKeyAndVisible];
}
//程序退出时保存数据
-(void)applicationWillTerminate:(UIApplication *)application { ❷
    NSError *error = nil;
    if (managedObjectContext != nil) {
        if ([managedObjectContext hasChanges]
```

```
            && ![managedObjectContext save:&error]) {
        NSLog(@"Unresolved error %@, %@",
            error, [error userInfo]);
                abort();
            }
        }
    }
//创建 Core Data 栈
#pragma mark-
#pragma mark Core Data stack
-(NSManagedObjectContext *) managedObjectContext { ❸
    if (managedObjectContext != nil) {
        return managedObjectContext;
    }
    NSPersistentStoreCoordinator *coordinator =
        [self persistentStoreCoordinator];
    if (coordinator != nil) {
        managedObjectContext =
        [[NSManagedObjectContext alloc] init];
        [managedObjectContext
        setPersistentStoreCoordinator: coordinator];
    }
    return managedObjectContext;
}
-(NSManagedObjectModel *)managedObjectModel { ❹
    if (managedObjectModel != nil) {
        return managedObjectModel;
    }
    managedObjectModel = [[NSManagedObjectModel
        mergedModelFromBundles:nil] retain];
    return managedObjectModel;
}
-(NSPersistentStoreCoordinator *) persistentStoreCoordinator { ❺
    if (persistentStoreCoordinator != nil) {
        return persistentStoreCoordinator;
    }
    NSURL *storeUrl = [NSURL fileURLWithPath:
```

```
        [[self applicationDocumentsDirectory]
    stringByAppendingPathComponent:
        @"NoteCoreData.sqlite"]];
    NSError * error = nil;
    persistentStoreCoordinator = [[NSPersistentStoreCoordinator
        alloc] initWithManagedObjectModel:[self
        managedObjectModel]];
    if(![persistentStoreCoordinator addPersistentStoreWithType:
        NSSQLiteStoreType configuration:nil
            URL:storeUrl options:nil error:&error]) {
        NSLog(@"Unresolved error %@, %@",
            error, [error userInfo]);
        abort();
    }
    return persistentStoreCoordinator;
}
//获取程序的档案目录
#pragma mark-
#pragma mark Application's Documents directory
-(NSString *)applicationDocumentsDirectory {
    return [NSSearchPathForDirectoriesInDomains
        (NSDocumentDirectory,
        NSUserDomainMask, YES) lastObject];              ❻
}
-(void)dealloc {
    [managedObjectContext release];
    [managedObjectModel release];
    [persistentStoreCoordinator release];
    [noteCoreDataController release];
    [window release];
    [super dealloc];
}
@end
```

在❶处声明了 NSManagedObjectContext 的对象 context，我们用它来获取、操作数据，并声明了第一个用于存储控制器的数组和标签视图控制器 ScanNoteViewController，因为这个视图是程序加载数据后显示出来的视图，所以在这里要对它进行初始化。

在❷处的 applicationWillTerminate 方法主要用于在程序结束时，利用 Core Data 保存所有的更改。

在❸处自定义 NSManagedObjectContext 对象，如果对象已经存在就直接返回，否则继续让对象获得下面声明的 NSPersistentStoreCoordinator 的对象 coordinator，它必须获得这个对象才能够获得数据。

在❹处自定义 Core Data 的数据模板，这里的模板从本地的 xcdatamodel 文件里获得。

在❺处定义 NSPersistentStoreCoordinator 对象，其实我们可以从这里看出这个类只是获取一个数据存储位置，因为这个类用来管理数据存储。

在❻处的方法用来返回程序的档案目录。

到这里我们就大体了解了 Core Data 各类的作用，而且已经搭建好整个项目的框架，接下来我们就来实现将数据通过 Core Data 存储起来。

**2. 利用 Core Data 写入数据**

全部的写入操作都在 Record 标签视图下的 RootViewController 文件中实现，不过在开始写入数据之前，我们应当规划数据存储的模型，也就是在 xcdatamodel 文件中创建实体。

在 Resources 文件夹下面打开 NoteCoreData. xcdatamodel 文件，如图 9-25 所示，这是一个图形化的实体模型构建文件。它共分为四部分，第 1 部分实体列表，用于展示、新建实体；第 2 部分属性列表，用来列出每一个实体的属性描述；第 3 部分 Atttribute 列表，用来编辑每个属性的详细信息如类型、默认值等；第 4 部分图形化展示视图，如果有多个实体，都会显示在这个视图里，而且它们之间的绑定关系等都可以直观地显示出来。

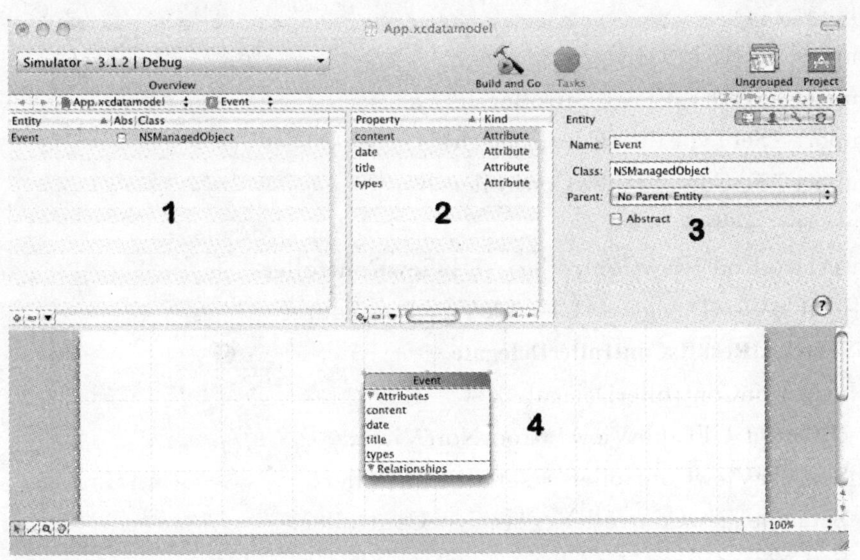


图 9-25　数据模型编辑文件

下面开始创建实体。首先，在第 1 部分中单击下方的“+”按钮，选择 Add Entity 选项，如图 9-26 所示，设置实体名称为“Event”。

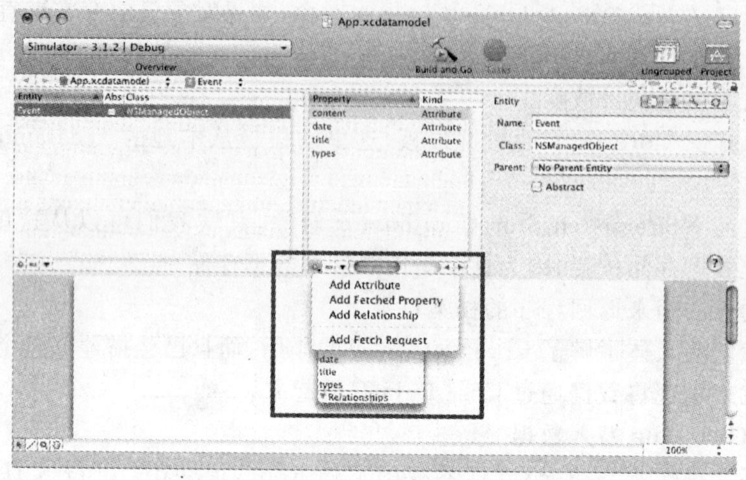


图 9-26 添加 Attribute

然后，在第 2 部分中，同样单击下方的“+”按钮，选择 Add Attribute，分别命名为 content、date、title、types；在第 3 部分里面将各自的类型更改为 String，更改完成后保存文件，开始进行写入操作。我们将在 Record 标签内的 RootViewController 文件内添加使用 Core Data 写入数据的代码。

选中 RootViewController.h 文件，修改代码：

**代码 9.21 RootViewController.h 文件**

```
#import <UIKit/UIKit.h>
#import "SetDateViewController.h"
#import "SetTypeViewController.h"
#import "SetContentViewController.h"
#import <sqlite3.h>
@interface RootViewController : UIViewController
<UITableViewDelegate, UITableViewDataSource
<NSFetchedResultsControllerDelegate,                ❶
UINavigationControllerDelegate> {
    IBOutlet UITableView *rootNoteView;
    SetDateViewController *setDatecontroller;
    SetTypeViewController *setTypeController;
    SetContentViewController *setContentController;
    NSString *myTitle;
    sqlite3 *database;
```

```
    NSFetchedResultsController *fetchedResultsController; ❷
    NSManagedObjectContext *managedObjectContext; ❸
}
@property (nonatomic, retain) UITableView *rootNoteView;
@property (nonatomic, retain) NSString *myTitle;
SetDateViewController *setDatecontroller;
@property (nonatomic, retain)
SetTypeViewController *setTypeController;
@property (nonatomic, retain)
SetContentViewController *setContentController;
@property (nonatomic, retain)
NSFetchedResultsController *fetchedResultsController;
@property (nonatomic, retain)
NSManagedObjectContext *managedObjectContext;
-(NSString *)dataFilePath;
-(void)myChange:(id)sender;
-(void)addRecord; ❹
@end
```

在❶处在程序中添加了两个协议 NSFetchedResultsControllerDelegate 和 UINavigationControllerDelegate,其中第一个协议是利用 Core Data 进行数据操作的协议。

在❷处声明了一个获取数据请求的控制器类,它用来获取实体请求,对数据进行操作。

在❸处声明了一个 NSManagedObjectContext 的对象,因为这里要进行数据的写入操作。

在❹处添加了一个方法,用于将数据写入到存储文件中。

选中 RootViewController. m 文件,修改代码:

**代码 9.22 RootViewController. m 文件**

```
#import "RootViewController.h"
#import "NoteCoreDataAppDelegate.h"
#import "CustomCell.h"
@implementation RootViewController
@synthesize rootNoteView;
@synthesize myTitle;
@synthesize setDatecontroller;
@synthesize setTypeController;
```

```
@synthesize setContentController;
@synthesize fetchedResultsController;
@synthesize managedObjectContext;
-(NSString *)dataFilePath {
    NSArray *paths = NSSearchPathForDirectoriesInDomains
(NSDocumentDirectory, NSUserDomainMask, YES);
    NSString *documentsDirectory = [paths objectAtIndex:0];
    return[documentsDirectory
    stringByAppendingPathComponent:@"NoteSQL.db"];
}
//添加记录方法,借助实体来将数据添加到 NSManagedObject 对象
#pragma mark-
-(void)addRecord {
    NSString *titleStr = [[(CustomCell *)
        [self.rootNoteView cellForRowAtIndexPath:
        [NSIndexPath indexPathForRow:0 inSection:0]]
        noteTextField] text];
    NSString *dateStr = [[(CustomCell *)
        [self.rootNoteView cellForRowAtIndexPath:
        [NSIndexPath indexPathForRow:1 inSection:0]]
        noteTextField] text];
    NSString *typesStr= [[(CustomCell *)
        [self.rootNoteView cellForRowAtIndexPath:
        [NSIndexPath indexPathForRow:2 inSection:0]]
        noteTextField] text];
    NSString *contentStr = [[(CustomCell *)
        [self.rootNoteView cellForRowAtIndexPath:
        [NSIndexPath indexPathForRow:3 inSection:0]]
        noteTextField] text];
    if (([titleStr length]<=0) || ([dateStr length]<=0) ||
        ([typesStr length]<=0) || ([contentStr length]<=0)) {
        UIAlertView *alert = [[UIAlertView alloc]
                                    initWithTitle:@"警告"
                                    message:@"请填写完整!"
                                    delegate:self
                                    cancelButtonTitle:@"确定"
                                    otherButtonTitles:nil];
```

```
    [alert show];
    [alert release];
}
else{
    //删除数据库操作代码
    if (sqlite3_open([[self dataFilePath]
            UTF8String], &database) != SQLITE_OK){
        sqlite3_close(database);
        NSAssert(0, @"Failed to open database");
    }
    NSString * insertSQL =
        [[NSString alloc] initWithFormat:
        @"INSERT INTO NoteSQL
            (title, date, types, content)
            VALUES('%@','%@','%@','%@');"
            ,titleStr, dateStr, typesStr, contentStr];
    char * errorMsg2;
    if (sqlite3_exec (database, [insertSQL UTF8String],
        NULL, NULL, &errorMsg2) != SQLITE_OK) {
        NSAssert1(0, @"Error updating tables:
        %s", errorMsg2);
        sqlite3_free(errorMsg2);
    }
    //将所获取的数据存入 NSManagedObjectContext 对象
    NSManagedObjectContext * context =
        [fetchedResultsController managedObjectContext];
    NSEntityDescription * entity =
        [[fetchedResultsController fetchRequest] entity];
    NSManagedObject * newManagedObject =
        [NSEntityDescription
        insertNewObjectForEntityForName:
        [entity name] inManagedObjectContext:context];
        [newManagedObject setValue:titleStr forKey:@"title"];
    [newManagedObject setValue:dateStr forKey:@"date"];
    [newManagedObject setValue:typesStr forKey:@"types"];
    [newManagedObject setValue:contentStr
                forKey:@"content"];                         ❶
```

```
            UIAlertView *alert = [[UIAlertView alloc]
                                  initWithTitle:@"信息"
                                  message:@"添加成功!"
                                  delegate:self
                                  cancelButtonTitle:@ "确定"
                                  otherButtonTitles:nil];
            [alert show];
            [alert release];
            NSError *error = nil;
            if (![context save:&error]) {
            NSLog(@"Unresolved error %@, %@", error,        ❷
            [error userInfo]);
                abort();
            }
        }
    }
    -(void)viewWillAppear:(BOOL)animated {
        [self.rootNoteView reloadData];
        [super viewWillAppear:YES];
    }
    -(id)initWithNibName:(NSString *)nibNameOrNil
        bundle:(NSBundle *)nibBundleOrNil {
        if (self = [super initWithNibName:nibNameOrNil
        bundle:nibBundleOrNil]) {
        self.title=@"添加记录";
        }
        return self;
    }
    -(void)navigationController:
        (UINavigationController *)navigationController
            willShowViewController:(UIViewController *)
            viewController animated:(BOOL)animated {
        [self.rootNoteView reloadData];
    }
    #pragma mark-
    #pragma mark View lifecycle
    //初始化操作
```

```
-(void)viewDidLoad {
    self.navigationController.delegate =self;
    NoteCoreDataAppDelegate * delegate =
    [[UIApplication sharedApplication] delegate];
    self.managedObjectContext =
        delegate.managedObjectContext;                ❸
    [super viewDidLoad];
    UIBarButtonItem * add = [[UIBarButtonItem alloc]
    initWithTitle:@"添加"
    style:UIBarButtonItemStyleBordered
                        target:self
        action:@selector(addRecord)];
    self.navigationItem.rightBarButtonItem = add;
    [add release];
    self.myTitle=nil;
    NSError * error = nil;
    if (![[self fetchedResultsController] performFetch:&error]){
    NSLog(@"Unresolved error %@, %@", error, [error userInfo]);
    abort();                                          ❹
    }
}
-(void)viewDidUnload {
}
-(void)myChange(id)sender{
    UITextField * myTextField=(UITextField *)sender;
    self.myTitle=myTextField.text;
}
#pragma mark -
#pragma mark Table view methods
……
//利用 NSFetchedResultsController 中的请求建立数据文件的连接
#pragma mark -
```

```
#pragma mark Fetched results controller
-(NSFetchedResultsController * )
    fetchedResultsController {
    if (fetchedResultsController != nil) {
        return fetchedResultsController;
    }
    NSFetchRequest * fetchRequest =                                   ❺
        [[NSFetchRequest alloc] init];
    NSEntityDescription * entity = [NSEntityDescription
        entityForName:@"Event"
    inManagedObjectContext:managedObjectContext];
    [fetchRequest setEntity:entity];
    [fetchRequest setFetchBatchSize:20];
    //根据键一值对应获取数据并对数据进行排序
    NSSortDescriptor * sortDescriptor1 =
        [[NSSortDescriptor alloc]
        initWithKey:@"title" ascending:NO];
    NSSortDescriptor * sortDescriptor2 =
        [[NSSortDescriptor alloc]
        initWithKey:@"types" ascending:NO];
    NSSortDescriptor * sortDescriptor3 =
        [[NSSortDescriptor alloc]
        initWithKey:@"date" ascending:NO];                            ❻
    NSSortDescriptor * sortDescriptor4 =
        [[NSSortDescriptor alloc]
        initWithKey:@"content" ascending:NO];
    NSArray * sortDescriptors = [[NSArray alloc]
        initWithObjects:sortDescriptor1,
        sortDescriptor2,sortDescriptor3,sortDescriptor4, nil];
    [fetchRequest setSortDescriptors:sortDescriptors];
```

```
    NSFetchedResultsController * aFetchedResultsController =
    [[NSFetchedResultsController alloc]
        initWithFetchRequest:fetchRequest
        managedObjectContext:managedObjectContext
                sectionNameKeyPath:nil
                cacheName:@"Root"];
    aFetchedResultsController.delegate = self;
    self.fetchedResultsController = aFetchedResultsController;      ❼
    [aFetchedResultsController release];
    [fetchRequest release];
    [sortDescriptor1 release];
    [sortDescriptor2 release];
    [sortDescriptor3 release];
    [sortDescriptor4 release];
    [sortDescriptors release];
    return fetchedResultsController;
}

-(void)controllerDidChangeContent:
(NSFetchedResultsController *)controller {                ❽
}
#pragma mark -
#pragma mark Memory management
-(void)didReceiveMemoryWarning {
    [super didReceiveMemoryWarning];
}
-(void)dealloc {
    [setContentController release];
    [setTypeController release];
    [setDatecontroller release];
    [rootNoteView release];
    [myTitle release];
    [fetchedResultsController release];
    [managedObjectContext release];
    [super dealloc];
}
@end
```

在 addRecord 方法中，前面部分代码仍然是获取表视图里的内容分别存储到相应的字符串中。

❶处的写入和读取操作都需要有 NSManagedObjectContext 对象和 NSFetchedResultsController 对象来监控数据变化和传送获取数据的请求，所以在这里用指定的实体名创建了 NSManagedObjectContext 的新对象，还声明了 NSEntityDescription 类的对象 entity，其中 NSEntityDescription 类主要是对于实体的属性描述，然后将字符串内的值写入 newManagedObject。在❷处如果写入数据出现错误，则返回错误信息。

❸处使用 UIApplication sharedApplication 实例获取该委托的引用，接着把 managedObjectContext 传给 NoteCoreDataAppDelegate。

在❹处如果操作错误，就返回错误信息，其中 abort ()为退出方法，当出现错误且不能恢复时，跳出一个警告面板提示用户按 Home 键退出程序。

在❺处是对于获取请求控制器的操作，主要用于建立连接。首先判断 fetchedResultsController 是否为空，如果不为空则实体创建一个获取请求类 NSFetchRequest 的对象，用来获取数据持久化对象，NSEntityDescription 用来确定获取哪一个实体。

在❻处我们利用 NSSortDescriptor 来对所获取的实体中的字段进行排序，并按照顺序存入一个数组中。

在❼处的代码是用来创建一个 NSFetchedResultsController 对象，并指定它的代理为应用程序本身。

在❽处添加的方法是利用 NSFetchedResultsController 监测数据内容的变化，在这里重新加载表视图。

**3. 利用 Core Data 读取数据**

到这里我们就已经通过 Core Data 实现了数据的写入操作，接下来开始数据的读取操作。选中 ScanNoteViewController. h 文件，修改代码：

**代码 9.23　ScanNoteViewController. h 文件**

```
#import <UIKit/UIKit.h>
#import <sqlite3.h>
@interface ScanNoteViewController :
        UIViewController <NSFetchedResultsControllerDelegate,
        UITableViewDelegate, UITableViewDataSource>{
        IBOutlet UITableView * mainTable;
        NSMutableArray * dateShow;
        sqlite3 * database;
NSFetchedResultsController * fetchedResultsController;
NSManagedObjectContext * managedObjectContext;          ❶
}
```

```
@property (nonatomic, retain) UITableView *mainTable;
@property (nonatomic, retain) NSMutableArray *dateShow;
@property (nonatomic, retain)
    NSFetchedResultsController *fetchedResultsController;
@property (nonatomic, retain)
    NSManagedObjectContext *managedObjectContext;
-(NSString *)dataFilePath;
-(NSMutableArray *)getArray;
@end
```

利用 Core Data 对数据写入和读取的操作方法相似。在❶处声明了一个获取数据请求的控制器类，用来获取实体，请求对数据进行操作；同样还声明了一个 NSManagedObjectContext 的对象，进行数据的读取操作。接下来修改 ScanNoteViewController.m 文件的代码来实现利用 Core Data 读取所存入的数据。

选中 ScanNoteViewController.m 文件，修改代码：

**代码 9.24 ScanNoteViewController.m 文件**

```
#import "ScanNoteViewController.h"
#import "CustomCell.h"
#import "recordInfo.h"
@implementation ScanNoteViewController
@synthesize mainTable;
@synthesize dateShow;
@synthesize fetchedResultsController;
@synthesize managedObjectContext;
-(NSString *)dataFilePath {
    NSArray *paths = NSSearchPathForDirectoriesInDomains
            (NSDocumentDirectory, NSUserDomainMask, YES);
    NSString *documentsDirectory = [paths objectAtIndex:0];
    return [documentsDirectory
stringByAppendingPathComponent:@"NoteSQL.db"];
}
//读取数据
-(NSMutableArray *)getArray {
    NSMutableArray *mutArray = [[NSMutableArray alloc] init];
        return mutArray;                                        ❶
//此处删除了源文件中的数据库读取代码。
}
```

```
-(id)initWithNibName:(NSString *)nibNameOrNil
bundle:(NSBundle *)nibBundleOrNil {
if (self = [super initWithNibName:nibNameOrNil
bundle:nibBundleOrNil]) {
        self.title = @"清单列表";
    }
    return self;
}
-(void)viewDidLoad {
    self.navigationItem.rightBarButtonItem = self.editButtonItem;
    [super viewDidLoad];
    NSError *error = nil;
    if (![[self fetchedResultsController] performFetch:&error]) {
    NSLog(@"Unresolved error %@, %@",                              ❷
        error, [error userInfo]);
        abort();
    }
}

-(void)viewWillAppear:(BOOL)animated {
        self.datashow=[self getArray];
        [self.mainTable reloadData];
}//重新加载数据
-(void)didReceiveMemoryWarning {
    [super didReceiveMemoryWarning];
}
-(void)viewDidUnload {
}
-(void)dealloc {
    [fetchedResultsController release];
    [managedObjectContext release];
    [dateShow release];
    [mainTable release];
    [super dealloc];
}
//数据源方法的实现,用读取的数据重新绘制表
#pragma mark -
```

```
#pragma mark Table Data Source Methods
-(NSInteger)numberOfSectionsInTableView:
    (UITableView *)tableView {
        return 1;
        return [[fetchedResultsController sections] count];       ❸
}

-(NSInteger)tableView:(UITableView *)tableView
    numberOfRowsInSection:(NSInteger)section {
        id <NSFetchedResultsSectionInfo> sectionInfo =
        [[fetchedResultsController sections]                       ❹
        objectAtIndex:section];
        return [sectionInfo numberOfObjects];
}
-(UITableViewCell *)tableView:(UITableView *)tableView
        cellForRowAtIndexPath:(NSIndexPath *)indexPath{
        static NSString *scanNoteViewControllerCell =
                @"scanNoteViewControllerCell";
    CustomCell *cell = (CustomCell *)[tableView
        dequeueReusableCellWithIdentifier:
        scanNoteViewControllerCell];
    if (cell == nil){
        NSArray *nib=[[NSBundle mainBundle]
                                loadNibNamed:@"CustomCell"
                                owner:self
                                options:nil];
        cell=[nib objectAtIndex:0];
    }
    //获取标题和日期绘制行
```

```
    NSManagedObject * managedObject =
        [fetchedResultsController
        objectAtIndexPath:indexPath];
    [cell.contentView addSubview:cell.noteLabel];
    cell.noteLabel.text=[[dateShow
          objectAtIndex:indexPath.row]title];
    cell.noteLabel.text = [[managedObject
        valueForKey:@"title"] description];                          ❺
    [cell.contentView addSubview:cell.noteTextField];
    recordInfo * tempRecord =
    [self.dateShow objectAtIndex:indexPath.row];
    cell.noteTextField.text=tempRecord.date;
    cell.noteTextField.text=[[managedObject
        valueForKey:@"date"] description];
    [cell.noteTextField setEnabled:NO];
    return cell;
}
// 利用 NSFetchedRequestController 中的请求建立数据文件的连接
#pragma mark -
#pragma mark Fetched results controller
-(NSFetchedResultsController *)fetchedResultsController { ❻
    if (fetchedResultsController != nil) {
        return fetchedResultsController;
    }
    NSFetchRequest * fetchRequest = [[NSFetchRequest alloc] init];
    NSEntityDescription * entity = [NSEntityDescription
                                     entityForName:@"Event"
        inManagedObjectContext:managedObjectContext];
    [fetchRequest setEntity:entity];
    [fetchRequest setFetchBatchSize:20];
    //根据键一值对应获取数据并对数据进行排序
    NSSortDescriptor * sortDescriptor1 = [[NSSortDescriptor alloc]
        initWithKey:@"title" ascending:NO];
    NSSortDescriptor * sortDescriptor2 = [[NSSortDescriptor alloc]
        initWithKey:@"types" ascending:NO];
    NSSortDescriptor * sortDescriptor3 = [[NSSortDescriptor alloc]
        initWithKey:@"date" ascending:NO];
```

```
    NSSortDescriptor * sortDescriptor4 = [[NSSortDescriptor alloc]
        initWithKey:@"content" ascending:NO];
    NSArray * sortDescriptors = [[NSArray alloc]
        initWithObjects:sortDescriptor1,sortDescriptor2,
        sortDescriptor3,sortDescriptor4, nil];
    [fetchRequest setSortDescriptors:sortDescriptors];
    NSFetchedResultsController * aFetchedResultsController =
    [[NSFetchedResultsController alloc]
    initWithFetchRequest:fetchRequest
    managedObjectContext:managedObjectContext
        sectionNameKeyPath:nil cacheName:@"Root"];
        aFetchedResultsController.delegate = self;
        self.fetchedResultsController = aFetchedResultsController;
        [aFetchedResultsController release];
        [fetchRequest release];
        [sortDescriptor1 release];
        [sortDescriptor2 release];
        [sortDescriptor3 release];
        [sortDescriptor4 release];
        [sortDescriptors release];
        return fetchedResultsController;
}
-(void)controllerDidChangeContent:
      (NSFetchedResultsController * )controller {
        [self.mainTable reloadData];                      ❼
}
@end
```

在 ScanNoteViewController.m 文件里，我们要利用 Core Data 实现数据的读取。对于视图的构建，在 9.5 节已经详细介绍过了，不再赘述。我们重点关注 Core Data 代码。

在❶处创建一个数组，用来存放要传递给表的数据。

在❷处 ViewDidLoad 方法中，利用 NSFetchedResultsController 与实体建立起来的连接，实现数据的初始化，注意如果出错就需要返回错误信息。

在❸处表视图的分区数按照所获取数据的分区数来确定。

在❹处确定表视图特定分区的行，这里返回的是实体里的行数。

在❺处绘制 Scan 标签表视图时，利用 managed object 的路径读取出标题(title)和日期(date)：

```
NSManagedObject * managedObject = [fetchedResultsController
                objectAtIndexPath:indexPath];
……
cell.noteLabel.text = [[managedObjectvalueForKey:
                @"title"] description];
```

在❻处是对 NSFetchedResultsController 的操作，与前面所讲的写入操作一样，同样是为了利用请求建立与实体的连接。

在❼处当数据内容确实改变之后，需要重新加载表视图的数据。

**4. 编译并运行程序**

到这里项目就完成了，编译并运行，填充数据之后就可以看到如图 9-23 所示的效果。

# 9.7 小结

在本章中首先介绍了 iPhone 应用程序的文件系统，重点讲解了如何实现应用程序数据的持久性存储。属性列表、归档、SQLite3、Core Data 是数据持久性存储中常用的四种方式，其中 Core Data 是在 SDK3.0 中新添加的特性。希望通过本章的学习，你能够对这些存储方式有深入的理解。

这里讲解的文件系统、数据持久性存储，是 iPhone 应用程序开发中必备的知识，很多 iPhone 程序中都涉及到数据处理，若想致力于 iPhone 应用程序的开发，务必要掌握本章内容。

# 第 10 章 用户设置

**本章内容**

- 什么是用户设置
- 在 Settings 中添加设置束
- 在应用程序中添加设置界面
- 如何在表行上添加控件

用户设置是很多系统中不可或缺的功能之一，Mac 系统下的系统偏好设置，iPhone/iPod touch 上的 Settings 应用程序（如图 10-1 所示）都是用户设置。在今后开发应用程序的过程中，尤其是当程序功能比较复杂时，为用户添加设置选项是一项很重要的工作。这一章我们就来学习如何为应用程序添加设置选项。

图 10-1 iPhone/iPod 的 Settings 应用程序

## 10.1 用户设置概述

iPhone 上的应用程序很多都带有设置选项，用户可以更改这些设置，以改变软件的界面风格或功能。如在一个备忘录中，用户可能希望调整输入字体的大小或者样式，使其符合自己的喜好，对此只需要简单设置一下就可以了，这些设置会被保存下来。

添加设置功能有两种方式：一种是在系统设置（Settings）中添加，另一种是在应用程序内部添加。

在 Settings 中添加设置选项非常简单，只需要在应用程序中增加一个设置束，Settings 中就会出现一个该应用程序的设置条目。进入此条目的下级视图，便可以看到应用程序的设置界面。使用这种方式，用户不需要使用代码或者 Interface Builder 设计设置界面，省去了很多繁琐的工作。而它的缺点也显而易见，那就是用户必须退出应用程序才可以进入 Settings 中设置的页面。因此在 Settings 中设置的选项最好是那些不经常更改的属性，如亮度、通讯方式、网络选择等。

上面我们提到了一个新名词：设置束（Settings Bundle），它是在应用程序中创建的一组文件，通过编写里面的 plist 文件来设计 Settings 中的设置界面，稍后我们来学习如何使用它。

另一种添加设置功能的方式为：直接在应用程序中增加设置界面。虽然这需要我们自己来设计界面，但它避免了必须退出应用程序才可更改设置的缺点。因此，在这里设置的选项应该是那些经常使用的操作，通常放在主视图的背面。

下面我们通过一个小程序来学习上述两种方式是如何实现的。

## 10.2 创建 NoteSetting 应用程序

图 10-2 为我们将要设置的主界面，未添加设置选项前，Text View 中正文字体大小为 18，字体样式为 AmericanTypewriter，背景图片是不能更换的。当我们对字体大小、字体样式、背景图片都做了修改后，界面显示如图 10-3 所示。

图10-2 主视图默认设置的显示　　图 10-3 主视图更改设置后的显示

仔细观察可以发现,上面两个视图的右下角有一个小按钮ⓘ,它是该项目所使用的模板 Utility Application 自带的控件,点击该按钮会转到主视图的背面(伴随着一个翻转的动画效果)。前面已经提到设置选项一般放置在主视图的背面,这就是我们使用这个模板的目的。

现在开始创建这个项目。打开 Xcode,选择 File→NewProject,弹出 New Project 窗口,如图 10-4 所示,选择 Utility Application 模板,将项目命名为 NoteSetting。

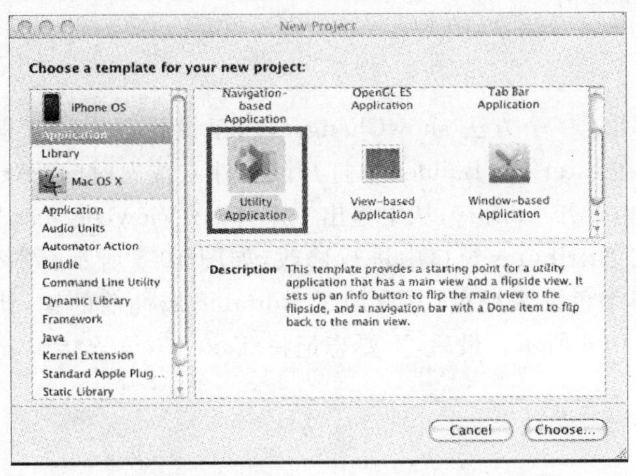


图 10-4 Utility Application 模板

也许你已经发现，这个模板与我们以前所用的模板有些不同，它把定义两个视图的文件分别放在不同的文件夹下。其中 Main View 中存放的是定义主视图（也就是上面图 10-2 和 10-3）的文件，Flipside View 中的文件用于定义主视图背面的设置视图，稍后我们就要在这里面定义设置选项。

## 10.3 设计主视图

首先，我们需要导入五张背景图片，在随书光盘的 Chapter 10 NoteSetting 文件夹中找到：黑白印画.png、青竹通幽.png、那时茶香.png、花生彼岸.png 和开到荼蘼.png，将它们拖到 Xcode 中 Resources 文件夹下。本例中，加入背景图片需要用到 Image View，界面的文本信息是在 Text View 中显示的，因此我们要在 MainViewController.h 文件中做以下声明：

**代码 10.1 MainViewController.h 文件**

```
#import "FlipsideViewController.h"
@interface MainViewController : UIViewController
                    <FlipsideViewControllerDelegate> {
    IBOutlet UITextView *mainTextView;
    UIImageView *backgroundImage;
}
@property (nonatomic, retain)IBOutlet UITextView *mainTextView;
@property (nonatomic, retain)UIImageView *backgroundImage;
-(IBAction)showInfo;
-(void)showChanges;
@end
```

头文件中还声明了一个方法 showChanges，用于设置字体和背景图片。接下来，双击 MainView.xib，在 Interface Bulider 中打开该文件。首先，打开 Attributes 窗口，调节视图的背景色。然后打开 Library，从中拖出一个 Text View 到 MainView，调节其大小。选中 Text View，在 Attributes 窗口中进行修改，如图 10-5 所示。本例中我们更改了文本显示内容，将文本颜色设为绿色，并取消了 Editable 属性前的钩，让该文本不可编辑，修改后的效果如图 10-6 所示。最后，不要忘记将 Text View 连接到 mainTextView 输出口。

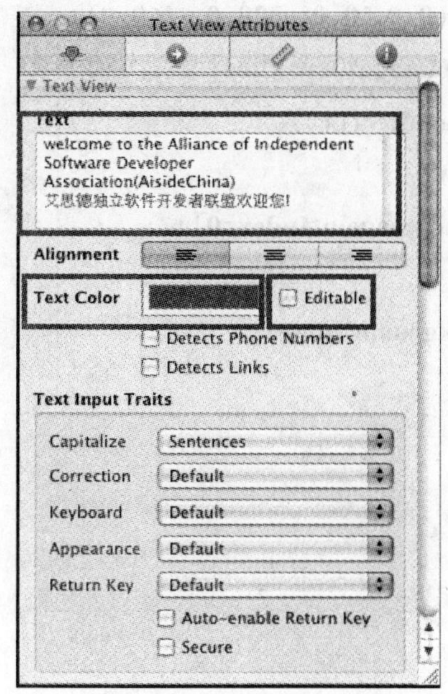


图 10-5 Text View 的属性检测器

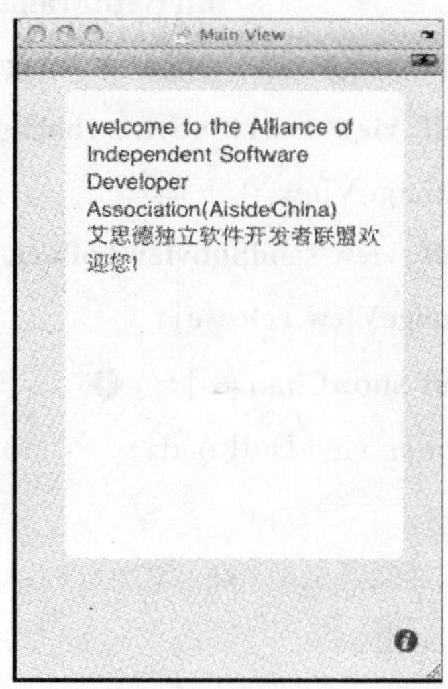


图 10-6 Main View 界面设计效果

保存并关闭 Interface Builder，返回 Xcode，选择 MainViewController. m 文件，添加如下黑体字所示的代码：

**代码 10.2 MainViewController. m 文件**

```
#import "MainViewController.h"
#import "MainView.h"
@implementation MainViewController
@synthesize mainTextView;
@synthesize backgroundImage;
-(void)showChanges {
    self.mainTextView.font =[UIFont
        fontWithName:@"AmericanTypewriter"      ❶
            size:18];
    UIImage * image = [UIImage imageNamed:@"黑白印画.png"];
    self.backgroundImage.image = image;    ❷
}
-(void)viewDidLoad {
```

```
    CGRect imageRect = CGRectMake(0.0, 0.0, 320.0, 460.0);  ┐
    UIImageView * imageView = [[UIImageView alloc]             │
                initWithFrame:imageRect];                       │
    self.backgroundImage = imageView;                           │
    [self.view insertSubview:backgroundImage atIndex:0];        │ ❸
    //imageView 在下面                                          │
    [self.view sendSubviewToBack:backgroundImage];              │
    [imageView release];                                        ┘
    [self showChanges];   ❹
    [super viewDidLoad];
}
......
-(void)dealloc {
    [backgroundImage release];
    [mainTextView release];
    [super dealloc];
}
@end
```

为了方便更改背景图片，我们在❸处用代码在主视图中定义了一个 Image View。使用 sendSubviewToBack:方法将 Image View 放在了视图的底层，如果不这样做，加入的图片会覆盖 Text View。

showChanges 方法中，在❶处使用 fontWithName: size:方法定义了显示文本的字体样式和大小，在❷处将背景图片设为"黑白印画.png"。

我们希望在程序载入时，可以显示已更改后的设置。为此，在 viewDidLoad 方法中的❹处调用了 showChanges 方法。

编译并运行该程序，如果未做其他更改，程序运行后主界面应该与前面图 10-2 相同。

## 10.4 在 Settings 中添加设置选项

应用程序启动后，Settings 会自动检测该应用程序中是否含有设置束，如果含有，会根据其内容显示应用程序的设置选项。因此，我们应先创建一个设置束。

### 10.4.1　创建一个设置束

在 Xcode 的 Groups & Files 窗格中，右键单击 NoteSetting，选择 Add→New File，弹出 New File 窗口，点击左侧 iPhone OS 层级下的 Resource，在右侧会显示一个 Settings Bundle 图标，如图 10-7 所示，这就是我们要创建的设置束文件。

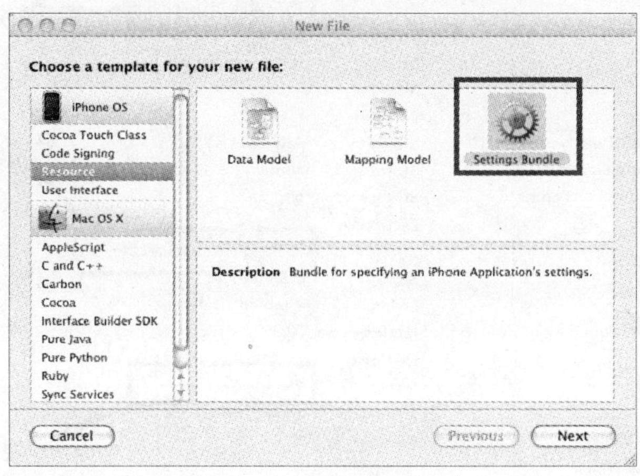


图 10-7　新建设置束文件

选中 Settings Bundle 图标，单击 Next。我们不妨使用默认文件名 Settings. Bundle，这样在 Xcode 中便多了一个新项 Settings. Bundle。展开会看到两个文件：Root. plist 和 en. lproj，这一章我们只涉及 Root. plist 文件。

单击 Root. plist 文件，在右侧将显示相应的内容，展开所有的节点，如图 10-8 所示。Root 节点下有两项，第一项为 StringsTable，这不是我们现在关心的内容，先不用管它。下一项是 PreferenceSpecifiers 数组，其中的内容便是偏好设置选项。默认情况下有 4 个 Dictionary 节点，每个节点下都有一个 Type 键，它们的取值分别为 PSGroupSpecifier、PSTextFieldSpecifier、PSToggleSwitchSpecifier、PSSliderSpecifier，分别指定一个设置选项。除了 PSGroupSpecifier 类型的项，每项都会有一个主键 Key，用于传递设置选项的值。

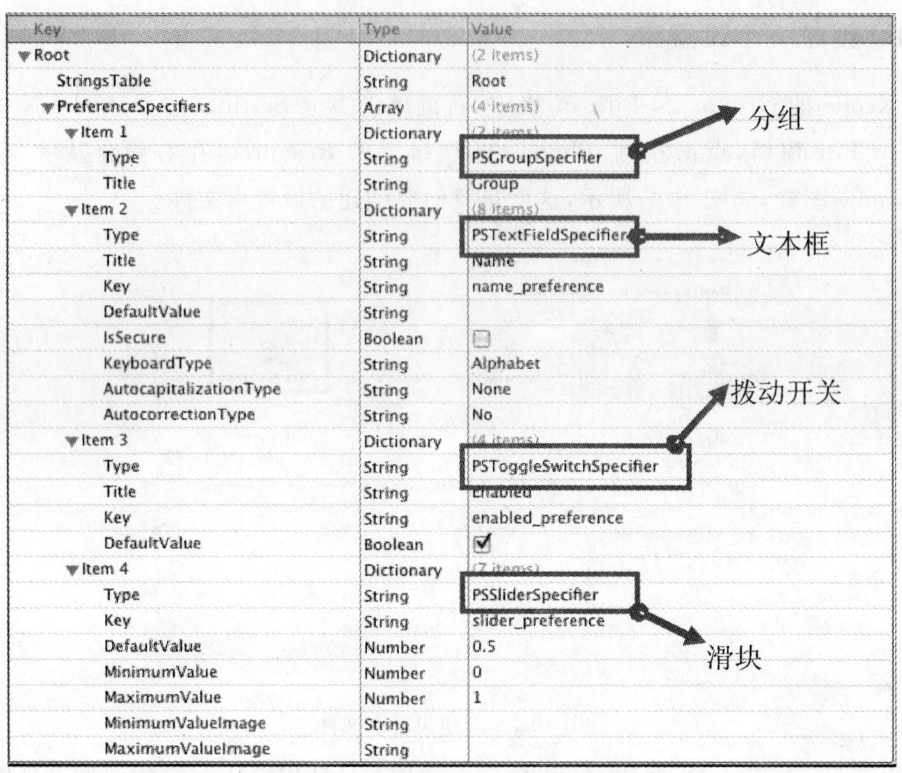

| Key | Type | Value |
|---|---|---|
| ▼Root | Dictionary | (2 items) |
| StringsTable | String | Root |
| ▼PreferenceSpecifiers | Array | (4 items) |
| ▼Item 1 | Dictionary | (2 items) |
| Type | String | PSGroupSpecifier |
| Title | String | Group |
| ▼Item 2 | Dictionary | (8 items) |
| Type | String | PSTextFieldSpecifier |
| Title | String | Name |
| Key | String | name_preference |
| DefaultValue | String | |
| IsSecure | Boolean | ☐ |
| KeyboardType | String | Alphabet |
| AutocapitalizationType | String | None |
| AutocorrectionType | String | No |
| ▼Item 3 | Dictionary | (4 items) |
| Type | String | PSToggleSwitchSpecifier |
| Title | String | Enabled |
| Key | String | enabled_preference |
| DefaultValue | Boolean | ☑ |
| ▼Item 4 | Dictionary | (7 items) |
| Type | String | PSSliderSpecifier |
| Key | String | slider_preference |
| DefaultValue | Number | 0.5 |
| MinimumValue | Number | 0 |
| MaximumValue | Number | 1 |
| MinimumValueImage | String | |
| MaximumValueImage | String | |



图 10-8　Root. plist 文件默认内容

属性列表中节点的类型除了 Dictionary 类型外，还可以是 Array、String、Boolean、Data、Number 等类型，但只有 Dictionary 和 Array 类型是可以存储多个节点的有序列表。

想知道现在会出现什么样的界面吗？编译并运行一下吧。我们还未对程序进行修改，因此程序界面并无变化，这不是我们想要的。退出应用程序，找到 Settings 图标，点击进入，是不是多了一条以 NoteSetting 命名的设置项？如图 10-9 所示，这就是我们刚刚创建的设置束。进入程序设置主视图，如图 10-10 所示。

图 10-9 Settings 中增加的设置束

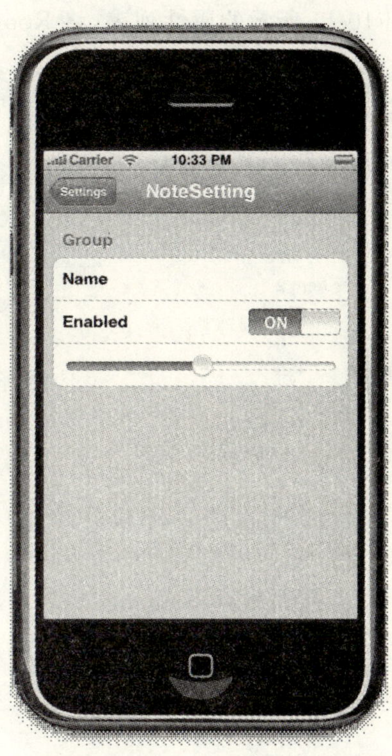


图 10-10 Settings 中的默认设置主界面

现在与 Root. plist 文件中的节点类型对应一下：PSGroupSpecifier 创建一个分组，PSTextFieldSpecifier 创建一个文本框，PSToggleSwitchSpecifier 创建一个拨动开关，PSSliderSpecifier 创建一个滑块。它们按照 Root. plist 文件中节点的先后顺序摆放着，下面我们就对 Root. plist 文件稍加修改，使设置界面变为我们想要的效果。

### 10.4.2 编写 Root. plist 文件

我们打算用拨动开关控制主视图背景图片是否可变，还希望出现一个可以进入的子列表，能够选择字体样式。因此，属性列表中默认的四种节点类型只有两个是可以利用的：PSGroupSpecifier 和 PSToggleSwitchSpecifier。

那么，有下级列表的设置项是如何定义的呢？这时侯就需要增加另一种节点类型了：PSMultiValueSpecifier。它在默认的属性列表中并未给出，但也是很常见的设置项样式。PSMultiValueSpecifier 是一个添加多值字段的项，它指定一个带有扩展指示器的行，点击此行进入子列表，选中一行，行右侧会出现一个钩，表示已选中此行，这个值将被传到设置主视图，这正是我们所希望实现的效果。

对照图 10-11 所示的属性列表，将 Root. plist 文件做如下修改：

| Key | Type | Value |
| --- | --- | --- |
| ▼Root | Dictionary | (2 items) |
| StringsTable | String | Root |
| ▼PreferenceSpecifiers | Array | (3 items) |
| ▼Item 1 | Dictionary | (2 items) |
| Type | String | PSGroupSpecifier |
| Title | String | 设置选项 |
| ▼Item 2 | Dictionary | (6 items) |
| Type | String | PSToggleSwitchSpecifier |
| Title | String | 是否改变背景图片 |
| Key | String | imageSwitch |
| TrueValue | String | Engaged |
| FalseValue | String | Disabled |
| DefaultValue | String | Engaged |
| ▼Item 3 | Dictionary | (6 items) |
| Type | String | PSMultiValueSpecifier |
| Title | String | 字体样式 |
| Key | String | fontStyle |
| ▼Values | Array | (4 items) |
| Item 1 | String | AmericanTypewriter |
| Item 2 | String | Marker Felt |
| Item 3 | String | Georgia |
| Item 4 | String | Zapfino |
| ▼Titles | Array | (4 items) |
| Item 1 | String | AmericanTypewriter |
| Item 2 | String | Marker Felt |
| Item 3 | String | Georgia |
| Item 4 | String | Zapfino |
| DefaultValue | String | AmericanTypewriter |

图 10-11　更改后的 Root. plist 文件内容

(1) 删除不需要的条目 Item 2 和 Item 4；

(2) 分组(PSGroupSpecifier)：Title 的值改为“设置选项”；

(3) 拨动开关(PSToggleSwitchSpecifier)：Title 的值改为“是否改变背景图片”，Key 的值为 imageSwitch。接下来增加两个子项，选中 Key 行，点击行末端的添加同级节点的按钮[+]，出现一个新行，其默认类型为 String，这里不需要更改它，将其键和值设为 TrueValue/Engaged。用同样方法再增加一个新行，键和值设为 FalseValue/Disabled。最后将 DefaultValue 改为 String 类型，其值设为 Engaged。

(4) 多值字段(PSMultiValueSpecifier)：增加一个新项，用来选择字体的样式。这一项操作有点复杂，可将 Item 2 折叠并选中，点击行末端的按钮[+]，出现一个新行 Item 3，将其 Type 改为 Dictionary。展开小三角，选中此行，在行末端出现一个按钮[≡]，这个按钮与前面我们用到的按钮有些不同，它代表的是增加子结点。点击三次按钮，增加三行，其键和值分别设为 Type/PSMultiValueSpecifier、Title/字体样式、Key/fontStyle。

我们还未定义子列表中的值。向 Item 3 中添加另外一个子项，命名为 Values，将其类型改为 Array，此行前端便出现了一个小三角，这表示将要定义一个数组。使该行处于展开状态，点击右侧的添加按钮，增加四个子项，其值分别为 AmericanTypewriter、Marker Felt、Georgia、Zapfino。这些值代表的是字体样式，用户也可以添加自己喜欢的字体样式。

折叠 Values 并选中，复制粘贴，便生成了 Values 的一个副本，将此新项名称改为 Titles，这一项是在子列表中显示的行标题。而 Values 中定义的是用户设置中实际存储的一组值。这两项必须相互对应，这样才能让 MainView 中的字体样式更改为我们选中的值。

这个多值字段还应该有一个默认值。再增加 Item 3 的一个子项，将其键和值分别设为 DefaultValue/AmericanTypewriter。将字体样式默认值设为 American Typewriter。

至此，属性列表编辑完成了。遍译并运行一下。进入系统设置看到的设置主界面应该如图 10-12 所示。第一行是一个控制是否改变背景图片的拨动开关，第二行带有一个扩展指示器。点击字体样式一行进入子列表，如图 10-13 所示；点击某行，出现选中标记；返回主设置视图，选中行的文本被显示出来。在前面章节的学习中，这样的操作用代码实现起来有些麻烦，在属性列表中定义是不是简单了很多？

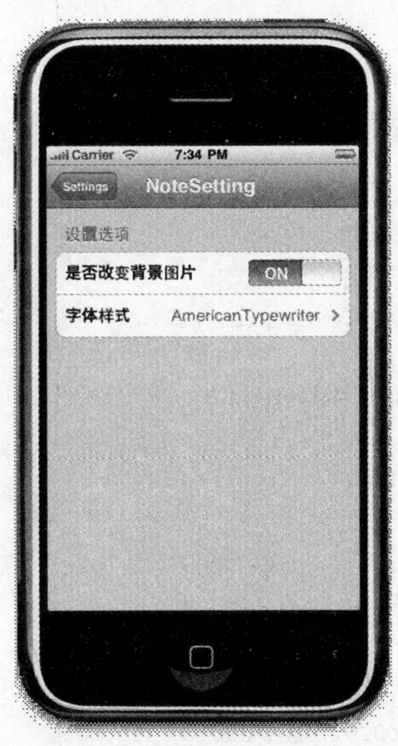

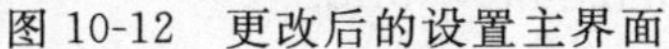
图 10-12 更改后的设置主界面

图 10-13 字体样式选择列表

**Tips**

设置束不仅包括以下五种：PSGroupSpecifier/PSTextFieldSpecifier/PSToggleSwitchSpecifier/PSSliderSpecifier/PSMultiValue-Specifier，还可以有 PSChildPaneSpecifier 等类型。PSChildPaneSpecifier 类型的设置项能够添加一个子设置视图，此子视图是在另一个属性列表中定义的，通过名为 File 的键一值对将其引入。

### 10.4.3 更改字体样式

设置界面已经完成了，如何才能让 Main View 读取到设置值，并作相应的改变呢？或者说，用户设置如何才能控制主视图中的显示呢？

前面提到过 Root.plist 文件中的每个项都会有一个主键 Key，它的作用就是将用户设置的各选项值传递到所控制的视图。在 Xcode 中选中 MainViewController.h 文件，在头部添加代码：

**#define kFontStyle @"fontStyle"**

打开 MainViewController.m，在 showChanges 方法中作如下更改：

**代码 10.3 MainViewController.m 文件**

```
-(void)showChanges {
    self.mainTextView.font =[UIFont
            fontWithName:@"AmericanTypewriter"
            size:19];
    NSUserDefaults *defaults =
            [NSUserDefaults standardUserDefaults];
    self.mainTextView.font =[UIFont
            fontWithName:[defaults objectForKey:kFontStyle]
            size:18];                                          ❶
    UIImage *image = [UIImage imageNamed:@"黑白印画.png"];
    self.backgroundImage.image = image;
}
-(void)viewDidLoad {
    CGRect imageRect = CGRectMake(0.0, 0.0, 320.0, 460.0);
    UIImageView *imageView = [[UIImageView alloc]
                                  initWithFrame:imageRect];
```

```
    self.backgroundImage = imageView;
    [self.view insertSubview:backgroundImage atIndex:0];
    //imageView 在下面
    [self.view sendSubviewToBack:backgroundImage];
    [imageView release];
    NSUserDefaults *defaults =
                        [NSUserDefaults standardUserDefaults];   ❷
    NSDictionary *styleDefaults = [NSDictionary
            dictionaryWithObject:@"AmericanTypewriter"        ❸
                      forKey:kFontStyle];
    [defaults registerDefaults:styleDefaults];   ❹
    [self showChanges];
    [super viewDidLoad];
}
```

NSUserDefaults 类通过键值存取用户设置信息。在 viewDidLoad 方法中，首先在❷处定义了一个指向用户默认设置值的指针 defaults；然后使用 dictionaryWithObject:forKey:方法在❸处将字体样式属性 fontStyle 初始化为 AmericanTypewriter，并赋给 NSDictionary 类的一个实例对象 styleDefaults；最后，在❹处使用 registerDefaults:将 styleDefaults 赋给 defaults。这样就在属性列表中完成了字体样式的初始化。

objectForKey:方法用于获取默认的用户设置，它将返回一个 Objective-C 对象，如 NSString、NSDate、NSNumber。在 showChanges 方法❶处返回的是 Settings 中选择的字体样式，这样 MainView 就获得了 Settings 中当前的用户设置。

编译并运行程序，主视图中字体样式为初始值 AmericanTypewriter。打开 Settings 应用程序，将字体样式选择为 Zapfino，退出 Settings，再返回 NoteSetting 应用程序，主视图的改变了？如图 10-14 所示。

图 10-14　修改字体样式后的效果

到目前为止，还未对拨动开关设置选项定义任何操作，我们将在本章的 10.6 节讲到它。

# 10.5 在应用程序中添加设置

你已经学会了第一种方法：在 Settings 中添加设置项并读取设置值，接下来将要介绍如何在应用程序中添加设置项。

## 10.5.1 定义设置视图

这一次仍将设置选项放置在 TableView 中，如图 10-15 所示。在 FlipView 中将要添加两个设置选项：

（1）在表视图第一组中添加一个 Slider 调节字体大小，并在其左边加入一个 Label 显示滑块的值；

（2）第二组是一个可勾选的列表，用来选择背景图片。

在前面的章节中并未提到表行上可以添加控件，这其实完全可以做到，而且并不复杂。

图 10-15 在应用程序中设置界面

打开 FlipsideViewController.h 文件，添加如下黑体字所示的代码：

**代码 10.4 FlipsideViewController.h 文件**

```
#define kFontSlider @"fontSlider"
#define kImageKey @"backgroundImage"
@protocol FlipsideViewControllerDelegate;
@interface FlipsideViewController : UIViewController {
    id <FlipsideViewControllerDelegate> delegate;
    IBOutlet UITableView *flipTableView;
    IBOutlet UISlider *flipSlider;
    IBOutlet UILabel *labelSize;
    NSArray *imageSelectList;
    NSIndexPath *lastIndexPath;
    NSString *imageSelect;
    int selectRow;
}
```

```
@property (nonatomic, retain) IBOutlet UITableView *flipTableView;
@property (nonatomic, retain) IBOutlet UISlider *flipSlider;
@property (nonatomic, retain) IBOutlet UILabel *labelSize;
@property (nonatomic, retain) NSArray *imageSelectList;
@property (nonatomic, retain) NSIndexPath *lastIndexPath;
@property (nonatomic, retain) NSString *imageSelect;
@property int selectRow;
@property (nonatomic, assign) id
            <FlipsideViewControllerDelegate> delegate;
-(IBAction)done;
-(void)valueChanged:(id)sender;
@end
@protocol FlipsideViewControllerDelegate
-(void)flipsideViewControllerDidFinish:
                    (FlipsideViewController *)controller;
@end
```

代码中声明了两个常量 kFontSlider 和 kImageKey，作为存取设置选项的键值。还声明了 7 个全局变量：

(1) flipTableView：表视图的输出口；

(2) flipSlider：滑块；

(3) labelSize：标签；

(4) imageSelectList：存储图片名的数组；

(5) lastIndexPath：跟踪最后选中的行；

(6) imageSelect：被选中的图片名；

(7) selectRow：被选中的行数。

valueChanged:方法用于跟踪滑块的值，并在标签文本中显示。打开 FlipsideViewController.xib 文件，拖入一个 Table View 并连接到 flipTableView 输出口，同时连接代理和数据源，将表更改为分组表样式，相信这些操作你已经很熟悉，不再赘述。

接下来，保存并返回 Xcode，选中 FlipsideViewController.m，添加如下黑体字所示的代码：

**代码 10.5 FlipsideViewController.m 文件**

```
#import "FlipsideViewController.h"
@implementation FlipsideViewController
@synthesize delegate;
@synthesize flipTableView;
```

```
@synthesize flipSlider;
@synthesize labelSize;
@synthesize imageSelectList;
@synthesize lastIndexPath;
@synthesize imageSelect;
@synthesize selectRow;
-(void)viewDidLoad {
    self.view.backgroundColor =
        [UIColor viewFlipsideBackgroundColor];
    NSArray * array = [[NSArray alloc]
            initWithObjects:@"黑白印画",@"青竹通幽",
            @"那时茶香",@"花生彼岸",@"开到荼蘼",nil];
    self.imageSelectList = array;
    [array release];
    [super viewDidLoad];
}
//label 中显示滑块的值
-(void)valueChanged:(id)sender {
    NSString * sizeLabel = [[NSString alloc]
            initWithFormat:@"%.f",flipSlider.value];      ❶
    self.labelSize.text = sizeLabel;
    [sizeLabel release];
}
//视图出现前读取上次设置值
-(void)viewWillAppear:(BOOL)animated {
    NSUserDefaults * defaults =
            [NSUserDefaults standardUserDefaults];
    flipSlider.value = [defaults floatForKey:kFontSlider]; ❷
    self.imageSelect = [defaults objectForKey:kImageKey];
    selectRow=[defaults integerForKey:@"intFont"];
    NSIndexPath * temp=[NSIndexPath
            indexPathForRow:selectRow inSection:1];
    self.lastIndexPath=temp;
    [temp release];
}
//视图消失前保留本次设置值
-(void)viewWillDisappear:(BOOL)animated {
    NSUserDefaults * defaults =
            [NSUserDefaults standardUserDefaults];
```

```
    [defaults setFloat:flipSlider.value forKey:kFontSlider]; ❸
    [defaults setObject:imageSelect forKey:kImageKey];
    [defaults setInteger:selectRow forKey:@"intFont"];
    [super viewWillDisappear:animated];
}
……
-(void)dealloc {
    [flipTableView release];
    [flipSlider release];
    [labelSize release];
    [imageSelectList release];
    [lastIndexPath release];
    [imageSelect release];
    [super dealloc];
}
#pragma mark -
#pragma mark Table Data Source Methods
-(NSInteger)numberOfSectionsInTableView:(UITableView *)tableView
{
    return 2;
}
-(NSInteger)tableView:(UITableView *)tableView
    numberOfRowsInSection:(NSInteger)section {
    if (section == 0) {
        return 1;
    }
    return 5;
}
//定义组头标题
-(NSString *)tableView:(UITableView *)tableView
titleForHeaderInSection:(NSInteger)section{
    if (section == 0) {
        NSString *sectionName = @"字体大小";
        return sectionName;
    }
    NSString *sectionName = @"背景图片";
    return sectionName;
}
-(UITableViewCell *)tableView:(UITableView *)tableView
```

```
        cellForRowAtIndexPath:(NSIndexPath *)indexPath {
    static NSString *SettingCellIdentifier = @"SettingCellIdentifier";
    UITableViewCell *cell =
    [tableView
dequeueReusableCellWithIdentifier:SettingCellIdentifier];
    if (cell == nil) {
        cell = [[[UITableViewCell alloc]
                    initWithFrame:CGRectZero
                    reuseIdentifier:SettingCellIdentifier]
  autorelease];
    }
    if (indexPath.section == 0) {
        NSUserDefaults *defaults =
                [NSUserDefaults standardUserDefaults];
        //表行上添加 Slider 控件
        UISlider *fontSizeSlider = [[UISlider alloc]
          initWithFrame:CGRectMake(90.0f,100.0f, 261.0f, 23.0f)]; ❹
        [fontSizeSlider addTarget:self
                      action:@selector(valueChanged:)          ❺
            forControlEvents:UIControlEventValueChanged];
        fontSizeSlider.minimumValue = 11;
        fontSizeSlider.maximumValue = 25;
        fontSizeSlider.value = [defaults floatForKey:kFontSlider];
        self.flipSlider = fontSizeSlider;
        cell.accessoryView = self.flipSlider; ❻
        [fontSizeSlider release];
        //表行上添加 Label 控件
        UILabel *fontSizeLabel = [[UILabel alloc]
                initWithFrame:CGRectMake(10, 10, 75, 25)];
        fontSizeLabel.textAlignment = UITextAlignmentLeft;
        fontSizeLabel.font = [UIFont boldSystemFontOfSize:14];
        NSString *sizeLabel = [[NSString alloc] initWithFormat:      ❼
                @"%.f",[defaults floatForKey:kFontSlider]];
        fontSizeLabel.text = sizeLabel;
        self.labelSize = fontSizeLabel;
        [cell.contentView addSubview:fontSizeLabel];
        [fontSizeLabel release];
    }
    //检索第二组中每行
```

```
    if (indexPath.section ==1 ) {
        NSUInteger row = [indexPath row];
        NSUInteger oldRow = [lastIndexPath row];
        //定义每行文本信息
        cell.textLabel.text = [imageSelectList objectAtIndex:row];
        //标记某行选中
        cell.accessoryType = (row == oldRow &&
lastIndexPath !=nil)?
            UITableViewCellAccessoryCheckmark:
                                UITableViewCellAccessoryNone;
    }
    return cell;
}
//标记某行选中
#pragma mark Table Delegate Methods
-(void)tableView:(UITableView *)tableView
didSelectRowAtIndexPath:(NSIndexPath *)indexPath {
    int newRow = [indexPath row];
    int oldRow = [lastIndexPath row];
    if (newRow != oldRow) {
        UITableViewCell *newCell =
                    [tableView
cellForRowAtIndexPath:indexPath];
        newCell.accessoryType =
UITableViewCellAccessoryCheckmark;
        UITableViewCell *oldCell =
                    [tableView
cellForRowAtIndexPath:lastIndexPath];
        oldCell.accessoryType = UITableViewCellAccessoryNone;
        lastIndexPath = indexPath;
        self.imageSelect = newCell.textLabel.text;
        selectRow=[indexPath row];
    }
    //选择第一行有效
    if (newRow ==0 && oldRow == 0) {
        UITableViewCell *newCell =
            [tableView
```

```
cellForRowAtIndexPath:indexPath];
            newCell.accessoryType
UITableViewCellAccessoryCheckmark;
            lastIndexPath = indexPath;
            self.imageSelect = newCell.textLabel.text;
            selectRow=[indexPath row];
        }
        //选中行不显示为蓝色
        [tableView deselectRowAtIndexPath:indexPath animated:YES];
    }
    @end
```

在该代码中定义了一个两组的表视图：第一组有一行，组标题为"字体大小"；第二组有五行，组标题为"背景图片"。代码讲解见下面的10.5.2和10.5.3小节。

### 10.5.2 在表行上添加标签和滑块

表视图单元中添加控件是在tableView:cellForRowAtIndexPath:方法中实现的。❹处我们在第一组中的第一行（唯一的行）创建了一个滑块并定义了它的位置。在❺处使用addTarget:action:forControlEvents:方法来实现滑块的值改变时调用❶处的valueChanged方法，即标签中会显示滑块值的变化。然后定义了它的最小值和最大值。最后在❻处将创建的滑块赋给第一行单元的accessoryView属性，这样一个滑块便添加到表行上了。其中下列语句是与用户设置相关的：

```
fontSizeSlider.value = [defaults floatForKey:kFontSlider];
```

它的作用是在应用程序再次加载时将上次存储的滑块值读出并显示在滑块上。由于fontSizeSlider是个局部变量，因此将它赋给了全局变量flipSlider，这样就可以在其他方法中存取滑块的值了。

在❼处使用同样的方式在表行上添加标签控件，文本显示为滑块的当前值。本例中设置了标签的部分属性：字体和对齐方式，并将fontSizeLabel赋给全局变量labelSize。

setFloat:方法用于存储滑块的值，它在viewWillDisappear方法❸处被调用，使得在视图转换之前完成存储工作，保证了当我们翻转视图到主视图时，读取到的是最新的设置值。在viewWillAppear方法❷处调用floatForKey:方法，目的是在FlipView出现前获取滑块的值并显示在滑块上。

### 10.5.3 可勾选列表

剩下的很大篇幅的代码用于定义第二组表视图，一个可勾选的列表。在本书前面章节已经介绍过，这里不做过多的解释。不同的是，我们存储了被选中行的行数，以便于再次打开时仍被选中着。

与滑块值的存取相似，使用 setObject：和 objectForKey：两个方法来存取图片的名字。

### 10.5.4 主视图初始化并获取设置值

设置视图定义好之后，接下来的工作就是让 MainView 读取设置值了。在 MainViewController.m 文件中做如下更改：

**代码 10.6 MainViewController.m 文件**

```
-(void)showChanges {
    NSUserDefaults * defaults =
                        [NSUserDefaults standardUserDefaults];
    self.mainTextView.font =[UIFont
            fontWithName:[defaults objectForKey:kFontStyle]
            size:18];
    UIImage * image = [UIImage imageNamed:@"黑白印画.png"];
    self.mainTextView.font =[UIFont
        fontWithName:[defaults objectForKey:kFontStyle]
                size:[defaults floatForKey:kFontSlider]];
    NSString * imageName = [[defaults objectForKey:kImageKey]
                    stringByAppendingString:@".png"];
    UIImage * image = [UIImage imageNamed:imageName];
    self.backgroundImage.image = image;
}
-(void)viewDidLoad {
    CGRect imageRect = CGRectMake(0.0, 0.0, 320.0, 460.0);
    UIImageView * imageView = [[UIImageView alloc]
                                initWithFrame:imageRect];
    self.backgroundImage = imageView;
    [self.view insertSubview:backgroundImage atIndex:0];
    //imageView 在下面
    [self.view sendSubviewToBack:backgroundImage];
    [imageView release];
    NSUserDefaults * defaults =
            [NSUserDefaults standardUserDefaults];
    NSDictionary * styleDefaults=[NSDictionary
            dictionaryWithObject:@"AmericanTypewriter"
                        forKey:kFontStyle];
```

```
    NSDictionary *sliderDefaults = [NSDictionary
            dictionaryWithObject:@"18"forKey:kFontSlider];❶
    NSDictionary *imageDefaults = [NSDictionary
          dictionaryWithObject:@"黑白印画" forKey:kImageKey];❷
    [defaults registerDefaults:styleDefaults];
    [defaults registerDefaults:sliderDefaults];
    [defaults registerDefaults:imageDefaults];
    [self showChanges];
    [super viewDidLoad];
}
-(void)viewDidAppear:(BOOL)animated{
    [self showChanges];❸
    [super viewDidAppear:animated];
}
```

在❶中,将滑块的值设为18;视图的背景图片在❷处初始化为“黑白印画”。与更改字体大小的原理相同,在showChanges方法中,使用floatForKey:方法获取滑块的值,设置字体大小,使用objectForKey:方法获取图片名。

新增了一个viewDidAppear:方法,并在❸处调用了showChanges方法。这样,在由FlipView转换到MainView前,设置值就已经被读出来了。

你也许会疑惑,为什么不使用viewWillAppear而选择viewDidAppear呢?那是因为:程序在FlipView的viewWillDisappear方法中保存设置,在MainView的viewDidAppear方法中读取这些设置,而viewWillAppear方法是在viewWillDisappear之前被调用的,因此,使用它将无法读取到最新的设置值。

编译并运行程序,效果如图10-3所示。

## 10.6 开关控制背景图片

该让开关派上用场了。本例中开关的作用是控制背景图片是否可以改变,有一个简单的方法可以实现:控制FlipView的表视图中是否创建第二个组。当开关处于打开状态时,表视图创建两个组,可以对背景图片进行更改;如果开关处于关闭状态,第二组不能被创建。

在FlipsideViewController.h文件头部添加如下代码:

```
#define kImageSwitch @"imageSwitch"
```

常量kImageSwitch用于传递开关的状态。在FlipsideViewController.m文件中做如下更改:

**代码 10.7 FlipsideViewController.m 文件**

```
-(NSInteger)numberOfSectionsInTableView:(UITableView *)tableView {
    NSUserDefaults * defaults =
            [NSUserDefaults standardUserDefaults];
    if ([[defaults objectForKey:kImageSwitch]
                isEqualToString: @"Disabled"]) {
        return 1;
    }
    return 2;
}
```

编译并运行一下,进入 Settings 应用程序,在 NoteSetting 设置项中将开关关闭,如图 10-16 所示,返回 NoteSetting 应用程序,在 FlipView 中表视图只显示一个组,如图 10-17所示。

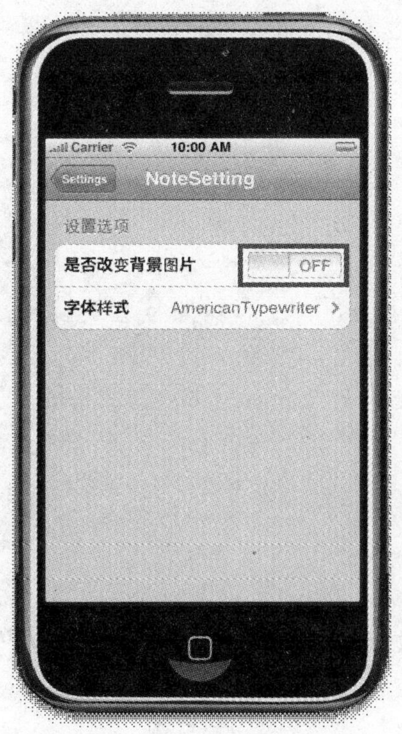

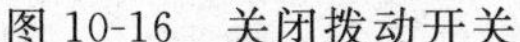

图 10-16 关闭拨动开关　　　　图 10-17 不显示选择背景图片的组

## 10.7 小结

这一章我们使用了一个新项目模板，创建了一个新文件 Settings Bundle。学习了添加用户设置的两种方式：在 Settings 中和在应用程序内部，并学习了如何在表行上添加控件。是不是很有趣？

想知道如何用手指改变一张图片的大小和位置吗？赶快翻到下一页吧！

# 第11章　触摸、手势和事件

**本章内容**

- 触摸、手势和事件的概念
- 轻击放大
- 拖拽定位
- 轻扫翻页
- 捏合缩放

众所周知，iPhone 是没有键盘的，所有操作都来自手指触摸。作为苹果公司两度申请的专利技术，强大的多触摸功能为 iPhone 用户带来了无可比拟的用户体验。浏览网页、删除邮件、放大缩小图片，这些通常人们习惯用鼠标来实现的操作，在 iPhone 上用手指就完全可以做到。不得不说，iPhone 开创了人机交互的新纪元。

iPhone 本身有很多默认的触摸手势，同时也为开发人员自定义更加丰富的手势提供了很好的技术支持。这一章我们就来为你揭晓触摸的底层机制。

## 11.1　了解相关术语

下面借助于在音乐播放器中浏览专辑的过程（如图 11-1 所示），具体了解一下触摸的相关术语。

图 11-1　轻扫浏览专辑

根据一次放到屏幕上手指的数量，将触摸分为单点触摸和多点触摸(即多触摸)。iPhone 能同时检测到任意数量的触摸点，只要你的手指足够多、点间距足够大。

手势是为触发某一操作而进行的一系列事件。iPhone 中常用的触摸手势有：轻击、拖拽、轻扫、捏合等。

(1) 手指触碰到屏幕并立即离开的过程称为轻击，如在上一章图 10-2 和图 10-3 中轻击屏幕右下角的小按钮，会翻转到视图的背面进行某些设置。

(2) 手指在屏幕上移动时，使小图标、图片等随触摸点位置的改变而改变的过程称为拖拽。

(3) 浏览专辑时，从手指接触屏幕到在屏幕上移动，最后离开屏幕的过程称为轻扫。

(4) 捏合是一个多触摸手势，它需要两个手指同时操作，通过改变两个触摸点间的距离实现对图片、网页等的放大、缩小。

由于会受到手指的大小、形状的不确定性因素的影响，iPhone 屏幕检测到触摸点的精确度要比用鼠标低很多。因此，手势的定义应存在一个较大的可接受范围，如果不这样做，iPhone 响应的操作也许会偏离实际想要的结果。

事件也是经常被提到的术语，有触摸开始、移动、取消和结束四个事件。当手指触发某个事件时，会调用相应的方法，见表 11-1。

**表 11-1　事件及其对应的方法**

| 事件 | 方法 |
|---|---|
| 触摸开始 | touchesBegan:withEvent: |
| 触摸移动 | touchesMoved:withEvent: |
| 触摸结束 | touchesEnded:withEvent: |
| 触摸取消 | touchesCancelled:withEvent: |

手指触摸到屏幕时，程序会调用 touchesBegan:withEvent:方法；手指在屏幕上移动时，调用 touchesMoved:withEvent:方法；手指脱离屏幕时，调用 touchesEnded:withEvent:方法；某个事件导致触摸中断时，调用 touchesCancelled:withEvent:方法；一般是在有电话打进来时触摸被取消。我们可以在不同的方法中定义不同的操作，当某个事件发生时，便会触发相应的操作。

接下来我们将通过几个简单的小程序来讲解手势是如何定义的，图 11-2 描述了触摸体系结构。

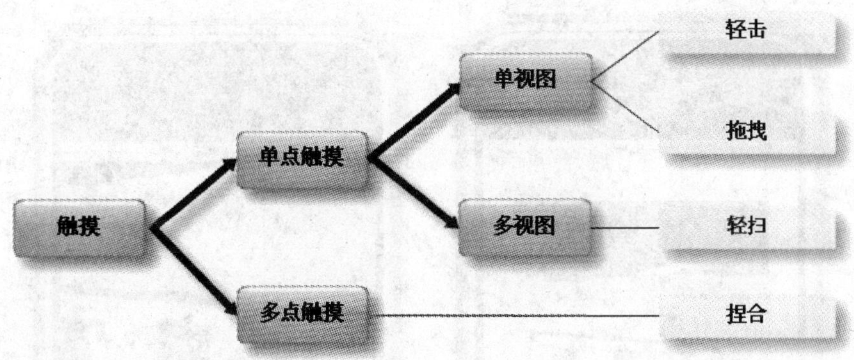


图 11-2　触摸体系结构

本章后面的案例将实现轻击、拖拽、轻扫、捏合的如下功能：

(1) 轻击一张图片，改变图片大小；

(2) 拖拽一张图片，改变图片的位置；

(3) 轻扫两个视图，实现视图间转换；

(4) 捏合一张图片，使图片缩放。

# 11.2　轻击和拖拽

轻击操作在 iPhone 的使用中无处不在，如轻击小图标进入程序主界面，轻击列表某行进入下级菜单，轻击添加按钮增加记录等等，方便又实用。我们也会经常用到拖拽手势，比如更改 iPod touch 界面上程序图标的位置，如图 11-3 所示。

作为单点触摸中常用的手势，轻击和拖拽实现起来比较简单，一起来学习一下吧。

## 11.2.1　构建应用程序 NoteTaps

在这个小程序中，我们定义一个操作，根据轻击的次数改变一张图片的大小。如：单击图片，其大小保持不变，如图 11-4 所示；双击图片，图片大小变为原来的两倍，如图 11-5 所示；三次轻击图片，变为原来的三倍等。

图 11-3　拖拽 App Store 图标

图 11-4 单击图片大小不变

图 11-5 双击图片放大

当然，在 iPhone 中通过捏合手势完全可以改变图片的大小（在 11.4 节中我们会讲到）。学完这一节，你就可以根据自己的需要定义轻击所触发的操作了。

首先，使用 View-based Application 模板创建一个新项目，将其命名为 NoteTaps。将随书光盘 Chapter 11 NoteTaps 文件夹中的 coverImage.png 文件拖至 Xcode 中的 Resources 文件夹。选中 NoteTapsViewController.h，添加如下黑体字所示的代码：

**代码 11.1 NoteTapsViewController.h 文件**

```
#import <UIKit/UIKit.h>
@interface NoteTapsViewController: UIViewController {
    UIImageView *imageView;
    UIImage *image;
    CGRect imageRect;
}
@property (nonatomic, retain) UIImageView *imageView;
@property (nonatomic, retain) UIImage *image;
@property CGRect imageRect;
@end
```

保存并选中 NoteTapsViewController. m，添加如下黑体字所示的代码：

**代码 11.2 NoteTapsViewController. m 文件**

```
#import "NoteTapsViewController.h"
@implementation NoteTapsViewController
@synthesize imageView;
@synthesize image;
@synthesize imageRect;
……
-(void) viewDidLoad {
    image = [UIImage imageNamed:@"coverImage.png"];
    imageRect = CGRectMake(0.0, 0.0, 260, 0.0);
    imageRect.size.height=260 * image.size.height / image.size.width;❶
    imageView = [[UIImageView alloc]initWithFrame:imageRect];
    imageView.image = image;
    imageView.center = CGPointMake (160,230);❷
    [self.view addSubview:imageView];
    [super viewDidLoad];
}
……
- (void)dealloc {
    [image release];
    [imageView release];
    [super dealloc];
}
@end
```

本例中，为了可以灵活地改变图片的大小，我们用代码添加一张图片，而不是用 Interface Builder。图片的初始化是在 viewDidLoad 方法中实现的。在❶处让图片的大小按比例缩放，显示在屏幕中。在❷处定义了图片的中心为视图的中点，这样看起来会协调一些。

现在，编译并运行一下程序，实际效果如图 11-4 所示。轻击图片，不会有任何反应，因为我们还未对触摸手势定义任何的操作。

## 11.2.2 轻击放大图片

返回 Xcode，将以下代码添加到 NoteTapsViewController. m 文件的@end 之前：

代码 11.3　NoteTapsViewController.m 文件

```
- (void) touchesBegan :( NSSet * ) touches
        withEvent :( UIEvent * ) event {
        UITouch * touch = [touches anyObject];
        int i = [touch tapCount]; ❶
        imageRect = CGRectMake(10.0, 30.0 , 260 * i, 0.0);  ┐
        imageRect.size.height =                              │ ❷
                260 * i * image.size.height / image.size.width;┘
        imageView.frame = imageRect;
        imageView.center =CGPointMake (160,230);
}
```

当手指开始触摸屏幕时，系统自动调用 touchesBegan:withEvent:方法。其中❶处定义了一个 UITouch 对象 touch，tapCount 是它的一个属性，用来记录轻击次数；❷处规定，根据轻击次数改变图片的大小。

> **Tips**
>
> tapCount 属性只对连续的轻击计数，如果再次轻击相隔时间过长，或者屏幕检测到多点触摸，tapCount 的值会重置为 0。

很多时候 touchesBegan:withEvent:方法与 touchesEnded:withEvent:方法不加区别地使用，例如在本例中我们也可以在 touchesEnded:withEvent:方法中定义这些操作，效果完全相同，只是一个在触摸开始时触发，一个在触摸结束时触发。

### 11.2.3　拖拽图片

要实现拖拽图片，我们需要使用 touchesMoved:withEvent:方法，在 NoteTapsViewController.m 文件中删除 touchesBegan:withEvent:方法并添加以下代码：

代码 11.4　NoteTapsViewController.m 文件

```
- (void) touchesMoved :( NSSet * ) touches
        withEvent :( UIEvent * ) event {
        [UIView beginAnimations:@"touchesMoved" context: nil];
        [UIView setAnimationDuration: 0.01];
        [UIView setAnimationBeginsFromCurrentState: YES];
        UITouch * touch = [touches anyObject];
```

```
    imageView.center = [touch locationInView: self.view]; ❶
    [UIView commitAnimations];
}
```

当手指在屏幕上移动时调用 touchesMoved:withEvent:方法。我们仍然定义了一个UITouch 对象,然后用 ❶处的 locationInView:方法获取某一视图上当前的触摸点。本例中我们采用了这样一种方法实现拖拽效果:让图片的中心随触摸点位置的改变而改变。当手指在屏幕上移动时,图片也会跟着移动,这样便有了用手拖拽着图片移动的感觉。为了让拖拽看起来更加流畅,我们还加入了动画效果。

## 11.3 轻扫翻页

轻扫也是 iPhone 中常见的操作,如删除表行、邮件,浏览网页等,都可以用轻扫手势实现。下面我们仍然通过一个简单的小程序来解析屏幕是如何识别轻扫手势并做出相应的操作的。

与上节中的 NoteTaps 程序相同,轻扫也只是单点触摸,不同的是涉及到了两个视图间的转换。一般而言,如果通过手势对一个视图执行某项操作,触摸手势是在该视图中实现的;但如果手势影响到多个视图,触摸手势应该在根视图控制器中实现。我们要实现的是:在第一个视图中向左轻扫翻到下一页,如图 11-6 所示;在第二个视图中向右轻扫返回上一页,如图 11-7 所示。

图 11-6 向左轻扫进入下一页

图 11-7 向右轻扫返回上一页

使用 View-based Application 模板创建一个新项目并命名为 NoteSwipe。将随书光盘 Chapter 11 NoteSwipe 文件夹中的 coverImage. png 和 aboutImage. png 文件拖入 Xcode 中的 Resources 下。我们需要在两个视图间进行转换，因此新建两个 UIView Contrloller subclass 文件，分别命名为 CoverView 和 AboutView，注意不要忘了同时创建相应的. xib 文件。

双击 CoverView. xib 文件，打开 Interface Builder，从 Library 中拖出一个 Image View 控件到视图。打开 Image View 的 Attributes 窗口，在 image 属性的下拉列表中选择 coverImage. png。以同样的方式，为 AboutView 视图添加 aboutImage. png 图片。

保存 CoverView. xib 和 AboutView. xib 并返回 Xcode。因为是在根视图控制器 NoteSwipeViewController 中实现触摸手势的，所以我们不需要对两个视图的. h 和. m 文件做任何修改。

选中 NoteSwipeViewController. h 文件，添加如下黑体字所示的代码：

**代码 11.5　NoteSwipeViewController. h 文件**

```
#define kMinimumGestureLength  25
#import <UIKit/UIKit.h>
@class CoverView;
@class AboutView;
@interface NoteSwipeViewController: UIViewController {
    CoverView  *firstView;
    AboutView  *secondView;
    CGPoint startPoint;
}
@property (nonatomic, retain) CoverView   *firstView;
@property (nonatomic, retain) AboutView   *secondView;
@property CGPoint startPoint;
@end
```

本例中，将 NoteSwipeViewController 作为 CoverView 和 AboutView 的根视图控制器，在 NoteSwipeViewController. h 中声明了两个子类的对象。还声明了一个 CGPoint 类型的变量，作为触摸的起始点。

接下来，选中 NoteSwipeViewController. m 文件，添加如下黑体字所示的代码：

**代码 11.6　NoteSwipeViewController. m 文件**

```
#import "NoteSwipeViewController.h"
#import "CoverView.h"
#import "AboutView.h"
@implementation NoteSwipeViewController
```

```
@synthesize firstView;
@synthesize secondView;
@synthesize startPoint;
……
- (void) viewDidLoad {
    CoverView * first = [[CoverView alloc]
                initWithNibName:@"CoverView" bundle:nil];
    self.firstView = first;                                    ❶
    [self.view insertSubview:first.view atIndex:0];
    [first release];
    AboutView * second = [[AboutView alloc]
                initWithNibName:@"AboutView" bundle:nil];
    self.secondView = second;
    [self.view insertSubview:second.view atIndex:0];
    [second release];
    [super viewDidLoad];
}
……
- (void) dealloc {
    [firstView release];
    [secondView release];
    [super dealloc];
}
-(void) touchesBegan :( NSSet *)
                touches withEvent :( UIEvent *) event {
    UITouch * touch = [touches anyObject];
    startPoint = [touch locationInView:self.view];  ❷
}
- (void)touchesMoved:(NSSet *)touches withEvent:(UIEvent *)event {
    UITouch * touch = [touches anyObject];
    CGPoint currentPoint = [touch locationInView:self.view];   ❸
    CGFloat deltaX = currentPoint.x - startPoint.x;   ❹

    [UIView beginAnimations:@"View Flip" context:nil];
    [UIView setAnimationDuration:1.25];
    [UIView setAnimationCurve:UIViewAnimationCurveEaseInOut];
```

```
    if (deltaX <= -kMinimumGestureLength &&
                        self.firstView.view.superview != nil) {
        [UIView setAnimationTransition:
                    UIViewAnimationTransitionCurlUp
                            forView:self.view
                              cache:YES];                    ❺
        [firstView.view removeFromSuperview];
        [self.view insertSubview:secondView.view atIndex:0];
    }
    else if(deltaX >=kMinimumGestureLength &&
                        self.firstView.view.superview == nil) {
        [UIView setAnimationTransition:
                    UIViewAnimationTransitionCurlDown
                            forView:self.view
                              cache:YES];                    ❻
        [secondView.view removeFromSuperview];
        [self.view insertSubview:firstView.view atIndex:0];
    }
    [UIView commitAnimations];
}
@end
```

在❶处 viewDidLoad 方法中将 CoverView 视图设为进入程序的初始画面；在❷处 touchesBegan: withEvent:方法中获得手指触摸的起始点位置；在❸处 touchesMoved: withEvent:方法中获得手指移动时的当前位置；然后在❹处将两点之间的横轴距离赋给 deltaX 变量。

本例中，我们规定如果手指移动的横轴跨度大于25 像素，则算作轻扫。根据判断轻扫的方向做出相应操作：在❺处定义向左轻扫进入 AboutView 视图，在❻处定义向右轻扫返回 CoverView 视图。这次仍然加入了动画，在轻扫时就有翻页的效果了。

# 11.4 捏合缩放图片

上面两个例子都只用到单点触摸，这在其他平台上也很常见，这一节我们就来真正接触 iPhone 的核心技术：多点触摸。

捏合手势是最简单常用的多点触摸的应用。只要两个手指轻轻一动，图片便任意缩放了。图 11-8 演示了当两个触摸点之间的距离增大时，图片会放大；图 11-9 演示了两个触摸点的距离变小时，图片被缩小了。下面我们就来看看 iPhone 是如何检测到捏合手势并控制图片大小的。

图 11-8 捏合使图片放大

图 11-9 捏合使图片缩小

同样使用一个 View-based Application 模板，新建一个名为 NotePinch 的项目。首先，需要定义一个函数来计算两点之间的距离。本例中，我们使用 C 语言来定义。因此，新建一个 C 文件，虽然这在本书中是第一次用到，但是不必担心，你会觉得它非常简单。

选择 Classes 文件夹，右键选择 Add→New File，在弹出窗口左边 Mac OS X 窗格中选中 C and C++，在右边栏中选择 C File，如图 11-10 所示。点击右下角的 Next，将其命名为 NoteMath。

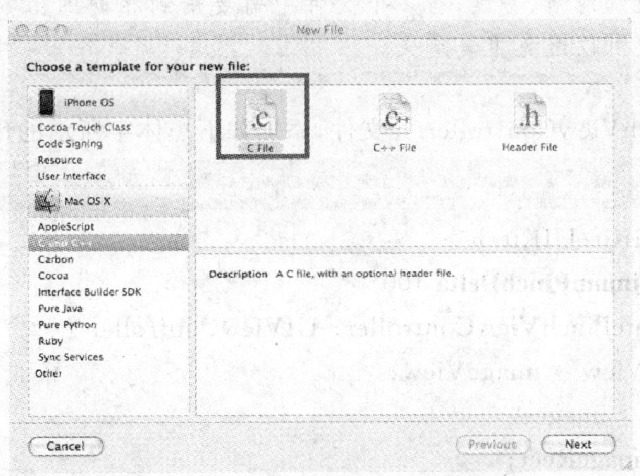


图 11-10 新建 C 文件

这样，在 Classes 文件夹中就新增了两个以 NoteMath 命名的文件，分别选中 NoteMath.h 文件和 NoteMath.m 文件，添加如下黑体字所示的代码：

**代码 11.7　NoteMath.h 文件**

```
#import <CoreGraphics/CoreGraphics.h>❶
CGFloat distanceBetweenPoints (CGPoint first, CGPoint second);❷
```

**代码 11.8　NoteMath.m 文件**

```
#include "NoteMath.h"
#include <math.h>❸
CGFloat distanceBetweenPoints (CGPoint first, CGPoint second) {
    CGFloat deltaX = second.x - first.x;
    CGFloat deltaY = second.y - first.y;
    return sqrt(deltaX * deltaX + deltaY * deltaY );❹
}
```

在修改之前，这是两个空文件，因此首先需要在❶处引入 CoreGraphics 框架。在 NoteMath.h 文件❷处声明了一个方法，在 NoteMath.m 文件中实现了这个方法。在❸处定义了两点之间的距离：两点间 x 坐标差和 y 坐标差的平方和的开方。因为用到 sqrt 方法，需在❹处引入头文件 math.h。上述方法也完全可以在 NotePinchViewController 文件中实现。

> **Tips**
>
> 最好是在一个单独的文件中定义所需要的函数。当程序中需要用到多个函数时，这会使你的程序结构更加清晰、易读；当多个文件中需要用到相同的函数，还可以避免重复定义。

选择 NotePinchViewController.h 文件，添加如下黑体字所示的代码：

**代码 11.9　NotePinchViewController.h 文件**

```
#import <UIKit/UIKit.h>
#define kMinimumPinchDelta 100
@interface NotePinchViewController: UIViewController {
    UIImageView *imageView;
    UIImage *image;
    CGRect imageRect;
    CGFloat initialDistance;
```

```
}
@property (nonatomic, retain) UIImageView *imageView;
@property (nonatomic, retain) UIImage *image;
@property CGRect imageRect;
@property CGFloat initialDistance;
@end
```

在头文件中定义了一个常量 kMinimumPinchDelta 和一个用于在触摸开始时获取两点间距离的变量 initialDistance。

将随书光盘 Chapter 11 NotePinch 文件夹中的 coverImage.png 图片拖入 Xcode 中的 Resources 文件夹，使用代码将图片加入界面，这一步你已经很熟悉了，不再赘述。

选择 NotePinchViewController.m 文件，添加如下黑体字所示的代码：

**代码 11.10 NotePinchViewController.m 文件**

```
#import "NotePinchViewController.h"
#include "NoteMath.h"
@implementation NotePinchViewController
@synthesize imageView;
@synthesize image;
@synthesize imageRect;
@synthesize initialDistance;
……
- (void)viewDidLoad {
    image = [UIImage imageNamed:@"coverImage.png"];
    imageRect = CGRectMake(0.0, 0.0, 260, 0.0);
    imageRect.size.height = 260 * image.size.height / image.size.width;
    imageView = [[UIImageView alloc]initWithFrame:imageRect];
    imageView.image = image;
    imageView.center = CGPointMake(160,230);
    [self.view addSubview:imageView];
    [imageView release];
    [super viewDidLoad];
}
……
- (void)dealloc {
    [image release];
    [imageView release];
```

```
    [super dealloc];
}
- (void)touchesBegan:(NSSet *)touches withEvent:(UIEvent *)event {
    if ([touches count] == 2) {  ❶
        NSArray *twoTouches = [touches allObjects];
        UITouch *first = [twoTouches objectAtIndex:0];
        UITouch *second = [twoTouches objectAtIndex:1];          ❷
        initialDistance =
        distanceBetweenPoints ([first locationInView:self.view],
            [second locationInView:self.view]);
    }
}
- (void) touchesMoved :( NSSet *) touches
        withEvent :( UIEvent *) event {
            if ([touches count] == 2) {
            NSArray *twoTouches = [touches allObjects];
            UITouch *first = [twoTouches objectAtIndex:0];
            UITouch *second = [twoTouches objectAtIndex:1];          ❸
            CGFloat currentDistance =
            distanceBetweenPoints([first locationInView:self.view],
                        [second locationInView:self.view]);
            int i = imageRect.size.width+
                        (currentDistance - initialDistance)/10;
            if (initialDistance == 0)
                initialDistance = currentDistance;
            else if (i> kMinimumPinchDelta) {
                    imageRect = CGRectMake(10.0, 30.0 , i, 0.0);
                    imageRect.size.height =
                        i * image.size.height / image.size.width;   ❹
                    imageView.frame = imageRect;
                    imageView.center =CGPointMake (160,230);
                }
    }
}
- (void) touchesEnded :( NSSet *) touches
    withEvent :( UIEvent *) event {
    initialDistance = 0;
```

```
}
@end
```

在 touchesBegan: withEvent:方法的❶处,touches 的 count 属性返回当前视图中检测到的触摸点数量;❷处定义如果计数为 2,则获取触摸开始时两手指间的距离。

同样地,在 touchesMoved: withEvent:方法中,❸处定义如果检测到两个触摸点,则计算手指移动时两点间的距离;在❹处根据两点间距离的变化控制图片的大小。

你也可以根据自己的计算确定捏合手势改变图片大小的幅度。应尽量使用户体验良好,这需要不断尝试才可以做到。

## 11.5 小结

本章通过三个小程序认识了几个常用的手势:轻击、拖拽、轻扫和捏合;学会了如何通过事件调用不同方法,进而触发相应的操作。现在,尽情地发挥你的想象力,自定义一个自己的手势吧!不过要记住,不要与 iPhone 中的常用手势混淆,这会让用户感到非常困惑。

# 第12章　国际化和本地化

**本章内容**

- NSLocalizedString 的应用
- .lproj 文件如何命名
- 本地化.xib 文件
- 本地化图像
- 本地化字符串
- 本地化应用程序名称

每天都有大量用户在 App Store 下载应用程序，他们来自不同的国家和地区，使用着不同的语言。用户在选择软件时，首先考虑的是软件支持的语言环境。如果软件仅支持一种语言，无形中就缩小了客户群，这不得不说是一笔巨大的损失。支持多语言环境能够使应用程序不受语言、地区的限制。

学习本章之前，也许你编写的应用程序会出现这样的状况：不论 iPhone 或模拟器的语言环境是什么，应用程序始终显示的是开发基础语言。更糟糕的是，在编写程序时没有注意到这个细节，出现多种语言混用的情况。不用担心，学完这一章，我们就能让我们的应用程序走向国际化！

## 12.1　了解国际化和本地化

为满足不同用户的需求，我们可以重写代码，开发出不同语言版本的软件，但这并不是高效的解决办法。借助 Mac OS X 和 iPhone 对国际化强大的技术支持，完全可以省去很多重复的工作。

国际化是一个设计和构建应用程序使之更加容易本地化的过程。相应地，本地化是将一个国际化后的应用程序翻译成不同的语言版本的过程，二者相辅相成。

### 12.1.1　需要本地化的资源

想一想应用程序中什么资源会受到地区和语言限制呢？下面列举了常见的几种：

(1) 静态文本：如提示信息、应用程序名称等；

(2) 图像：如界面中有文字信息的图片和应用程序小图标；

(3) 声音：如教育类软件的语音教学，有地方特色的背景音乐等；

(4) 动态文本:如时间、日期等。

当我们进行本地化时,应用程序会为每种支持的语言创建一个. lproj 文件夹,里面存放着支持该语言的相应文件,如. png 文件、. xib 文件、. strings 文件等。因此,如果应用程序支持两种语言,则会有两个. lproj 文件夹。应用程序启动时,会根据用户的语言和地区设置,检测应用程序中是否有与设置相匹配的. lproj 文件夹。如果存在,则会载入该语言版本的应用程序。

提供不同的语言版本的图片、声音等文件非常简单,而对于字符串文件,手动输入每个字符串,然后翻译成不同的语言,是一件相当耗时的工作。如果应用程序可以自动搜索程序中的字符串并生成字符串文件,然后再将字符串文件翻译成不同的语言版本,是不是简单了很多?通过对应用程序进行国际化,完全可以做到这一点。

下面是传统的字符串声明:

```
NSString * title = @"show message";
```

国际化后的字符串声明是这样的:

```
NSString * title = NSLocalizedString (@"show message",@"alert title");
```

NSLocalizedString 有两个参数,第一个参数是开发基础语言表述的字符串,第二个参数是对该字符串的注释。如果应用程序未做本地化,则只返回第一个参数。

通过运行 genstrings 命令行,程序会生成一个名为 Localizable. strings 的文件,自动搜索代码中所有的 NSLocalizedString 宏,提取出所有的字符串,并嵌入到该文件中,运行后显示如下:

```
/* alert title */
"show message" = "show message";
```

若将该字符串翻译为中文,可以这样做:

```
/* alert title */
"show message" = "显示信息";
```

### 12.1.2 .lproj 文件的命名规则

若选择汉语作为语言,选择中国作为地区,应用程序会查找是否有 zh_CN. lproj 文件夹。. lproj 文件夹应遵循国际标准化组织(ISO)的规定来命名,见表 12-1。文件夹名称的前两位字母是语言代码,后两位字母是国家(地区)代码。所有的语言代码都是小写的,地区代码是大写的。虽然 Mac OS 文件系统不区分大小写,但 iPhone 文件系统是区分大小写的,因此应严格按照 ISO 的规定命名。所有的语言至少有一个三位代码;有些语言有两个三位代码,一种是该语言的英语拼写,一种是本地拼写。部分语言有二位代码,当一种语言既有二位代码又有三位代码时,应使用二位代码。

**表 12-1 部分语言和国家的 ISO 代码**

| 语言 | ISO 二位代码 | ISO 三位代码 | ISO 地区代码 |
|---|---|---|---|
| 英语 | en | eng | US(美国) |
| 法语 | fr | fre/fra | FR(法国) |
| 汉语 | zh | chi/zho | CN(中国) |
| 德语 | de | ger/deu | DE(德国) |
| 日语 | ja | jpn | JP(日本) |

以选择法语作为 iPhone 语言、选择法国作为地区的用户为例，应用程序查找本地化文件夹的流程如图 12-1 所示。

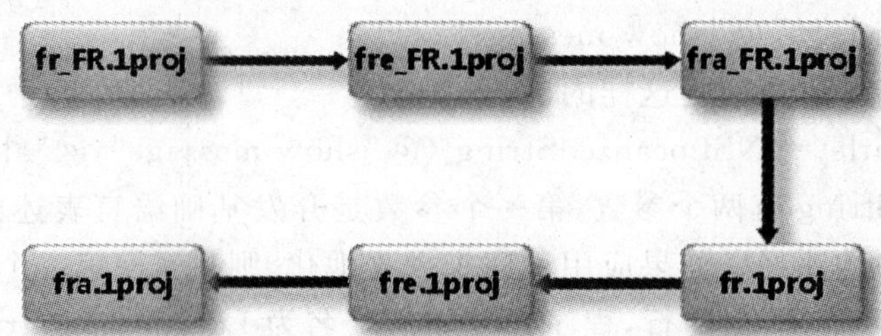


图 12-1 .1proj 文件夹查找顺序

fr_FR. lproj 是最规范的命名方式，如果没有精确匹配的文件夹，应用程序会查找只有语言代码匹配的文件夹 fr. lproj。如果这些都没有，应用程序会载入开发基础语言版本。

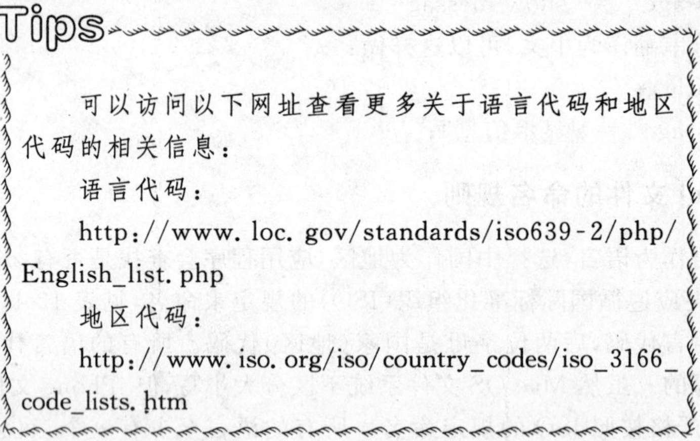
Tips

可以访问以下网址查看更多关于语言代码和地区代码的相关信息：

语言代码：

http://www.loc.gov/standards/iso639-2/php/English_list.php

地区代码：

http://www.iso.org/iso/country_codes/iso_3166_code_lists.htm

# 12.2 创建一个国际化的项目

下面通过创建一个小应用程序，切身感受一下 iPhone 对国际化的强大支持。本例中，假设我们的开发基础语言是英语，图 12-2 是未本地化的效果图。Image View 嵌入在整个视图下面，中间左边是三个 Label，右边是三个 Text Field，右下方是一个 Round Rect Button。我们在文本框中输入标题、类别、内容，点击保存按钮，会弹出显示输入信息的警告。本地化后，将系统中的语言环境改为中文，效果如图 12-3 所示。

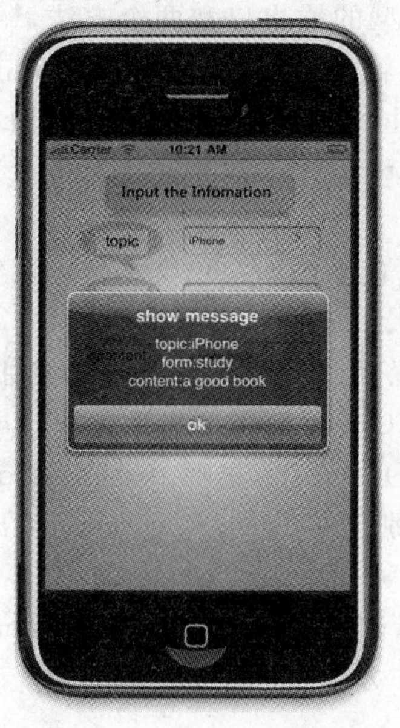

图 12-2　英文环境下的界面　　　图 12-3　中文环境下的界面

使用 View-based Application 模板创建一个项目，命名为 NoteGlobality。首先，在随书光盘 Chapter 12 NoteGlobality 文件夹中，找到 backgroundImage. png，拖入 Resources 文件夹。

### 12.2.1　声明输出口并连接

在 Xcode 中选中 NoteGlobalityViewController. h 文件，添加如下黑体字所示的代码：

**代码 12.1　NoteGlobalityViewController. h 文件**

```
#import <UIKit/UIKit.h>
@interface NoteGlobalityViewController: UIViewController {
    IBOutlet UITextField *topic;
    IBOutlet UITextField *form;
    IBOutlet UITextField *content;
}
@property (nonatomic, retain) UITextField *topic;
@property (nonatomic, retain) UITextField *form;
@property (nonatomic, retain) UITextField *content;
- (IBAction) save :( id) sender;
```

```
- (IBAction) textFieldDoneEditing :( id) sender;
@end
```

在该文件中声明了三个 UITextField 类型的输出口和两个方法。save:方法在点击按钮时调用,textFieldDoneEditing:方法用于输入完成时取消键盘。双击 NoteGlobalityViewController. xib 文件,打开 Interface Builder,参照图 12-4 设计界面,步骤如下:

(1) 拖入一个 Image View 控件,打开 Attributes 窗口,在其 image 属性的下拉列表中选择 backgroundImage. png 文件。

(2) 拖入三个 Label 控件,分别放置在黄色小图的白色文本区。调整 Label 的大小,将 layout 属性设为居中,分别键入文本 topic、form、content。

(3) 拖入三个 Text Field 控件,调整位置。点击 File's Owner,按住鼠标右键拖动到第一个 Text Field,连接 topic 输出口。打开 Connections 窗口,选择 Did End On Exit 事件旁的小圆圈,按住鼠标左键拖动到 File's Owner 上,选择 textFieldDoneEditing:方法。重复上面的操作,将下面两个 Text Field 分别连接 form 和 content 输出口及 textFieldDoneEditing:方法。

(4) 拖入一个 Round Rect Button 控件,调整它的位置。打开 Connections 窗口,将它的 Touch Up Inside 事件连接到 save:操作方法。

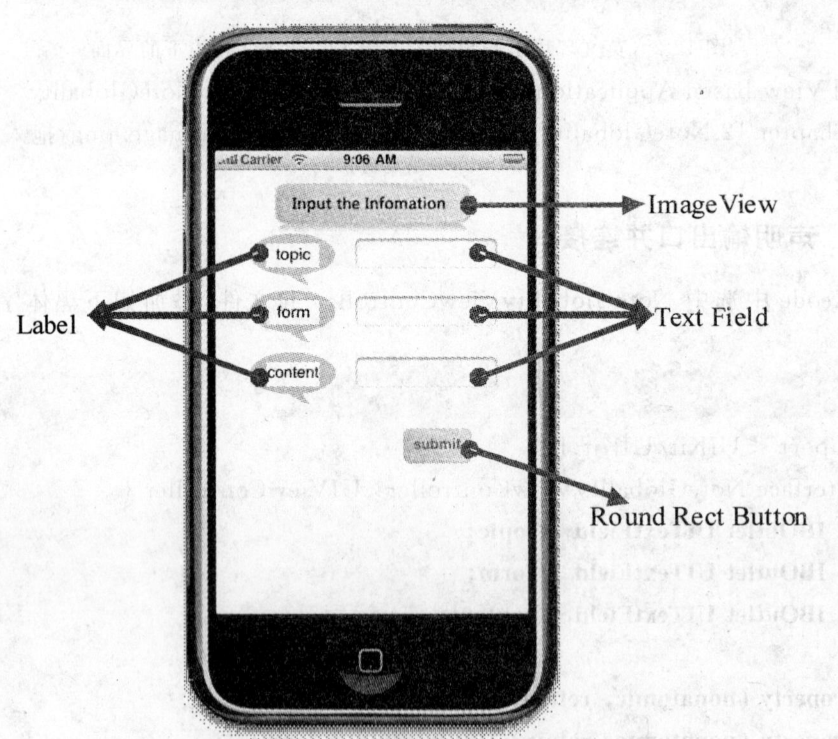


图 12-4 显示界面

关闭 Interface Builder，编译并运行一下程序。点击右下角按钮，由于还未定义 save：方法触发的操作，程序并无任何反应。

### 12.2.2 定义操作

选择 NoteGlobalityViewController.m，添加如下黑体字所示的代码：

**代码 12.2 NoteGlobalityViewController.m 文件**

```
#import "NoteGlobalityViewController.h"
@implementation NoteGlobalityViewController
@synthesize topic;
@synthesize form;
@synthesize content;

- (IBAction) save :( id) sender {
    NSString *topicString = topic.text;
    NSString *formString = form.text;
    NSString *contentString = content.text;
    NSString *title =
    NSLocalizedString(@"show message",@"alert title");
    NSString *cancelButton =
    NSLocalizedString(@"ok", @"alert cancelbutton");            ❶
    NSString *msg = [[NSString alloc]
    initWithFormat:NSLocalizedString(@"topic:%@\n
    form:%@\ncontent:%@",@"alert message"),
    topicString, formString, contentString];
    UIAlertView *alert = [[UIAlertView alloc] initWithTitle: title
                                        message:msg
                                        delegate:nil
                                cancelButtonTitle:cancelButton
                                otherButtonTitles:nil];

     [alert show];
     [alert release];
     [msg release];
```

```
}
- (IBAction) textFieldDoneEditing :( id) sender {
        [sender resignFirstResponder];
}
```

save:方法用于读取三个文本框中输入的信息,并显示在点击按钮弹出的警告中。我们在❶处每个输出字符串的语句中,使用了 NSLocalizedString 类。现在还未做本地化,因此当程序运行到含 NSLocalizedString 的语句时,只返回第一个参数,也就是开发基础语言。

编译并运行程序,效果如图 12-2 所示。尝试修改一下系统设置中的语言环境,打开 Settings→General→International→Language,选择简体中文,如图 12-5 所示。点击右上角的 Done 按钮,等待几秒钟,系统语言环境便会修改为简体中文。打开 NoteGlobality 应用程序,你会发现仍然是英文版本,为了让应用程序支持中文版本,我们还需做下面的工作。

图 12-5　设置语言环境

## 12.3　本地化应用程序

在本地化应用程序之前,我们首先分析一下本例中哪些资源需要本地化,见表 12-2。

表 12-2　本地化文件分析

| 需本地化的资源 | 对应文件的类型 |
| --- | --- |
| 标签和按钮上显示的文本 | . xib 文件 |
| 背景图片中的文本信息 | . png 文件 |
| 警告框中的显示信息 | . strings 文件 |
| 应用程序的名字 | . strings 文件 |

下面我们就来分步进行,将应用程序本地化为中文版本。

### 12.3.1 本地化.xib文件

右键单击 NoteGlobalityViewController.xib 文件，在出现的菜单中选择 Get Info，打开如图 12-6 所示的窗口。单击 General 选项卡，点击左下方的 Make File Localizable 按钮，窗口切换到 Targets 选项卡。

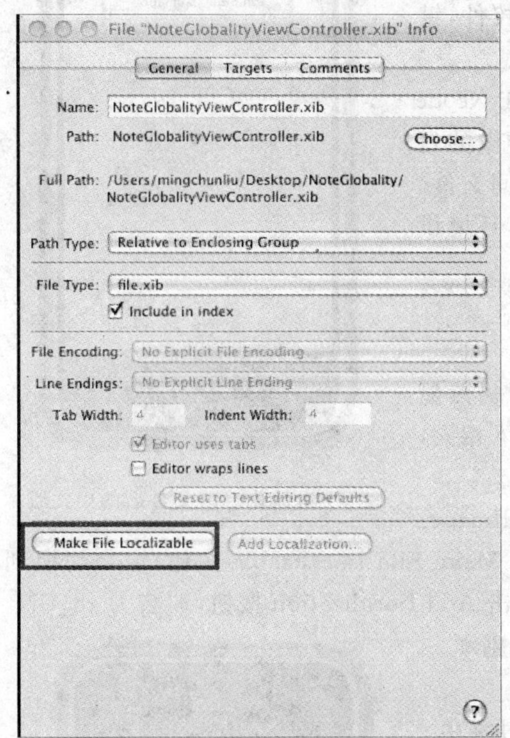


图 12-6 Info 窗口 General 选项卡

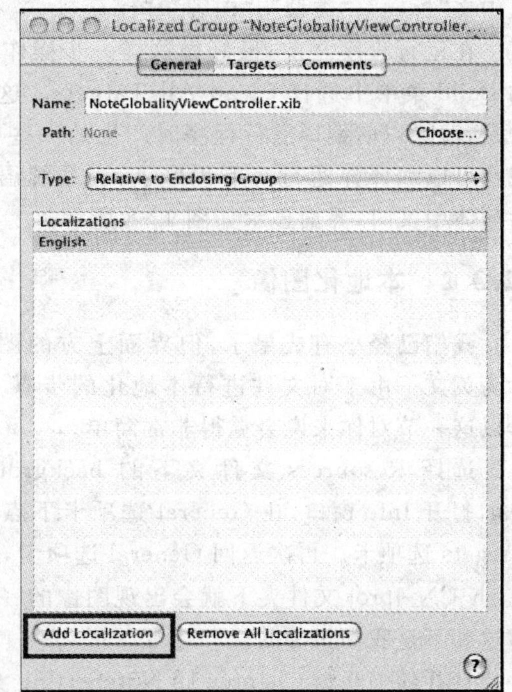


图 12-7 增加本地化文件夹

关闭此窗口，返回 Xcode，NoteGlobalityViewController.xib 文件有变化了：左边多了一个小三角。这表示该文件还有下级目录，展开后出现一个名为 English 的文件，双击打开此文件，这与我们原先的.xib 文件没什么不同。

与此同时，应用程序在项目文件夹中创建了一个 English.lproj 子文件夹，英文版本的 NoteGlobalityViewController.xib 文件就在里面。按照前面 12.1.2 小节提到的.lproj文件夹的命名规则，应使用语言的二位代码，存放英语版本相关文件的文件夹应取名为 en.lproj。English.lproj 是旧式的项目命名方式，但由于其运行良好，我们没必要更改它。

回到 Xcode，再次打开 NoteGlobalityViewController.xib 文件的 Info 窗口。在 General选项卡下，点击左下角的 Add Localization 按钮，如图 12-7 所示。这将会增加一个新的.lproj文件，用于存储中文版本应用程序的相关文件。将文件夹命名为 zh_CN.

lproj，前面已经讲解了这样命名的原因，不再赘述。

关闭 Info 窗口，查看项目文件夹，确实增加了一个 zh_CN. lproj 文件夹，里面存放的是 NoteGlobalityViewController. xib 文件的一个副本，在 Xcode 中的 NoteGlobalityViewController. xib 下显示为 zh_CN 文件。双击打开 Interface Builder，将 Xib 文件翻译为中文版本，即在 Attributes 窗口中更改标签和按钮上的文本显示。也可以依次双击控件，键入中文“标题”、“类型”、“内容”、“保存”。

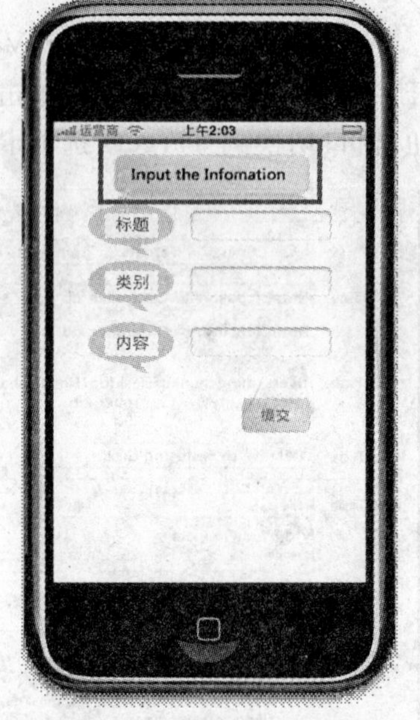

12-8　本地化. xib 文件后的中文版本

在运行程序之前，最好做这样一步操作：在 Xcode 的 Build 菜单中选择 Clean All Targets。这样做的目的是，清除以前编译运行程序时产生的一些中间文件。现在可以编译并运行一下程序了，当系统语言环境设为简体中文时，界面显示如图 12-8 所示。

## 12.3.2　本地化图像

我们已经小有成果了，但界面上方的图像仍然显示为英文。由于对文件进行本地化的步骤大体相同，所以这一节对你来说会觉得非常简单。

选择 Resources 文件夹下的 backgroundImage. png，打开 Info 窗口，在 General 选项卡下点击 Make File Localizable 按钮，窗口切换到 Targets 选项卡。再次返回 General 选项卡，点击 Add Localization 按钮，取名为 zh_CN。在 zh_CN. lproj 文件夹下就会出现图像的一个副本，但这并不是我们想要的。

打开随书光盘 Chapter 12 NoteSetting 文件夹中的 zh_CN. lproj 文件夹，同时，将 backgroundImage. png 拖入项目文件夹下的 zh_CN. lproj 中，替换图像的副本。查看 Xcode 中 Resources 文件夹下的 backgroundImage. png 文件，展开组，同样会有两个文件，分别命名为 English 和 zh_CN。此时两个文件已经是一个显示英文、一个显示中文的图片了。

图 12-9　中文环境下未本地化的警告信息

在选择 Clean All Targets 后，编译并运行程序看一下效果。在中文的语言环境下，图像改变了。在文本框中输入信息，点击保存按钮。我们又发现新问题了：警告框中提示信息还是英文的，如图 12-9 所示。不用担心，学完下一小节我们就能够解决这个问题。

如果应用程序中加入了小图标，并且此图标是受语言和地区限制的，完全可以按照上述方法对其进行本地化。

### 12.3.3 本地化警告信息

前面提到，使用 NSLocalizedString 类的目的是当运行 genstrings 命令行时，程序能够自动获取需要本地化的字符串并生成字符串文件，下面就来完成这个操作。

打开终端应用程序，以下为参考目录：Mac OS X→应用程序→实用工具→终端。键入 cd 和一个空格，然后将此项目文件夹拖到终端窗口，命令行中显示文件夹的路径后按回车键，接下来输入命令行：

```
genstrings ./Classes/*.m
```

此命令行的目的是查找本程序 Classes 文件夹中扩展名为.m 文件中出现的所有 NSLocalizedString 类。最后按下回车键，关闭终端应用程序。打开项目文件夹，发现新增了一个名为 Localizable.strings 的新文件，将其拖入 Xcode 中 Resources 文件夹下。确保文本编码为 Unicode(UTF-16)，如图 12-10 所示，否则该文件中会出现乱码。

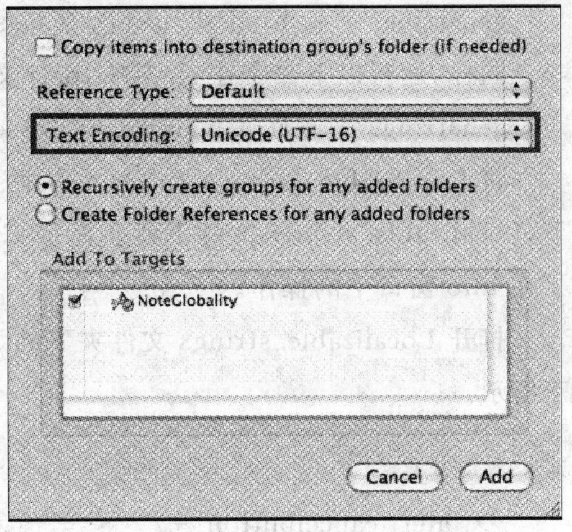


图 12-10 选择文本编码

在 Xcode 中选中 Localizable.strings 文件，内容如下：

**代码 12.3 Localizable.strings 文件**

```
/* alert cancelbutton */
"ok" = "ok";

/* alert title */
"show message" = "show message";

/* alert message */
"topic:%@\nform:%@\ncontent:%@" =
                    "topic:%1$@\nform:%2$@\ncontent:%3$@";
```

在生成该文件时，系统自动对字符串按照字母顺序进行了排序。NSLocalizedString 的第二个参数生成为注释语句，第一个参数生成为等号左边的字符串，也就是开发基础语言。等号右边是翻译的本地化语言，由于还未做本地化，因此等号两边的字符串相同。

接下来，在 Info 窗口中为 Localizable.strings 文件添加本地化文件。这一操作相信你已经非常熟悉了，不再赘述。

获得字符串文件并本地化还有另一种更简单的方法。由于已经建立了 English. lproj 和 zh_CN. lproj 文件夹，因此可以运行以下命令行，直接在 English. lproj 文件夹中创建 Localizable. strings 文件：

```
genstrings -o English.lproj ./Classes/*.m
```

同样，运行以下命令在 zh_CN. lproj 文件夹中创建该文件：

```
genstrings -o zh_CN.lproj ./Classes/*.m
```

将 Localizable. strings 文件分别从两个文件夹中拖入 Xcode 中 Resources 文件夹下，Localizable. strings 文件下就会显示 English 和 zh_CN 两个子目录。这种方法省去了在 Info 窗口中的操作。

打开 Localizable. strings 文件夹下的 zh_CN 文件，将字符串翻译为中文并保存，如下所示：

**代码 12.4 Localizable. strings 文件**

```
/* alert cancelbutton */
"ok" = "确定";

/* alert title */
"show message" = "显示信息";

/* alert message */
"topic:%@\nform:%@\ncontent:%@" =
                              "标题:%1$@\n类型:%2$@\n内容:%3$@";
```

至此，用户界面上的本地化问题就解决了。编译并运行一下程序，你应该会看到前面图 12-3 的效果。

### 12.3.4 本地化应用程序名称

应用程序的名称，即 iPhone 主菜单上的应用程序图标下的显示名称，如图 12-11 中第三个小图标下的文字，由于未做本地化，在任何语言下都只显示 NoteGlobality。

图 12-11 应用程序的名称

这一步就更加简单了。在 Resources 文件夹下手动添加一个.strings文件，如图12-12所示，将其命名为InfoPlist.strings。

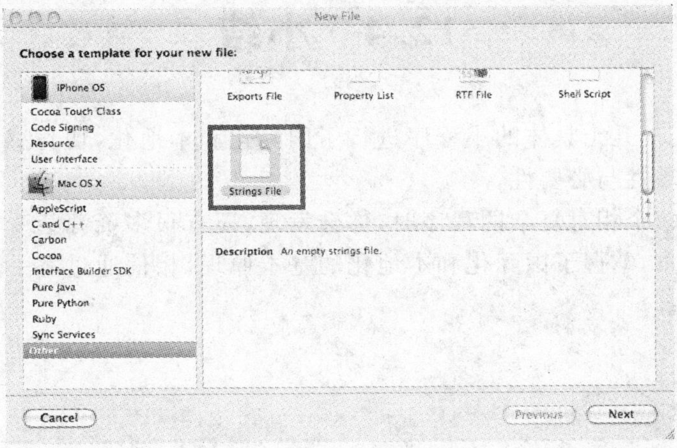

图 12-12 新建字符串文件

在该文件中只需输入一句代码：

**CFBundleDisplayName ＝"NoteGlobality"；**

打开InfoPlist.strings文件的Info窗口，重复同样的操作。返回Xcode，查看该文件是否已有English和zh_CN两个版本。点击zh_CN，修改代码为：

**CFBundleDisplayName** ＝"国际化"；

在中文环境下运行程序，上述应用程序的名称就显示为"国际化"了。

图 12-13 中文环境下应用程序的名称

## 12.4　小结

本章中，我们对几个具体的小程序进行了国际化和本地化，切身感受了使应用程序支持多语言的简单性与必要性。

当真正实现一个相对复杂的程序时，你会发现，还有很多资源会受到语言和地区的限制。学完这一章，掌握了国际化和本地化的基本原理，相信通过举一反三对你来说那已经不再是问题。

# 第4篇　扩展篇

本篇介绍了加速计、CoreLocation 定位和 Google 地图、多媒体和蓝牙等内容，是对应用程序进行功能和用户体验上的改进，使读者能开发出更加完善的软件。

扩展篇包括 4 章：

第 13 章加速计，详细讲解了加速计的概念以及在实际应用中如何使用，包括获取设备的方向、获取和过滤加速计的数据等。

第 14 章使用 CoreLocation 和 Map Kit，讲解了如何使用 CoreLocation 定位，以及如何通过 MapKit 框架调用 Google Map 显示地图信息。

第 15 章多媒体，介绍了如何访问 iPhone/iPod 照片库，以及音频和视频的处理。

第 16 章 Bluetooth，包括两部分内容：如何在设备间实现蓝牙连接和如何通过蓝牙传输数据。

# 第13章　加速计

**本章内容**

- 什么是加速计
- 确定设备的方向
- 获取加速计原始数据
- 过滤加速计数据
- 检测 iPhone 的摇动

iPhone/iPod 都有一个内置的加速计，这是苹果公司设计的亮点之一，也是 SDK 程序员津津乐道的一个独特组件。通过感知重力和检测用户的各种动作，加速计可以使应用程序更好地提升用户体验。例如，当 iPhone/iPod 被横置时，可以通过获取加速计的数据旋转应用程序界面，使用户操作更方便；当 iPhone/iPod 被摇晃时，可以使应用程序触发一次取消或擦除操作等。这些细微之处的考虑和设计，让用户实际操作时，感觉更加美妙和神奇。

本章我们将主要介绍加速计的基本概念，深入地了解加速计的工作原理，同时介绍一下如何使用加速计来确定设备的方向，如何从加速计上获取原始数据和处理这些数据。

## 13.1　什么是加速计

什么是加速计呢？从根本上说它是检测 iPhone 受力的一种设备。iPhone 内置的加速计是一个三轴加速计，这种加速计的组合，能够检测设备在三维空间中任何方向上的运动和重力（万有引力）；此外，它常用来检测人们手持 iPhone/iPod 的方式。图 13-1 是 iPhone 三轴加速计的三维示意图。

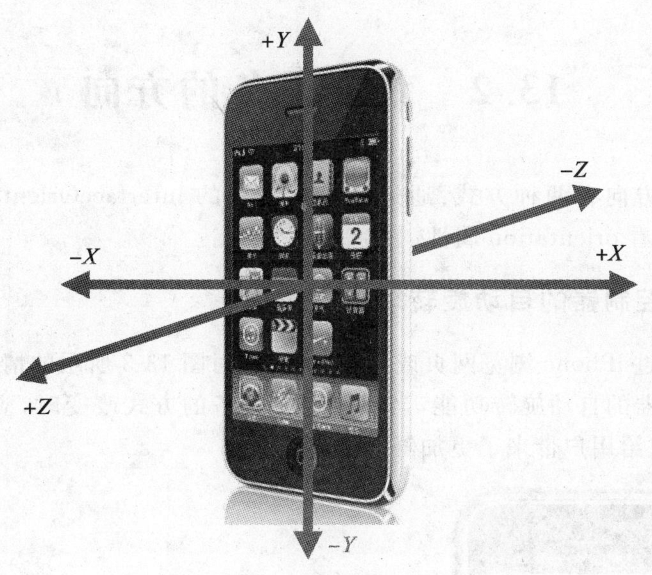


图 13-1 iPhone 三轴加速计的三维示意图

从图 13-1 可以看出，当正常握持设备时，$X$ 轴沿着设备的水平方向，$Y$ 轴沿着设备的垂直方向，$Z$ 轴沿着设备的正面方向（即由里向外的方向）。通常，各个方向上检测值的单位都用“g”表示，例如：静止纵向竖直握持设备时，会检测到 $Y$ 轴负方向受到大约 1 g 的重力。如果以一定的角度握持设备，那么会把 1 g 的重力分散到其他各个轴上，这和我们中学学过的物理知识中的受力分析是一致的。图 13-2 表示竖直握持设备成 45°角时，加速计检测到的 $X$ 和 $Y$ 轴受力情况。理想状态（iPhone 加速计受到的重力为 1 g）下 $+X$ 和 $-Y$ 轴受力均为 $\sqrt{2}$ g/2。如果检测到设备的某个方向上的受力远大于 1 g，那么就可以确定设备突然运动了。

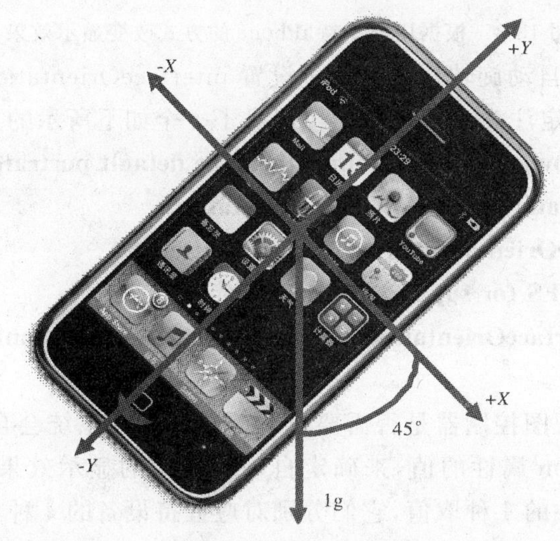


图 13-2 手持 iPhone 成 45°时的受力情况

# 13.2　获取设备的方向

获取设备的方向有两种方式:通过视图控制器的 interfaceOrientation 属性和通过 UIDevice 对象查看 orientation 属性。

## 13.2.1　视图控制器的自动旋转功能

如果你使用过 iPhone 浏览网页或者观看影片,对图 13-3 所示的情况一定非常熟悉。这就是视图控制器的自动旋转功能,当我们握持设备的方式改变时,应用程序能够自动旋转用户界面,这给用户带来了更加舒适的体验。

图 13-3　根据用户手持 iPhone 的方式改变显示效果

iPhone/iPod 实现自动旋转功能是通过设置 interfaceOrientation 属性实现的。在以前的程序代码中,你一定注意到模板中已经提供了一个如下所示的方法:

```
// Override to allow orientations other than the default portrait orientation.
- (BOOL)shouldAutorotateToInterfaceOrientation:
        (UIInterfaceOrientation)interfaceOrientation {
        // Return YES for supported orientations
        return (interfaceOrientation == UIInterfaceOrientationPortrait);
}
```

该方法用来确定视图控制器是否需要自动调整视图。系统会自动调用上述方法,并根据 intefaceOrientation 属性的值,来确定自动旋转后的显示效果。表 13-1 列出了 interfaceOrientation 属性的 4 种取值,它们分别对应握持设备的 4 种方式。

**表 13-1 interfaceOrientation 的属性值及描述**

| InterfaceOrientation 的属性值 | 描述 |
|---|---|
| UIInterfaceOrientationPortrait | 设备竖置、Home 键在下 |
| UIInterfaceOrientationPortraitUpsideDown | 设备倒置、Home 键在上 |
| UIInterfaceOrientationLandscapeLeft | 设备横置、Home 键在右 |
| UIInterfaceOrientationLandscapeRight | 设备横置、Home 键在左 |

为了更形象地说明这 4 种方式的不同之处，我们通过一个简单的程序来演示一下，它将要完成的是根据手持 iPhone 方式的改变调整程序界面。

我们将按照以下步骤完成本程序：

- 创建 Rotate 项目，添加图片资源；
- 构建程序界面；
- 修改 RotateViewController. m 文件，实现自动旋转功能。

**1. 创建 Rotate 项目，添加图片资源**

打开 Xcode，使用 View-based Application 模板创建一个项目，命名为 Rotate；在随书光盘 Chapter 13 Rotate 文件夹中找到 coverImage. png，拖入 Resources 文件夹中，以作为程序背景。

**2. 构建程序界面**

双击 RotateViewController. xib 文件，添加一个 Image View；并将它的 image 属性设置为刚才添加的图片资源，调整视图模式使图片能完整地显示；保存并关闭 RotateViewController. xib 窗口，返回 Xcode。

**3. 修改 RotateViewController. m 文件，实现自动旋转功能**

接下来，选中 RotateViewController. m 文件，我们要对文件中的方法 shouldAutorotateToInterfaceOrientation：做以下修改。

**代码 13.1 RotateViewController. m 文件**

```
- (BOOL)shouldAutorotateToInterfaceOrientation:
    (UIInterfaceOrientation)interfaceOrientation {
// Return YES for supported orientations
return (interfaceOrientation == UIInterfaceOrientationPortrait ||
interfaceOrientation == UIInterfaceOrientationLandscapeLeft ||
interfaceOrientation == UIInterfaceOrientationLandscapeRight ||
interfaceOrientation==UIInterfaceOrientationPortraitUpsideDown);   ❶
}
```

上述代码❶处，可以用于处理 4 种不同的 iPhone/iPod 握持方式，通过检查 interfaceOrientation 属性，并将它与当前 iPhone/iPod 的握持方式对比来调整视图方向。点击 Build and Go 按钮，编译并运行程序，当我们横置或者倒置 iPhone/iPod 时，将出现图 13-4 所示的几种效果。

图 13-4 改变握持 iPhone 的方式出现不同的效果

### 13.2.2 设备的 orientation 属性

我们还可以通过查看设备的 orientation 属性，来获取设备的方向信息。要访问 orientation 属性，必须声明一个 UIDevice 类的对象，具体方法如下：

```
UIDevice * device = [UIDevice currentDevice];
```

它的返回值是一个 UIDeviceOrientation 常量，该常量的取值与上一小节的 interfaceOrientation 的取值非常类似，见表 13-2。

**表 13-2 UIDeviceOrientation 的 7 种取值**

| UIDeviceOrientation 的取值 | 描述 | 三轴加速计受力(单位:g) |
|---|---|---|
| UIDeviceOrientationPortrait | 设备竖置、Home 键在下 | X:0.0 Y:-1.0 Z:0.0 |
| UIDeviceOrientationPortraitUpsideDown | 设备倒置、Home 键在上 | X:0.0 Y:1.0 Z:0.0 |
| UIDeviceOrientationLandscapeLeft | 设备横置、Home 键在右 | X:-1.0 Y:0.0 Z:0.0 |
| UIDeviceOrientationLandscapeRight | 设备横置、Home 键在左 | X:1.0 Y:0.0 Z:0.0 |
| UIDeviceOrientationFaceUp | 设备平躺、Home 键朝上 | X:0.0 Y:0.0 Z:-1.0 |
| UIDeviceOrientationFaceDown | 设备平躺、Home 键朝下 | X:0.0 Y:0.0 Z:1.0 |
| UIDeviceOrientationUnknown | 状态未知 | X、Y、Z 未知 |

它的取值是根据加速计的 X、Y 和 Z 轴来确定的，表中前四种取值对应 interfaceOrientation 的 4 种不同取值，后三个取值是 UIDeviceOrientation 特有的。注意表 13-2 中加速计各轴的受力数据是理想值。

当设备的方向发生改变时，可以调用 UIDevice 类的下面两个特殊方法：

(1) -(void)beginGeneratingDeviceOrientationNotifications，用来产生设备方向改变的通知 UIDeviceOrientationDidChangeNotification，它将启动设备的加速计硬件，并且分发加速事件给接收者。以下代码将创建一个通知接收者，当它接收到 UIDeviceOrientationDidChangeNotification 通知时，将调用相应的处理方法 deviceDidRotate：更新 orientation 属性值。

```
[[NSNotification Center defaultConter] addObserver:self
                    selector:@selector(deviceDidRotate:)
name:@"UIDeviceOrientationDidChangeNotification"
                    object:nil];
```

(2) -(void)endGeneratingDeviceOrientationNotifications，用于结束设备方向改变通知，它将通知设备可以关闭加速硬件，以减少耗电量，保证电池寿命。

上述方法只是获取设备的大致方向，但是在实际应用中我们可能需要使用加速计三个轴上检测到的原始数据。关于如何获取加速计的原始数据将在下一节详细介绍。

## 13.3 获取加速计的数据

为了获取加速计的原始数据，我们首先需要了解 UIAccelerometer 类和 UIAcceleration 类，以及 UIAccelerometerDelegate 协议。

### 13.3.1 访问加速计

要从加速计获取信息，首先必须创建一个 UIAccelerometer 类的对象来访问加速计，具体方法如下：

```
UIAccelerometer * accelerometer=[UIAccelerometer sharedAccelerometer];
```

创建完对象之后，还需要为加速计设置一个适当的更新间隔，即加速计的更新频率，具体方法如下：

```
accelerometer.updateInterval = 1.0f/60.f;
```

上述代码设置的更新频率为每秒钟更新 60 次。更新频率应当按照实际需求来设置，实际应用时在满足需求的情况下，应尽量使用最低限度的更新频率，不要频繁更新，以减少设备耗电量，从而保证电池的寿命。系统提供的更新范围是 10～100 Hz，一般情况下，游戏的更新频率设置为 30～60 Hz，检测方向的更新频率通常设置为 10～20 Hz。

同时，我们需要为加速计指定一个委托，具体方法如下：

```
accelerometer.delegate = self;
```

最后，当程序不再使用加速计时，需要调用下面这行代码将加速计的委托重置为 nil，告知系统它可以关闭加速计硬件设备了，从而维持电池寿命。

```
accelerometer.delegate = nil;
```

Tips

加速计更新频率及适用描述

| 频率 | 适用描述 |
| --- | --- |
| 10～20 | 适用于检测设备的当前矢量方向 |
| 30～60 | 适用于游戏中需要实现实时用户输入的应用程序 |
| 70～100 | 适用于需要检测高频率的手势(动作)的应用程序 |

## 13.3.2 获取加速计原始数据

在委托类的头文件中添加 UIAccelerometerDelegate 协议，通过实现 accelerometer：didAccelerate：方法便可获取加速计的原始数据，代码如下：

```
- (void)accelerometer:(UIAccelerometer *)accelerometer
        didAccelerate:(UIAcceleration *)acceleration {
        //do something here
}
```

该方法的第一个参数是 UIAccelerometer 类型的对象，即上一小节我们声明的访问加速计对象；第二个参数是 UIAcceleration 类型的对象。每个 UIAcceleration 类型的对象都有 $x$、$y$ 和 $z$ 三个属性，它们的取值是一个带符号的浮点数，分别对应三轴加速计的 $X$、$Y$ 和 $Z$ 轴方向的受力情况，表 13-2 说明了一些简单状态下的受力情况。该方法的调用频率取决于刚才设置的加速计 updateInterval 值，每隔 updateInterval 时间，该方法将被调用一次。

下面我们通过一个构建"指南针"的实例来说明如何获取加速计的原始数据。首先获取加速计的信息，然后根据此信息计算出一个角度，使得"指南针"的指针指向一个方向。程序的最终效果如图 13-5 所示。

图 13-5 "指南针"程序的效果

完成本程序需要以下步骤：

· 创建项目 Compass，导入图片资源；

· 修改 CompassViewController. h 文件，导入 math. h 文件，并声明显示图像的对象；

· 修改 CompassViewController. m 文件,获取加速计的数据,并计算旋转角度。

**1. 创建项目 Compass,导入图片资源**

打开 Xcode,使用 View-based Application 模板创建一个项目,将其命名为 Compass。在随书光盘 Chapter 13 Compass 文件夹中,找到 compass. png 和 pointer. png 图片,拖入 Resources 文件夹中。

**2. 修改 CompassViewController. h 文件,导入 math. h 文件,并声明显示图像的对象**

展开 Classes 文件夹,选中 CompassViewController. h 文件,添加如下黑体字所示的代码。

**代码 13.2 CompassViewController. h 文件**

```
#import <UIKit/UIKit.h>
#import <math.h>   ❶
@interface CompassViewController : UIViewController
<UIAccelerometerDelegate>{
    UIImageView *contentView;
    UIImageView *compassView;
    float accelX;                  ❷
    float accelY;
}
@end
```

在 CompassViewController. h 文件中,我们需要在❶处引入 math. h 文件进行数学计算;在❷处添加了 UIAccelerometerDelegate 协议,同时还声明了两个 UIImageView 对象以及两个用来接收加速计 *X* 轴和 *Y* 轴数据的浮点对象 accelX 和 accelY。

**3. 修改 CompassViewController. m 文件,获取加速计的数据,并计算旋转角度**

选中 CompassViewController. m 文件,并添加如下黑体字所示的代码:

**代码 13.3 CompassViewController. m 文件**

```
#import "CompassViewController.h"
@implementation CompassViewController

-(void)viewDidLoad{
    contentView = [[UIImageView alloc] initWithFrame:
                                    [[UIScreen mainScreen]
                                    applicationFrame]];        ❶
    contentView.image = [UIImage imageNamed:
                                @"compass.png"];
```

```
    self.view = contentView;
    [contentView release];
    compassView = [[UIImageView alloc] initWithImage:
                   [UIImage imageNamed:@"pointer.png"]];
    [compassView setCenter:CGPointMake(160.0f,240.0f)];
    [contentView addSubview:compassView];
    [compassView release];
    UIAccelerometer *accelerometer = [UIAccelerometer
                                      sharedAccelerometer];  ❷
    accelerometer.updateInterval = 1.0f/20.0f;  ❸
    accelerometer.delegate = self;  ❹
}
#pragma mark-
#pragma UIAccelerometer Delegate Method
- (void)accelerometer:(UIAccelerometer *)accelerometer
        didAccelerate:(UIAcceleration *)acceleration {
    //none filter
    accelX = -[acceleration x];
    accelY = [acceleration y];                                    ❺
    float angle = atan2(accelY,accelX);
    [compassView setTransform:
                 CGAffineTransformMakeRotation(angle)];
}
- (void)dealloc {
    UIAccelerometer *accelerometer = [[UIAccelerometer
                    sharedAccelerometer] setDelegate: nil];❻
    [super dealloc];
}
@end
```

在 viewDidLoad 方法中，首先在❶处创建了两个 UIImageView 对象，并分别设置了它们的大小和位置；同时，在❷处创建了一个访问加速计的对象，在❸处设置它的更新频率为每秒更新 20 次，并在❹处指定委托为 self。❺处代码是本程序的核心部分，它实现了委托方法，利用 UIAcceleration 实例对象，获取到了加速计 $X$ 和 $Y$ 轴的数据，并根据这些数据计算旋转角度，从而使“指南针”的指针旋转相应的角度。当程序运行结束时，需要调用❻处代码将加速计的委托重置为 nil，从而使加速计停止检测更新。

最后，点击 Build and Go 按钮，编译并运行程序，最终效果如图 13-5 所示。

# 13.4 过滤加速计数据

上面获取的加速计数据是未经过加工的原始数据，包含各个方面的受力，不能满足用户某些特殊情况下的需求。例如：有时候我们只需要考虑 iPhone 受到的重力，而忽略移动 iPhone/iPod 时它受到的外力；还有些情况下需要我们忽略重力对 iPhone/iPod 的影响，而只考虑其他外力对它的作用。在这些情况下，需要使用滤波器对加速计的原始数据进行过滤处理。

## 13.4.1 使用低通滤波器

低通滤波器用于过滤出加速计数据中的重力部分，以减少其他外力的影响，由此产生的过滤值对反映重力的影响也就更为准确，例如过滤 iPhone 移动时的外力。低通滤波器需要用到傅里叶变换知识，如果学习过这方面的知识，理解它会更加容易，不过即使没学过也没关系。一般情况下使用如下代码就能实现简单的过滤了，其中 kFilterFactor 表示过滤因子，它的取值范围是 0 到 1。

```
accelX = (acceleration.x * kFilterFactor)+(accelX * (1-kFilterFactor));
accelY = (acceleration.y * kFilterFactor)+(accelY * (1-kFilterFactor));
```

下面我们就使用低通滤波器对 Compass 程序获取到的加速计原始数据进行过滤。选中 CompassViewController.m 文件，对程序做以下修改：

**代码 13.4 CompassViewController.m 文件**

```
#define kFilterFactor   0.1 ❶
#import "CompassViewController.h"
@implementation CompassViewController
- (void)accelerometer:(UIAccelerometer *)accelerometer
        didAccelerate:(UIAcceleration *)acceleration {
    //none filter
    accelX = [acceleration x];
    accelY = [acceleration y];
    //Low-pass fliter
accelX=(acceleration.x * kFilterFactor)+(accelX * (1-kFilterFactor)); ❷
accelY=(acceleration.y * kFilterFactor)+(accelY * (1-kFilterFactor)); ❸
float angle = atan2(accelY,accelX);
[compassView setTransform:
                CGAffineTransformMakeRotation(angle)];
}
```

```
- (void)viewDidLoad{
        contentView = [[UIImageView alloc] initWithFrame:
                                            [[UIScreen mainScreen]
                                            applicationFrame]];
    contentView.image = [UIImage imageNamed:@" background.jpg"];
    self.view = contentView;
    [contentView release];
    compassView = [[UIImageView alloc] initWithImage:
                          [UIImage imageNamed:@"compass.png"]];
    [compassView setCenter:CGPointMake(160.0f,240.0f)];
    [contentView addSubview:compassView];
    [compassView release];
    UIAccelerometer *accelerometer = [UIAccelerometer
    sharedAccelerometer];
    accelerometer.updateInterval = 1.0f/20.0f;
    accelerometer.delegate = self;
}
- (void)dealloc {
    UIAccelerometer *accelerometer = [[UIAccelerometer
    sharedAccelerometer] setDelegate: nil];
    [super dealloc];
}
@end
```

在代码中，我们首先在❶处定义了一个常量 kFilterFactor 表示过滤因子，代码❷和❸分别用于过滤 $X$ 和 $Y$ 轴上除重力以外的其他外力。

点击 Build and Go 按钮，编译并运行程序，效果和图 13-5 一样，不过当你慢慢地移动 iPhone/iPod(为了保证 iPhone/iPod 受到的重力不受大的影响)时，会发现“指南针”指示的方向并未发生多大变化，这是因为低通滤波器已经过滤掉了大部分的移动外力。

## 13.4.2　使用高通滤波器

使用低通滤波器能过滤掉移动数据，但是有时候我们更希望能使用加速计检测设备的瞬时运动，因而需要过滤掉重力，以减少引力的影响，从而获取加速计的瞬时移动数据，这时就需要使用高通滤波器了。上一小节我们已经过滤出了重力，要获取移动的数据只需从原始数据中去除低通滤波数据即可。代码如下：

```
gravX = (acceleration.x * kFilterFactor) + (accelX * (1-kFilterFactor));
gravY = (acceleration.y * kFilterFactor) + (accelY * (1-kFilterFactor));
```

**accelX = acceleration.x - gravX;**

**accelY = acceleration.y - gravY;**

现在就来修改一下上一小节的代码，选中 CompassViewController.m 文件，对程序做以下修改：

**代码 13.5 CompassViewController.m 文件**

```
#define kFilterFactor 0.1
#import "CompassViewController.h"
@implementation CompassViewController

- (void)accelerometer:(UIAccelerometer *)accelerometer
        didAccelerate:(UIAcceleration *)acceleration {
    //none filter
    accelX = [acceleration x];
    accelY = [acceleration y];
    //Low-pass fliter
    accelX =(acceleration.x * kFilterFactor)+(accelX * (1-kFilterFactor));
    accelY = (acceleration.y * kFilterFactor)+(accelY * (1-kFilterFactor));
    //High-pass filter
    float gravX = (acceleration.x * kFilterFactor)
                    + (accelX * (1-kFilterFactor));
    float gravY = (acceleration.y * kFilterFactor)                ❶
                    + (accelY * (1-kFilterFactor));
    accelX = acceleration.x-gravX;
    accelY = acceleration.y-gravY;
    float angle = atan2(accelY, accelX);
    [compassView setTransform:
              CGAffineTransformMakeRotation(angle)];
}
- (void)viewDidLoad {
    contentView = [[UIImageView alloc]initWithFrame:
                                        [[UIScreen mainScreen]
                                        applicationFrame]];
    contentView.image = [UIImage imageNamed:@" background.jpg"];
    self.view = contentView;
    [contentView release];
    compassView = [[UIImageView alloc] initWithImage:
```

```
    [UIImage imageNamed:@"compass.png"]];
    [compassView setCenter:CGPointMake(160.0f,240.0f)];
    [contentView addSubview:compassView];
    [compassView release];
    UIAccelerometer * accelerometer = [UIAccelerometer
    sharedAccelerometer];
    accelerometer.updateInterval = 1.0f/20.0f;
    accelerometer.delegate = self;
}
- (void)dealloc {
    UIAccelerometer * accelerometer = [[UIAccelerometer
    sharedAccelerometer] setDelegate: nil];
    [super dealloc];
}
@end
```

在代码❶处，我们用加速计原始数据减去低通滤波器过滤得到的重力，就是最后需要的外力数据，这是一种非常简单快捷的方法。

编译并运行程序，初始运行效果也和图 13-5 一样，但是即使我们缓慢地移动 iPhone/iPod，“指南针”指示的方向也会发生明显的变化，因为加速计的感知是非常灵敏的，而且此时检测的主要就是移动时的数据。

## 13.5 检测摇动

摇动是 iPhone 具有的最让人赞叹的三维手势之一，它使得用户不需要触摸屏幕就可以执行一系列的操作，例如用户通过摇动设备擦除一段文字，或者是通过摇动来更改 iPhone 的背景颜色等。怎么样，听起来是不是很酷？其实这些是比较容易实现的，我们只需要设计一个简单的检查器，检测加速计某个轴上的绝对值是否大于设置的阈值（可以根据实际情况设置，下面程序中我们设置为 2.0 g），如果它大于阈值则表示发生了一次摇动。当然这可能是一个无意的动作，所以有时候为了使检测到的手势更加准确，我们可以约定摇动一定的次数才算是一次摇动。

下面我们将通过一个简单的时钟程序来演示摇动手势，程序运行效果如图 13-6 所示；当 iPhone/iPod 被摇动时，程序的背景图片将随机发生改变，如图 13-7 所示。

图 13-6 摇动前的效果

图 13-7 摇动后的效果

本程序大体需要以下几个步骤：

- 创建新项目 Clock，添加图片资源；
- 在 ClockViewController. h 中，声明 updateLabel 方法；
- 创建加速计对象，实现委托方法和 updateLabel 方法；
- 在 ClockAppDelegate. m 中，使用计时器调用 updateLabel 方法。

**1. 创建新项目 Clock，添加图片资源**

现在我们就来完成这个小的应用程序。打开 Xcode，选择 View-based Application 模板创建一个新项目，命名为随书光盘 Clock。在随书光盘 Chapter 13 Clock 文件夹中，找到 bg0. png、bg1. png、bg2. png 和 bg3. png 四幅图片，同时将它们拖入 Resources 文件夹中。当摇动 iPhone 时，时钟的背景在这几张图片之间随机切换。

**2. 在 ClockViewController. h 中，声明 updateLabel 方法**

添加图片完毕之后，选中 ClockViewController. h 文件，添加如下黑体字所示的代码：

**代码 13.6 ClockViewController. h 文件**

```
#import <UIKit/UIKit.h>
@interface ClockViewController : UIViewController
```

```
    <UIAccelerometerDelegate>{
    UILabel * timerabel;
    UIImageView * imageView;
}
@property (nonatomic, retain) UILabel * timerabel;
@property (nonatomic, retain) UIImageView * imageView;
- (void)updateLabel; ❷
@end
```

在此文件的❶处，我们添加了 UIAccelerometerDelegate 协议，声明了一个标签来显示时间，一个 UIImageView 对象用来显示时钟背景；在❷处还声明了一个 updateLabel 方法，用来更新标签上的文本。

**3. 创建加速计对象，实现委托方法和 updateLabel 方法**

接下来，选中 ClockViewController. m 文件，添加以下代码：

**代码 13.7　ClockViewController. m 文件**

```
#import "ClockViewController.h"
@implementation ClockViewController
@synthesize timerabel;
@synthesize imageView;
- (void)updateLabel {
    NSDateFormatter * formatter =
                [[NSDateFormatter alloc] init];
    [formatter setDateFormat:@"MM-dd-hh-mm-ss"];
    NSString * locationString = [formatter stringFromDate:
                                [NSDate date]];
    NSArray * timeArray =
    [locationString componentsSeparatedByString:@"-"];
    int hour = [[timeArray objectAtIndex:2] intValue];
    int min = [[timeArray objectAtIndex:3] intValue];
    int sec = [[timeArray objectAtIndex:4] intValue];
    timerabel.text = [NSString stringWithFormat:
                @"%02d:%02d:%02d", hour, min,sec];
    [formatter release];
}
- (void)viewDidLoad {
```

```
    UIAccelerometer * accelerometer =
    [UIAccelerometer sharedAccelerometer];                          ❸
    accelerometer.updateInterval = 1.0f/20.0f;
    accelerometer.delegate = self;

    imageView = [[UIImageView alloc] initWithFrame:
            [[UIScreen mainScreen] applicationFrame]];
    imageView.image = [UIImage imageNamed:@"bg0.png"];
    self.view = imageView;
    CGRect frame = CGRectMake(20,130,300,80);
    timerabel = [[UILabel alloc] initWithFrame:frame];
    timerabel.text = [NSString stringWithFormat:                    ❹
                        @"%@",[NSDate date]];
    [timerabel setBackgroundColor: [UIColor clearColor]];
    [timerabel setTextColor:[UIColor orangeColor]];
    [timerabel setFont: [UIFont fontWithName:
                        @"Zapfino" size:40]];
    [imageView addSubview:timerabel];
}
#pragma mark-
#pragma UIAccelerometer Delegate Method
- (void)accelerometer:(UIAccelerometer * )accelerometer
            didAccelerate:(UIAcceleration * )acceleration {
        if (fabsf(acceleration.x > 2.0) ||                          ❺
            fabsf(acceleration.y > 2.0) ||
            fabsf(acceleration.z > 2.0)) {

            srandom(time(NULL));
            int i= random() % 4;                                    ❻
            imageView.image = [UIImage imageNamed:
            [NSString stringWithFormat:@"bg%d.png",i]];
    }
}
- (void)dealloc {
    [timerabel release];
    [imageView release];
    [super dealloc];
```

```
}
@end
```

在 updateLabel 方法中，我们在❶处创建了一个 NSDateFormatter 类的实例对象 formatter 来设置时间格式；在❷处创建了一个 NSString 实例对象获取当前时间，此程序我们只需显示时、分、秒。

在 viewDidLoad 方法中，我们首先在❸处创建了一个访问加速计的对象，并设置它的更新频率为 20Hz，委托为 self。同时，在❹处创建了一个 UIImageView 实例对象和 UILabel 实例对象，初始化并调整它们的大小、字体和颜色，使得最终界面效果更美观。

最后，我们需要在❺处的 accelerometer:didAccelerate:方法中构建一个简单的检查器，检测设备是否被摇动，判断方法是检测加速计三个轴上的受力数据的绝对值是否大于2.0 g(阈值)，如果大于 2.0g 就定义为一次摇动，并触发某个操作，这里是触发❻处的方法来随机更改 Clock 程序的背景图片。

**4. 在 ClockAppDelegate.m 中，使用计时器调用 updateLabel 方法**

至此程序并没有完成，我们还需要调用 updateLabel 方法，动态更新标签文本。选中 ClockAppDelegate.m 文件，添加如下黑体字所示的代码。

**代码 13.5 ClockAppDelegate.m 文件**

```
#import "ClockAppDelegate.h"
#import "ClockViewController.h"
@implementation ClockAppDelegate
@synthesize window;
@synthesize viewController;
- (void)applicationDidFinishLaunching:(UIApplication *)application {
    // Override point for customization after app launch

    [NSTimer scheduledTimerWithTimeInterval:(1.0)
                                     target:self
                                   selector:@selector(onTimer)    ❶
                                   userInfo:nil
                                    repeats:YES];
    [window addSubview:viewController.view];
    [window makeKeyAndVisible];
}
- (void)dealloc {
    [viewController release];
    [window release];
```

```
    [super dealloc];
}
- (void)onTimer {
    [viewController updateLabel];  ❷
}
@end
```

在❶处，使用计时器对象，每隔 1 秒钟触发一次 onTimer 方法，该方法将调用❷处的 updateLabel 方法动态更新标签文本。

最后，编译并运行程序，效果如图 13-6 所示；当我们摇动 iPhone/iPod 时，时钟的背景图片将随机发生改变，效果如图 13-7 所示。

对于其他的一些三维手势，如倾斜、旋转等，需要构建更为严格缜密的摇动检查器。

## 13.6 小结

在本章中，我们讲解了 iPhone 内置加速计的三轴结构，以及获取 iPhone 方向的两种方式，并通过演示自动旋转功能展示了 iPhone 的神奇之处；通过简单的"指南针"程序，对如何处理加速计数据做了介绍。如果在程序中使用过滤器，可以分离出实际应用时需要的数据，这对于一些特殊应用来说是非常有效的。最后，通过一个更换"时钟"背景图片的程序，说明了如何检测摇动手势，相信这会使你更有兴趣去深入研究其他的三维手势。

在下一章，将要学习的是 iPhone 的另一个特性：使用 Core Location 定位，同时借助 Google 地图显示位置信息。

# 第14章 使用Core Location和MapKit

## 本章内容

- 定位的几种技术
- 如何进行定位
- 如何获取位置信息
- 使用Google地图

在上一章中,我们学习了iPhone/iPod内置加速计,对于它的一些简单应用已经有了一定的了解。本章将继续介绍iPhone/iPod另一个强大的功能——Core Location,它能够利用不同的技术实现精确的地理定位,我们将在本章简单说明这几种技术。当获得一个地理位置的准确的经、纬度之后,人们总是希望能够弄清楚它到底是什么地方,为此接下来,本章将介绍一个重要的框架——MapKit框架,使用它能够很轻松地加载Google地图。通过结合使用Core Location和MapKit,人们已经开发出很多有用的应用程序,如iPod touch自带的地图软件,如图14-1所示。

图14-1 iPod touch的地图软件

按照本书的惯例，我们仍然通过一个实例来学习本章内容。在这个例子中将使用 Core Location 知识读取当前地理位置的经度和纬度，然后利用 MapKit 框架实现调用 Google 地图来显示当前位置。

# 14.1　Core Location 介绍

CoreLocation.framework 是 iPhone SDK 中用来检测用户位置的框架，该框架包含两个重要的类：CLLocationManager 类和 CLLocation 类，此外还包括 CLLocationManagerDelegate协议，它们是 iPhone/iPod 实现定位功能的核心技术。在介绍它们之前，我们先来了解一下常用的几种定位技术。

## 14.1.1　定位的几种技术

Core Location 实现定位功能，主要应用了 GPS（全球定位系统）、蜂窝基站三角网定位和 Wi-Fi 定位服务（WPS）等三种技术，如图 14-2 所示。

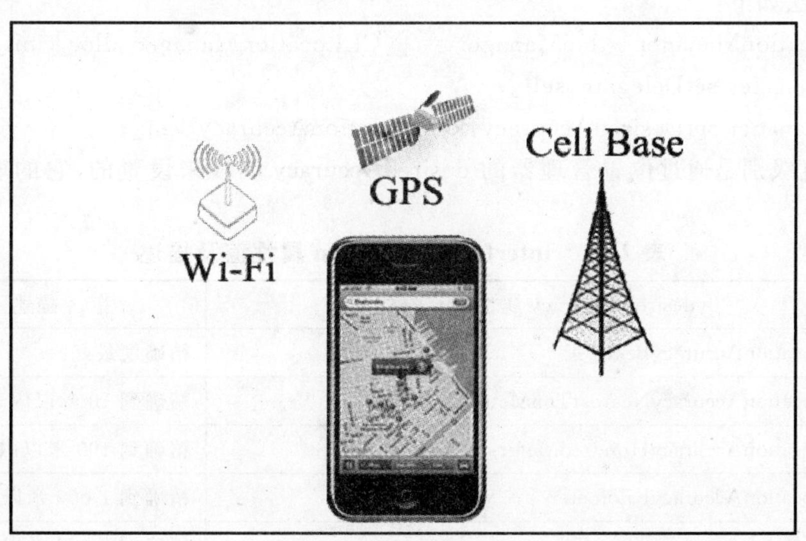


图 14-2　定位的三种技术

GPS 定位技术是三种技术中最精确的，它是通过获取地球同步卫星的微波信号来确定当前位置的。新型的 iPhone 3GS 内置了 GPS，但是第一代 iPhone 和 iPod 是没有的。我们日常使用的车载导航系统就是采用 GPS 定位技术的。

蜂窝基站三角网定位技术是根据 iPhone 所处的蜂窝基站的位置粗略确定当前位置，它在蜂窝基站比较密集的地方会比较精确，但是如果处在城市郊区，定位的准确性就不高了。该技术只用于 iPhone，在 iPod touch 上不能使用。

Wi-Fi 定位技术是使用 iPhone/iPod 上的 Wi-Fi 连接，通过 IP 地址找到它所使用的无线路由器，并利用 MAC 地址搜索服务提供商的大型数据库来定位。

Tips

| 定位方式 | iPhone | iPod touch |
| --- | --- | --- |
| GPS | √(第一代不具有) | × |
| 蜂窝基站 | √ | × |
| Wi-Fi | √ | √ |

注：Core Location 根据设备的实际需要选择合适的定位方式。

### 14.1.2　位置管理器

要实现定位功能，只需要使用 CoreLocation.framework 框架下的两个类 CLLocationManager、CLLocation 和一个 CLLocationManagerDelegate 协议。首先我们需要创建一个 CLLocationManager（位置管理器）对象，同时设置它的委托和精确度级别，具体方法如下：

```
CLLocationManager *locManager = [[CLLocationManager alloc] init];
[locManager setDelegate:self];
[locManager setDesiredAccuracy:kCLLocationAccuracyBest];
```

精确度级别是通过位置管理器的 desiredAccuracy 属性来设置的，它的取值可以参照表 14-1。

**表 14-1　interfaceOrientation 属性值及描述**

| desiredAccuracy 属性值 | 描述 |
| --- | --- |
| kCLLocationAccuracyBest | 精确度最好 |
| kCLLocationAccuracyNearestTenMeters | 精确到 10 米以内 |
| kCLLocationAccuracyHundredMeters | 精确到 100 米以内 |
| kCLLocationAccuracyKilometer | 精确到 1 000 米以内 |
| kCLLocationAccuracyThreeKilometers | 精确到 3 000 米以内 |

在实际使用时，应该选择最合适的精确度级别，因为选择 Core Location 的精确度级别越高，iPhone 消耗的电量会越多，不利于延长设备电池的寿命。

很多时候，并不希望 iPhone 时刻进行位置更新检测，因为这会给用户带来很多麻烦，就像刚才所说的会增加耗电量一样。用户总是希望在位置移动一定距离后再进行重新定位，这就需要设置位置管理器的 distanceFilter 属性，通常称它为“距离筛选器”，具体实现方法如下：

```
LocManager.distanceFilter = 300;
```

它表示只有当 iPhone 移动的距离大于 300 米时，才通知委托进行重新定位。当然也可以不使用距离筛选器，此时只需要设置 distanceFilter 的值为 kCLDistanceFilterNone 就可以了。接下来，可以使用下面这行代码，启动位置管理器来进行定位：

```
[locManager startUpdatingLocation];
```

如果不需要再进行位置更新，就应该使用下面这行代码停止检测位置更新：

```
[locManager stopUpdatingLocation];
```

### 14.1.3　获取位置信息

在上一小节中，我们已经学习了如何创建位置管理器对象以及如何使用它进行定位，这一小节主要介绍如何通过 Cllocation 对象的属性和方法获取位置信息。

CLLocation 对象主要包括 coordinate、altitude、horizontalAccuracy、verticalAccuracy 和 timestamp 等属性；此外，还有一个 getDistanceFrom：方法，用来获取更新前后两个位置的距离。下面我们对 CLLocation 的一些主要属性进行说明。

**coordinate** 用来存储地理位置的 latitude 和 longitude，分别代表该地理位置的纬度和经度，它们的取值都是浮点型。在地理学中，地球北半球的纬度为正值，南半球为负值；东半球的经度为正值，西半球为负值。获取经、纬度的具体方法如下：

```
float latitude = location.coordinate.latitude;
float longitude = location.coordinate.longitude;
```

**location** 是 CLLocation 类的一个实例对象，这里我们将获取到的经、纬度存储为浮点型；此外，还可以通过创建 CLLocationDegrees 类的实例对象，以度为单位存储经、纬度，代码如下：

```
CLLocationDegrees latitude = location.coordinate.latitude;
CLLocationDegrees longitude = location.coordinate.longitude;
```

**altitude** 属性表示某个位置的海拔高度，返回值为浮点型，实际定位时极不准确。可以用以下代码获取它的值：

```
float altitude = location.altitude;
```

**horizontalAccuracy** 属性表示水平准确度，返回值为浮点型。它是以 coordinate 为圆心的圆的半径，半径越小定位越精确，如果 horizontalAccuracy 为负值，表示 Core Location 定位失败。可以用以下代码获取它的值：

```
float horizontalAccuracy = location.horizontalAccuracy;
```

**verticalAccuracy** 属性表示垂直准确度，返回值也是浮点型。它的取值和海拔的取值 altitude 有关系，与实际情况相差很大。可以用以下代码获取它的值：

```
float verticalAccuracy= location.verticalAccuracy;
```

timestamp(时间戳)属性表示位置管理器确定位置的时间，返回的是一个 NSDate 实例。可以用以下代码获取它的值：

```
NSDate timestamp = location.timestamp;
```

此外，CLLocation 还有其他的几个属性，如 speed、course 等，这里就不再详细介绍了，可以通过 Xcode 的帮助文档进一步了解它们。

### 14.1.4 CLLocationManagerDelegate 协议

设置了位置管理器委托后，我们需要实现 CLLocationManagerDelegate 协议下的两个方法：

```
- (void)locationManager:(CLLocationManager *)manager
        didUpdateToLocation:(CLLocation *)newLocation
               fromLocation:(CLLocation *)oldLocation
```

这里更新位置的方法，当我们的移动范围大于距离筛选器的值时，位置管理器会调用此方法进行重新定位。

```
- (void)locationManager:(CLLocationManager *)manager
           didFailWithError:(NSError *)error
```

这是定位失败的处理方法，当 Core Location 无法获取到当前位置时，位置管理器将调用此方法。它返回的错误信息主要有两种：kCLErrorDenied 和 kCLErrorLocationUnknown。

程序开始运行时，会弹出如图 14-3 所示的对话框，询问用户是否允许使用当前位置信息。如果用户拒绝访问，将返回一个 kCLErrorDenied 错误信息；如果 Core Location 无法定位到用户当前的位置，将返回一个 kCLErrorLocationUnknown 错误信息，提示用户无法获取到当前位置，但此时位置管理器仍将在后台继续扫描。

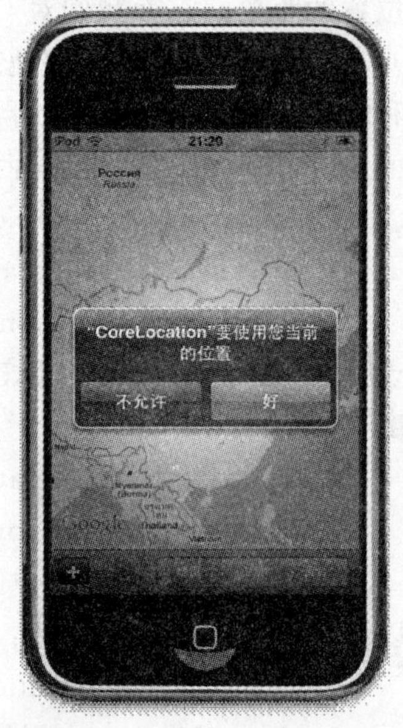


图 14-3 是否允许访问对话框

下面我们创建一个 CoreLocation 应用程序，它要完成的任务是获取当前地理位置信息并显示出来，在后续小节中我们将在此基础上继续完善它。

在开始程序之前，看看需要做哪些工作：

- 创建新项目，在 CoreLocationViewController. h 文件中引入框架、添加协议并声明位置管理器；
- 构建程序界面；
- 修改 CoreLocationViewController. m 文件，获取用户当前位置信息。

**1. 创建新项目，在 CoreLocationViewController. h 文件中引入框架、添加协议并声明位置管理器**

打开 Xcode，使用 View-based Application 模板创建一个项目，将其命名为 CoreLocation。展开 Frameworks 文件夹，添加 CoreLocation. framework 框架。添加完成后，展开 Classes 文件夹，选中 CoreLocationViewController. h 文件，添加如下黑体字所示的代码：

**代码 14.1 CoreLocationViewController.h 文件**

```
#import <UIKit/UIKit.h>
#import <CoreLocation/CoreLocation.h> ❶
@interface CoreLocationViewController : UIViewController
<CLLocationManagerDelegate> {
    CLLocationManager *locManager;
    CLLocationCoordinate2D loc;
    IBOutlet UITextView *textView;                         ❷
}
@property (nonatomic, retain) CLLocationManager *locManager;
@property (nonatomic, retain) IBOutlet UITextView *textView;
@end
```

在 CoreLocationViewController.h 中,添加了 CoreLocation/CoreLocation.h 头文件❶;同时,在代码段❷处添加了 CLLocationManagerDelegate 协议,声明了一个位置管理器实例对象和一个 UITextView 类型的输出口 textView,输出口用来显示获取到的位置信息。

**2. 构建程序界面**

这个程序的界面非常简单,我们只需要在视图中添加一个 Text View 就可以了。双击 CoreLocationViewController.xib 文件,从 Library 窗口中拖出一个 Text View 到视图中,调整它的大小和位置,并且连接输出口 textView;完成之后,保存并关闭 Xib 窗口。

**3. 修改 CoreLocationViewController.m 文件,获取用户当前位置信息**

接下来我们要完成的是使用位置管理器获取用户当前的位置信息,选中 CoreLocationViewController.m 文件,添加如下黑体字所示的代码:

**代码 14.2 CoreLocationViewController.m 文件**

```
#import "CoreLocationViewController.h"
@implementation CoreLocationViewController
@synthesize locManager;
@synthesize textView;
- (void)viewDidLoad {
    locManager = [[CLLocationManager alloc] init];
    locManager.delegate = self;
    locManager.desiredAccuracy = kCLLocationAccuracyBest;   ❶
    [locManager startUpdatingLocation];
    [super viewDidLoad];
}
- (void)dealloc {
    [locManager stopUpdatingLocation]; ❷
```

```
    [locManager release];
    [textView release];
    [super dealloc];
}
#pragma mark -
#pragma mark CoreLocation Delegate Methods
- (void)locationManager:(CLLocationManager *)manager
    didUpdateToLocation:(CLLocation *)newLocation
           fromLocation:(CLLocation *)oldLocation {                    ❸
    loc = [newLocation coordinate];
    float latitude = loc.latitude ;
    float longitude = loc.longitude;
    float altitude = newLocation.altitude;
    float horizontalAccuracy = newLocation.horizontalAccuracy;
    float verticalAccuracy = newLocation.verticalAccuracy;
    textView.text = [NSString stringWithFormat:@"latitude :%f\n
    ↳ longitude:%f\naltitude:%f\nhorizontalAccuracy:%f\n
    ↳ verticalAccuracy:%f",latitude ,longitude,altitude,
    ↳ horizontalAccuracy,verticalAccuracy];
    [textView setEditable:NO];
}
- (void)locationManager:(CLLocationManager *)manager
        didFailWithError:(NSError *)error{
    NSString *errorMessage;
    if ([error code] == kCLErrorDenied){                               ❹
          errorMessage = @"访问被拒绝!";
}
if ([error code] == kCLErrorLocationUnknown) {
      errorMessage = @"无法定位到你的位置!";
}
UIAlertView *alert = [[UIAlertView alloc]
                             initWithTitle:nil
                             message:errorMessage
                             delegate:self
                             cancelButtonTitle:@"确定"
                             otherButtonTitles:nil];
    [alert show];
    [alert release];
```

```
}
@end
```

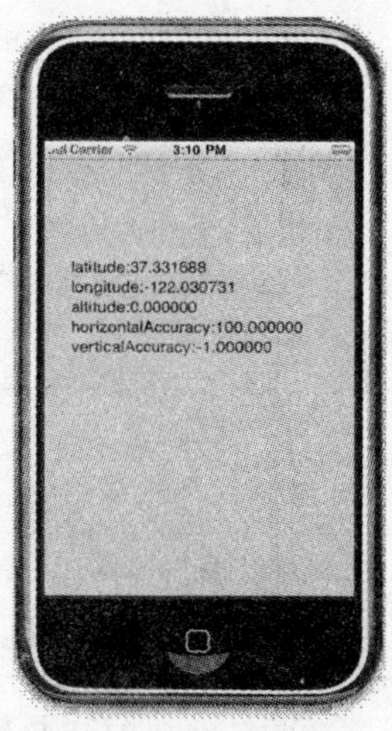


图 14-4　获取位置信息

至此，CoreLocation 程序就算完成了。在程序❶处我们首先创建了一个 CLLocationManager 实例对象，并设置它的委托为 self；同时将它的定位精确度级别设为最高，启动位置管理器后将调用程序❸处的委托方法获取当前位置的经度和纬度等信息，并把位置信息显示在 textView 中；本程序❹处的方法是用来处理定位失败和用户拒绝访问情况的；程序结束时，我们需要调用 stopUpdatingLocation 方法❷，停止位置更新。

编译并运行程序，最终效果如图 14-4 所示。

# 14.2　使用 MapKit 显示位置

通过上一节的学习，已经能够准确地定位到用户当前的位置，并获取到了相关的信息，不过我们还希望能够在地图中将它标识出来，这样即使我们处在一个陌生的环境，也不用担心找不到公交车站或是餐馆了。要完成上述功能，可以通过 MapKit. framework 框架调用 Google 地图。

## 14.2.1　使用 Google 地图

Google Map 是展示 Core Location 最理想的方式。使用 MapKit. framework 框架可以很轻松地在我们的应用程序中嵌入一幅功能完整的地图，同时具有很多独特的特征，例如缩放地图、在地图上添加注解等。嵌入地图时需要用到 MKMapView 类，这个类中包括很多重要的属性和方法，这里我们只介绍几个在程序中经常用到的。

**region** 属性用来设置地图的哪一部分被显示，它是 MKCoordinateRegion 类型的结构体对象，定义如下：

```
typedef struct {
    CLLocationCoordinate2D center;
    MKCoordinateSpan span;
} MKCoordinateRegion;
```

center 就是上一节中 coordinate 的取值，包含经、纬度信息，此处用来表示地图的中心位置。span 表示地图的一个跨度，它包括了该区域(region)的经度和纬度变化度信息，即缩放地图的比例，是 MKCoordinateSpan 类型的结构体对象，定义如下：

```
typedef struct {
    CLLocationDegrees latitudeDelta;
    CLLocationDegrees longitudeDelta;
} MKCoordinateSpan;
```

我们还可以通过 **MapType** 属性设置地图的类型，它的取值及描述见表 14-2。

**表 14-2　MapType 属性值及描述**

| MapType 属性值 | 描述 |
|---|---|
| MKMapTypeStandard | 表示标准的街道级地图 |
| MKMapTypeSatellite | 表示卫星图 |
| MKMapTypeHybrid | 表示以上两种类型的混合 |

下面将利用这些知识来修改 14.1 节中的 CoreLocation 例子。前面我们已经获取到了准确的地理位置信息，现在要完成的就是充分利用这些信息并调用 Google Map 显示当前位置的地图。修改后程序运行的效果如图 14-5 所示。

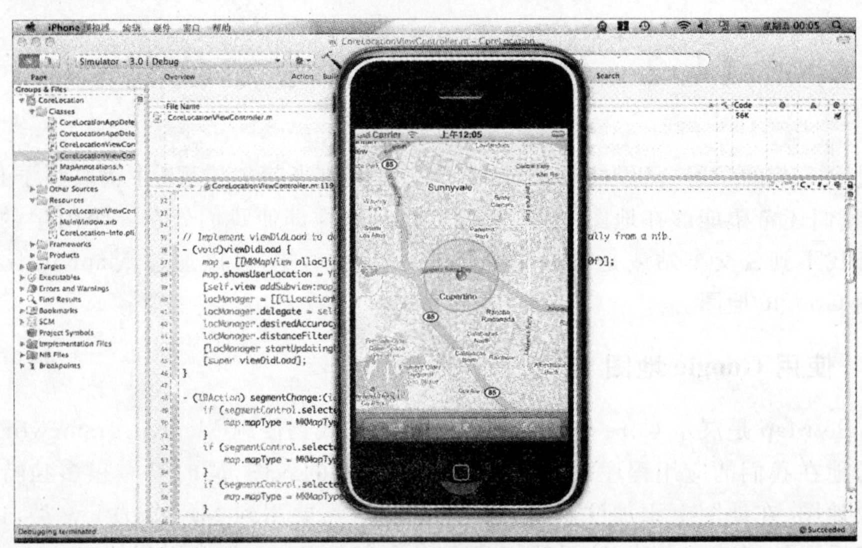

图 14-5　使用 Google Map 显示当前位置的地图

本程序大体包括以下步骤：

- 修改 CoreLocationViewController.h 文件，声明地图类实例对象；
- 修改程序界面；
- 修改 CoreLocationViewController.m 文件，添加 Google 地图。

**1. 修改 CoreLocationViewController.h 文件，声明地图类实例对象**

使用 Xcode 打开 CoreLocation 项目，我们需要在 Frameworks 文件夹下添加 MapKit.framework 框架；完成之后选中 CoreLocationViewController.h 文件，对它做以下修改：

**代码 14.3　CoreLocationViewController.h 文件**

```
#import <UIKit/UIKit.h>
#import <CoreLocation/CoreLocation.h>
#import <MapKit/MapKit.h> ❶
@interface CoreLocationViewController : UIViewController
    <CLLocationManagerDelegate, MKMapViewDelegate>{ ❷
    CLLocationManager * locManager;
    CLLocationCoordinate2D loc;
    IBOutlet UITextView * textView;
        IBOutlet UISegmentedControl * segmentControl;
        MKMapView * map;
        }
@property (nonatomic, retain) CLLocationManager * locManager;
@property (nonatomic, retain) IBOutlet UITextView * textView;          ❸
@property (nonatomic, retain) IBOutlet UISegmentedControl
 * segmentControl;
@property (nonatomic, retain) MKMapView * map;
- (IBAction) segmentChange:(id)sender;
@end
```

在程序的 CoreLocationViewController.h 文件中，首先需要在程序❶处引入 MapKit/MapKit.h 文件。然后，在❷处添加了 MKMapViewDelegate 协议。在❸处声明了一个 UISegmentControl 类型的输出口和一个 MKMapView 类型的对象，用于添加地图；最后，还声明了一个 segmentChange:操作方法。

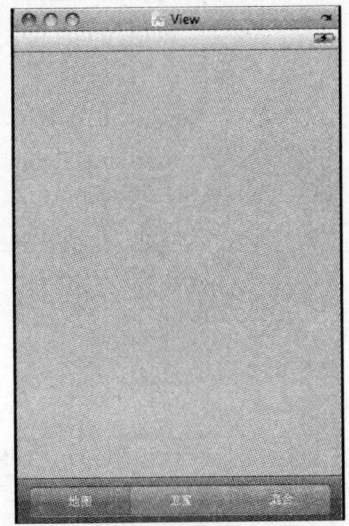

图 14-6　CoreLocation 程序界面

**2. 修改程序界面**

接下来对程序界面做一些修改，双击 CoreLocationViewController.xib 文件，往视图中添加一个 Tool Bar 控件，去掉上面的 Item；再添加一个分段控件 Segment Control，调整它的大小；当然你还可以往视图中添加一个 Map View，这里我们不采用这种方法，而是在程序中用代码实现。界面最终效果如图 14-6 所示。

然后，我们需要为分组控件连接到输出口 segmentControl，并将它的 Value Changed 事件与 segmentChange:操作方法连接。上述工作完成之后，保存并关闭 Xib 窗口，返回 Xcode。

**3. 修改 CoreLocationViewController. m 文件，添加 Google 地图**

选中 CoreLocationViewController. m 文件，我们需要对它做以下修改：

**代码 14.4　CoreLocationViewController. m 文件**

```
#import "CoreLocationViewController.h"
@implementation CoreLocationViewController
@synthesize locManager;
@synthesize segmentControl;
@synthesize map;
- (void)viewDidLoad {
    map = [[MKMapView alloc]
        initWithFrame:CGRectMake(0.0f, 0.0f, 320.0f, 411.0f)]; ❶
    map.showsUserLocation = YES;
    [self.view addSubview:map]; ❷
    locManager = [[CLLocationManager alloc] init];
    locManager.delegate = self;
    locManager.desiredAccuracy = kCLLocationAccuracyBest;
    locManager.distanceFilter = 100;
    [locManager startUpdatingLocation];
    [super viewDidLoad];
}
- (IBAction) segmentChange:(id)sender {
    if (segmentControl.selectedSegmentIndex == 0) {
        map.mapType = MKMapTypeStandard;
    }
    if (segmentControl.selectedSegmentIndex == 1) {
        map.mapType = MKMapTypeSatellite;                    ❸
    }
    if (segmentControl.selectedSegmentIndex == 2) {
        map.mapType = MKMapTypeHybrid;
    }
}
- (void)dealloc {
    [segmentControl release];
    [map release];
    [locManager release];
    [super dealloc];
```

```
}
#pragma mark -
#pragma mark Core Location Delegate Methods
- (void)locationManager:(CLLocationManager *)manager
    didUpdateToLocation:(CLLocation *)newLocation
        fromLocation:(CLLocation *)oldLocation {
    loc = [newLocation coordinate];
    float latitude = loc.latitude ;
    float longitude = loc.longitude;
    float altitude = newLocation.altitude;
    float horizontalAccuracy = newLocation.horizontalAccuracy;
    float verticalAccuracy = newLocation.verticalAccuracy;
    textView.text = [NSString stringWithFormat:
    @"latitude :%f\nlongitude:%f\naltitude:%f\nhorizontalAccuracy:%f\n
    ➥verticalAccuracy:%f",latitude ,longitude,altitude,horizontalAccuracy,
    ➥verticalAccuracy];
    [textView setEditable:NO];

    MKCoordinateRegion region;
    MKCoordinateSpan span;
    span.latitude Delta= 1;
    span.longitudeDelta= 1;
    region.span=span;                                    ❹
    region.center=loc;
    [map setRegion:region animated:YES];
    [map regionThatFits:region];
}

- (void)locationManager:(CLLocationManager *)manager
      didFailWithError:(NSError *)error{
    NSString *errorMessage;
    if ([error code] == kCLErrorDenied){
        errorMessage = @"访问被拒绝!";
    }
    if ([error code] == kCLErrorLocationUnknown) {
        errorMessage = @"无法确定你的位置!";
    }
```

```
    UIAlertView *alert = [[UIAlertView alloc]
                            initWithTitle:nil
                            message:errorMessage
                            delegate:self
                            cancelButtonTitle:@"确定"
                            otherButtonTitles:nil];
    [alert show];
    [alert release];
}
@end
```

图 14-7　卫星图效果

在程序中，首先在❶处创建了一个 MKMapView 类型的实例对象，指定了它的位置和大小，并设置它能显示用户当前的位置，在❷处将地图加载到视图中。

在❸处的 segmentChange：操作方法中，指定了分组控件的各分组分别代表 MapType 的三种不同的取值，并用下面两行代码指定区域的中心位置和跨度：

```
region.span=span;
region.center=loc;
```

同时设定地图放大时将变成原来的 1 倍：

```
span.latitude Delta=1;
span.longitudeDelta=1;
```

最后我们用下面的代码设置地图以动画的形式(程序刚开始运行时的效果)加载到新的区域：

```
[map setRegion:region animated:YES];
[map regionThatFits:region];
```

编译并运行程序，地图将以动画的形式加载，最后的效果如图 14-5 所示；当点击分组控件选择地图类型为卫星时，出现如图 14-7 所示的卫星图效果。

### 14.2.2　添加地图注解

仔细观察图 14-1 你会发现地图上面有两个“小别针”，它就是这里要讲的地图注解——Map Annotations。它主要有两个属性 title 和 subtitle，可以通过设置这两个属性来显示我们希望查看到的信息。当用户点击“小别针”时，这些信息将被显示出来。下面我们要做的就是为 CoreLocation 程序添加地图注解，这样就可以对某个位置做标记了，程序最终效果如图 14-8 所示。

本程序需要完成如下工作：

- 在项目中添加 MapAnnotations 类；
- 修改 CoreLocationViewController.h 文件，声明 addPin:方法；
- 修改程序界面；
- 修改 CoreLocationViewController.m 文件，实现 addPin:方法。

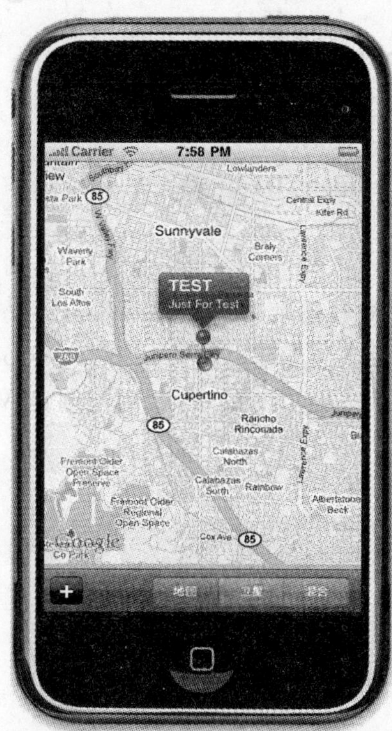


图 14-8　添加注解后的运行效果

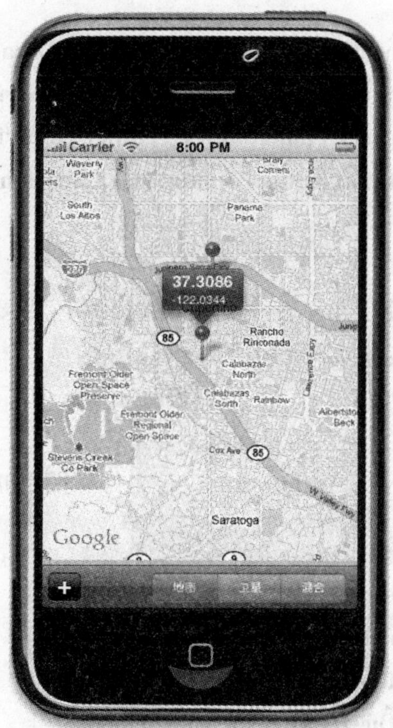


图 14-9　使用注解显示经、纬度信息

**1. 在项目中添加 MapAnnotations 类**

使用 Xcode 打开 CoreLocation 项目，我们继续在前一小节的基础上进行修改。展开 Classes 文件夹，右键选择 Add→New File…，出现新建文件向导时，选择 Objective-C Class 模板，并在 Subclass of 选项中选择 NSObject，将其命名为 MapAnnotations。选中 MapAnnotations.h 文件，添加如下黑体字所示的代码：

**代码 14.5　MapAnnotations.h 文件**

```
#import <Foundation/Foundation.h>
#import <CoreLocation/CoreLocation.h>
#import <MapKit/MapKit.h>
```

```
@interface MapAnnotations : NSObject <MKAnnotation>{
    CLLocationCoordinate2D coordinate;
    NSString * title;
    NSString * subtitle;
}                                                          ❷
@property(nonatomic, retain)CLLocationCoordinate2D coordinate;
@property (nonatomic, retain) NSString * title;
@property (nonatomic, retain) NSString * subtitle;
- (id)initWithCoordinate:(CLLocationCoordinate2D) c; ❸
@end
```

在❶处引入了两个头文件，同时还需要在❷处添加 MKAnnotation 协议，并声明两个字符串变量，分别用来存储注解的两个标题；还声明了一个 CLLocationCoordinate2D 类型的实例对象，它将作为❸处 initWithCoordinate:方法的参数，用于创建 MKAnnotations 类型的地图注解实例对象。

选中 MapAnnotations. m 文件，添加如下黑体字所示的代码：

**代码 14.6 MapAnnotations. m 文件**

```
#import "MapAnnotations.h"
@implementation MapAnnotations
@synthesize coordinate;
@synthesize title;
@synthesize subtitle;

- (id)initWithCoordinate:(CLLocationCoordinate2D) c{
    coordinate=c;
    return self;
}
- (void) dealloc {
    [title release];
    [subtitle release];
    [super dealloc];
}
@end
```

### 2. 修改 CoreLocationViewController. h 文件，声明 addPin:方法

上述工作完成之后，我们需要再对 CoreLocationViewController. h 文件做一点修改，

具体代码如下：

**代码 14.7　CoreLocationViewController.h 文件**

```
#import <UIKit/UIKit.h>
#import <CoreLocation/CoreLocation.h>
#import <MapKit/MapKit.h>
#import <MapKit/MKAnnotation.h>  ❶
#import "MapAnnotations.h"
@interface CoreLocationViewController : UIViewController
<CLLocationManagerDelegate, MKMapViewDelegate>{
CLLocationManager *locManager;
CLLocationCoordinate2D loc;
IBOutlet UITextView *textView;
IBOutlet UISegmentedControl *segmentControl;
MKMapView *map;
MapAnnotations *mapAnnotations; ❷
}
@property (nonatomic, retain) CLLocationManager *locManager;
@property (nonatomic, retain) IBOutlet UITextView *textView;
@property (nonatomic, retain) IBOutlet UISegmentedControl
 *segmentControl;
@property (nonatomic, retain) MKMapView *map;
@property (nonatomic, retain) MapAnnotations *mapAnnotations;
- (IBAction) segmentChange:(id)sender;
- (IBAction) addPin:(id) sender; ❸
@end
```

在 CoreLocationViewController.h 文件中，我们首先在❶处引入 MapKit/MKAnnotation.h 文件和 MapAnnotations.h 文件；然后，在❷处声明了一个 MKAnnotation类型的实例对象；最后，在❸处声明了一个 addPin:操作方法，用来响应用户添加地图注解的操作。

**3. 修改程序界面**

下面我们需要对 CoreLocation 程序的用户界面做一点修改。双击打开 CoreLocationViewController.xib 文件，我们需要在 Tool Bar 上添加一个 Bar Button Item，更改 Item 类型，并调整各个控件的位置和大小，最终界面如图 14-10 所示。

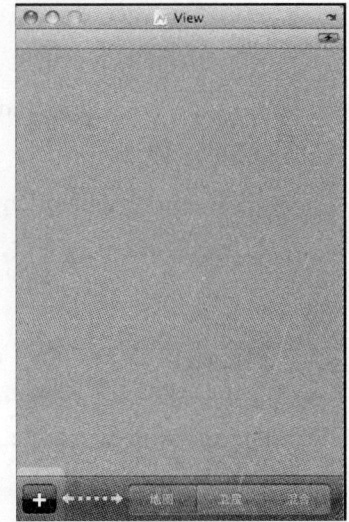


图 14-10　添加地图注解的用户界面

另外，还需要为 Item 连接上 addPin:操作方法。上述工作完成之后，保存并关闭 Xib 窗口，返回 Xcode。

**4. 修改 CoreLocationViewController.m 文件，实现 addPin:方法**

选中 CoreLocationViewController.m 文件，我们需要对它做以下修改：

**代码 14.8 CoreLocationViewController.m 文件**

```
#import "CoreLocationViewController.h"
@implementation CoreLocationViewController
@synthesize segmentControl;
@synthesize map;
@synthesize locManager;
@synthesize mapAnnotations;
- (void)viewDidLoad {
    map = [[MKMapView alloc]
            initWithFrame:CGRectMake(0.0f, 0.0f, 320.0f, 411.0f)];
    map.showsUserLocation = YES;
    [self.view addSubview:map];
    locManager = [[CLLocationManager alloc] init];
    locManager.delegate = self;
    locManager.desiredAccuracy = kCLLocationAccuracyBest;
    locManager.distanceFilter = 100;
    [locManager startUpdatingLocation];
    [super viewDidLoad];
}
- (IBAction) segmentChange:(id)sender {
    if (segmentControl.selectedSegmentIndex == 0) {
        map.mapType = MKMapTypeStandard;
    }
    if (segmentControl.selectedSegmentIndex == 1) {
        map.mapType = MKMapTypeSatellite;
    }
    if (segmentControl.selectedSegmentIndex == 2) {
            map.mapType = MKMapTypeHybrid;
    }
```

```
}
- (IBAction) addPin:(id) sender {
    mapAnnotations = [[MapAnnotations alloc]
                       initWithCoordinate:map.region.center]; ❶
    mapAnnotations.title = [NSString stringWithFormat:
                       @"%f",map.region.center.latitude];
    mapAnnotations.subtitle = [NSString stringWithFormat:
                       @"%f",map.region.center.longitude];          ❷
    [map addAnnotation:mapAnnotations];
    [mapAnnotations release];
}
- (void)dealloc {
    [segmentControl release];
    [map release];
    [locManager release];
    [mapAnnotations release];
    [super dealloc];
}
#pragma mark -
#pragma mark Core Location Delegate Methods
- (void)locationManager:(CLLocationManager *)manager
    didUpdateToLocation:(CLLocation *)newLocation
            fromLocation:(CLLocation *)oldLocation {
    loc = [newLocation coordinate];
    MKCoordinateRegion region;
    MKCoordinateSpan span;
    span.latitude Delta=0.1;
    span.longitudeDelta=0.1;
    region.span=span;
    region.center=loc;
    [map setRegion:region animated:YES];
    [map regionThatFits:region];
```

```
        mapAnnotations = [[MapAnnotations alloc]
                                initWithCoordinate:loc];
        mapAnnotations.title=@"TEST";                          ❸
        mapAnnotations.subtitle=@"Just For Test";
        [map addAnnotation:mapAnnotations];
        [mapAnnotations release];
}
- (void)locationManager:(CLLocationManager *)manager
        didFailWithError:(NSError *)error{
    NSString *errorMessage;
    if ([error code] == kCLErrorDenied){
        errorMessage = @"访问被拒绝!";
    }
    if ([error code] == kCLErrorLocationUnknown) {
        errorMessage = @"无法确定你的位置!";
    }
    UIAlertView *alert = [[UIAlertView alloc]
                                initWithTitle:nil
                                message:errorMessage
                                delegate:self
                                cancelButtonTitle:@"确定"
                                otherButtonTitles:nil];
    [alert show];
    [alert release];
}
@end
```

在此文件中我们实现了 addPin:操作方法，它的作用是往地图中添加注解。首先在❶处调用 MapAnnotation 类中定义的 initWithCoordinate:方法创建一个实例对象，并且在❷处设置 title 和 subtitle 的属性值为相应位置的纬度和经度，然后将注解加入地图中。

其次，程序开始进行定位时，在❸处我们为当前位置添加了一个地图注解，并设置了它要显示的注解信息，代码如下：

```
mapAnnotations.title=@"TEST";
```

mapAnnotations. subtitle＝@"Just For Test"；

编译并运行，最终效果如图 14-8 所示；当我们点击 Tool Bar 上的添加按钮时，地图上添加了一个红色“小别针”，点击它，将显示该位置的纬度和经度，如图 14-9 所示。

## 14.3 小结

在这一章中，我们主要学习了如何使用 Core Location 技术进行定位以及获取位置的经、纬度信息；利用这些信息我们可以使用 Google Map 显示相应的地图，这需要利用 MapKit. framework 框架中的 MKMapView 类；还可以使用该框架下的 MKAnnotations 类为地图添加注解。此外，MapKit. framework 框架中还包括很多其他的构建地图的类和方法，例如 MKPinAnnotationView 类和 MKAnnotation 类，它们可以为地图注解添加视图，并且可以在上面添加细节展示按钮以显示更多关于某个位置的信息，就像前面章节中介绍的对表操作一样。如果你对此感兴趣，可以借助相关的技术资料或者查看 Xcode 帮助文档进行深入的学习。

# 第 15 章 多媒体

## 本章内容

- 使用 iPhone 照片库资源
- 播放音频
- 播放视频

在前两章，主要学习了 iPhone 的两个特性：加速计和定位，本章将简单介绍 iPhone/iPod 中多媒体的相关知识。多媒体（Multimedia）可以理解为图像、音频、视频和动画等各种媒体的统称。首先，我们介绍如何使用图像选取器（UIImagePickerController）来读取 iPhone/iPod 照片库中的图片；然后，讲解有关 Core Audio 的知识以及音频播放的各种方式；最后，涉及到的是有关视频播放的问题。这些知识点在软件开发过程中是非常有用的，是否能在程序中充分利用多媒体资源，是软件成败的关键因素之一。

## 15.1 iPhone/iPod 照片库

众所周知，iPhone 应用程序采用的是沙盒机制，因此很难获取程序本身沙盒之外的数据，当然也包括 iPhone/iPod 照片库中的图片资源。那我们是不是就不能读取相册的内容了呢？当然不是，苹果公司已经为我们提供了一种非常简单的解决方法——图像选取器（UIImagePickerController）。

UIImagePickerController 是 ViewController 的一个子类，它是 iPhone SDK 中提供给用户访问 iPhone/iPod 照片库的接口。在访问 iPhone/iPod 照片库时，首先需要创建一个图像选取器对象，并设置它的委托，可以用以下代码实现：

```
UIImagePickerController *picker = [[UIImagePickerController alloc] init];
```

其次，在使用图像选取器时，需要指定 sourceType，即图片的来源，它的取值主要有以下 3 种类型：

- UIImagePickerControllerSourceTypeCamera，使用 iPhone 内置的照相机拍摄照片，并使用此照片作为图片来源；
- UIImagePickerControllerSourceTypePhotoLibrary，使用 iPhone/iPod 中的照片库；
- UIImagePickerControllerSourceTypeSavedPhotosAlbum，使用照相机胶卷，如果

没有照相机，将从设备保存图片的文件夹中获取。

最后，还要实现 UIImagePickerControlDelegate 协议以下的两个方法：

- imagePickerController：didFinishPickingImage：editingInfo：方法，用户点击 choose 按钮（如图 15-1 所示）时执行的方法；
- imagePickerControllerDidCancel：方法，用户取消选择时执行的方法。

接下来我们将通过实现一个简单的照片选取程序，进一步说明如何使用 UIImagePickerController 对象选取 iPhone/iPod 照片库中的照片。程序主要实现从照片库中选择照片、对照片进行简单编辑等功能，最后，只需轻轻点击“保存”就可以存储照片了。程序运行时，我们随意选择 iPod 相册中的一张照片并调整它的大小，如图 15-1 所示，点击 Choose，最后效果如图 15-2 所示。

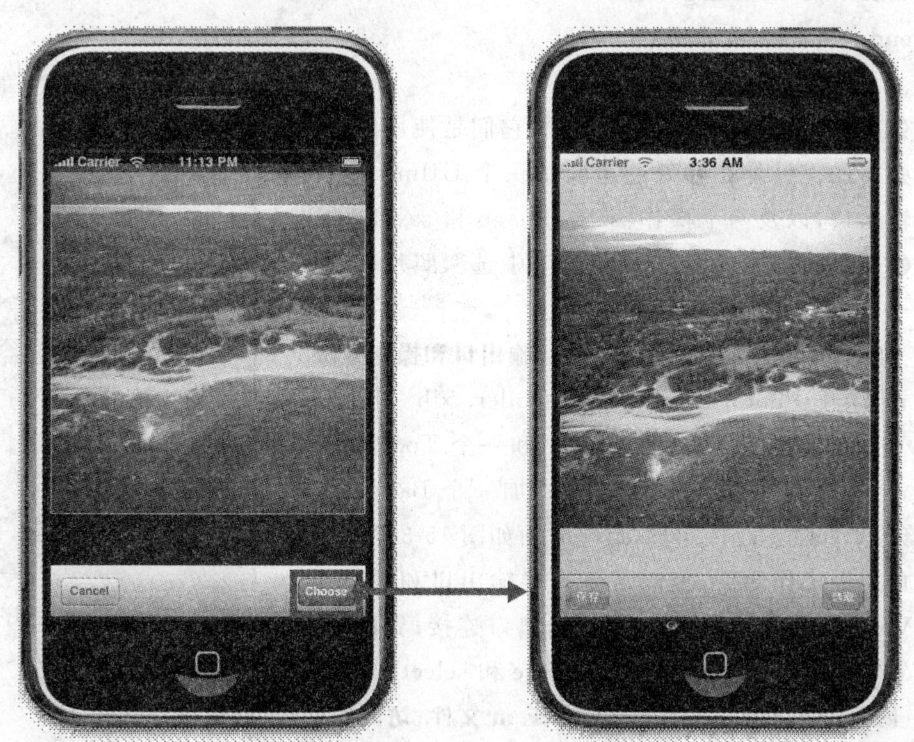


图 15-1 调整照片大小　　图 15-2 程序运行效果

要完成此程序大致需要做以下工作：

- 创建新项目 PhotoDemo，修改 PhotoDemoViewController. h 文件；
- 构建 PhotoDemo 程序界面，连接输出口和操作方法；
- 修改 PhotoDemoViewController. m 文件，访问照片库，获取照片。

**1. 创建新项目 PhotoDemo，修改 PhotoDemoViewController. h 文件**

打开 Xcode，使用 View-based Application 模板创建一个项目，将其命名为 PhotoDemo。选中 PhotoDemoViewController. h 文件，添加如下黑体字所示的代码：

**代码 15.1　PhotoDemoViewController.h 文件**

```
#import <UIKit/UIKit.h>
@interface PhotoDemoViewController : UIViewController
<UINavigationControllerDelegate, UIImagePickerControllerDelegate>{
    IBOutlet UIImageView *myImageView; ❶
}
@property (nonatomic, retain) IBOutlet UIImageView *myImageView;
- (IBAction)selectImage; ❷
- (IBAction)saveImage; ❸
@end
```

首先在头文件中添加了两个委托，它们是使用图像选取器时必需的。另外在❶处还声明了一个 UIImageView 类型的输出口，以及两个操作：selectImage 和 saveImage，如❷和❸所示，它们分别用来从照片库中选取照片和保存照片。

**2. 构建 PhotoDemo 程序界面，连接输出口和操作方法**

双击打开 PhotoDemoViewController.xib 文件，从 Library 窗口中拖出一个 Image View 和一个 Tool Bar 到视图中；同时还需要在 Tool Bar 上添加一个 Bar Button Item，修改 Item 的 Title 属性，最终界面如图 15-3 所示。

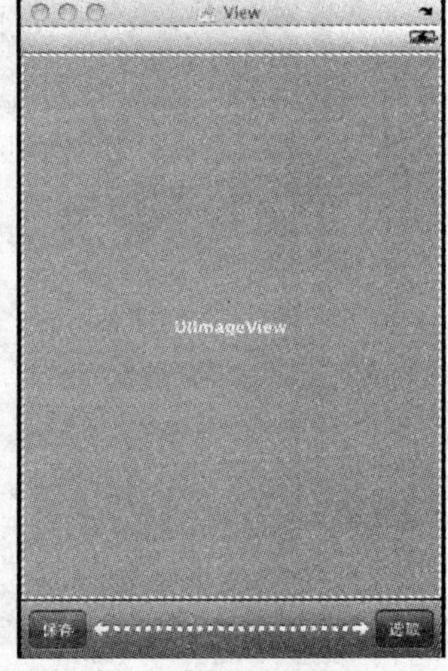

图 15-3　PhotoDemo 程序界面

完成以上工作之后，我们需要连接输出口和操作，将 Image View 视图与 myImageView 输出口连接；界面上的"保存"和"选取"按钮，分别与 saveImage 和 selectImage 操作方法连接。

**3. 修改 PhotoDemoViewController.m 文件，访问照片库，获取照片**

接下来选中 PhotoDemoViewController.m 文件，添加如下黑体字所示的代码：

**代码 15.2　PhotoDemoViewController.m 文件**

```
#import "PhotoDemoViewController.h"
@implementation PhotoDemoViewController
@synthesize myImageView;
```

```
- (IBAction)selectImage {
    UIImagePickerController * picker =
                        [[UIImagePickerController alloc]init];
    picker.delegate = self;
    picker.allowsImageEditing = YES;
    picker.sourceType =                                          ❶
        UIImagePickerControllerSourceTypePhotoLibrary;
    [self presentModalViewController:picker animated:YES];
    [picker release];
}
- (IBAction)saveImage {
if (myImageView.image != nil) {
    NSString * uniquePath = [[NSHomeDirectory() ❷
            stringByAppendingPathComponent:@"Documents"]
    stringByAppendingPathComponent:@"Image.png"];
    [UIImagePNGRepresentation(myImageView.image)
    writeToFile:    uniquePath atomically:YES]; ❸
    UIAlertView * alert = [[UIAlertView alloc]
                            initWithTitle:@"保存成功"
                            message:nil
                            delegate:nil
                            cancelButtonTitle:@"确定"          ❹
                            otherButtonTitles:nil];
    [alert show];
    [alert release];
    }
}

- (void)dealloc {
    [myImageView release];
    [super dealloc];
}
#pragma mark -
#pragma mark ImagePicker Delegate Methods
```

```
-(void )imagePickerController:(UIImagePickerController * )picker
          didFinishPickingMediaWithInfo:(NSDictionary * )Info {
    myImageView. image = image;
    [picker dismissModalViewControllerAnimated:YES ];
}                                                                      ❺

-(void)imagePickerControllerDidCancel:
                (UIImagePickerController * )picker {
  [picker dismissModalViewControllerAnimated:YES];                    ❻
}
@end
```

图 15-4　访问 iPhone 相册

在❶处创建了一个 UIImagePickerController 实例对象 picker,并设置它的委托为 self;同时将 allowsImageEditing 属性设置为 YES,默认为 NO,这样我们就可以对照片进行简单的缩放裁剪操作了。在此将 sourceType 设置为 UIImagePickerControllerSourceTypePhotoLibrary,表示将从设备相册中选取照片,也可以尝试设置为另外两种源类型。

saveImage 方法用来保存选取的照片。在此方法中,首先在❷处指定了图片存取路径和文件名称;并使用❸处的方法将图片写到 Documents 文件中,如果文件写入成功,将弹出一个警告框,提示图片保存成功,如❹处代码所示。

在协议的 imagePickerController:didFinishPickingImage:editingInfo:方法中,❺处我们要实现的是将选取的照片显示在 Image View 中;如果用户取消选取操作,将调用❻处的 imagePickerControllerDidCancel:方法,撤销图像选取器。

编译并运行程序,点击"选取",将访问到设备相册,如图 15-4 所示;从相册中任意选取一张照片调整它的大小,如图 15-1 所示,最后效果如图 15-2 所示。

# 15.2 iPhone 音频

音频是多媒体的一个重要组成部分，它让软件变得更加生动、形象，我们经常在按钮点击或弹出提示框时使用音频。有时候使用音频作为提示或警告信息，会比文字更有效果。一个拥有良好用户体验的软件总会充分利用音频使本身变得更加卓越。本节我们将介绍两种主要音频播放方式：使用 System Sound API 和 AVAudioPlayer 类。前者属于 System Sound Services 框架下的一个类；后者属于 AVFoundation. framework 框架下的一个类，该框架中的另外一个类是 AVAudioRecorder，主要提供了录制音频的一些方法。

## 15.2.1 System Sound API 播放短音频

System Sound API 播放音频具有以下特点：

- 不能循环播放；
- 不能控制音量；
- 不支持回放；
- 对音频格式有限制，支持. caf、. aif、. wav 等文件。

它主要用来播放一些较短的音频资源（一般小于 30s），如实现按钮点击时发出的声音。播放音频时，首先，需要声明一个 SystemSoundID 类型的实例；其次，需要使用 AudioServicesCreateSystemSoundID 方法注册声音；最后，使用 AudioServicesPlaySystemSound 方法播放音频。当我们不再使用该音频时，可以调用 AudioServicesDisposeSystemSoundID 方法释放先前注册的音频。它的具体使用方法将在下面的实例中说明。

## 15.2.2 AVAudioPlayer 播放长音频

AVAudioPlayer 播放音频时，与 System Sound API 相比具有以下特点：

- 一般用来播放较长的音频文件；
- 能循环播放；
- 能在暂停和播放之间切换；
- 支持更多的音频格式。

本节和 15.3 节，将逐步构建一个简易的多功能播放器程序。程序的最终效果如图 15-5 所示，在我们面前显示的是一个播放列表，列表中有音频和视频（将在 15.3 节完成）两个分组，选择相应的条目即可播放。这一节我们完成音频部分的内容，下一节继续完成视频部分。

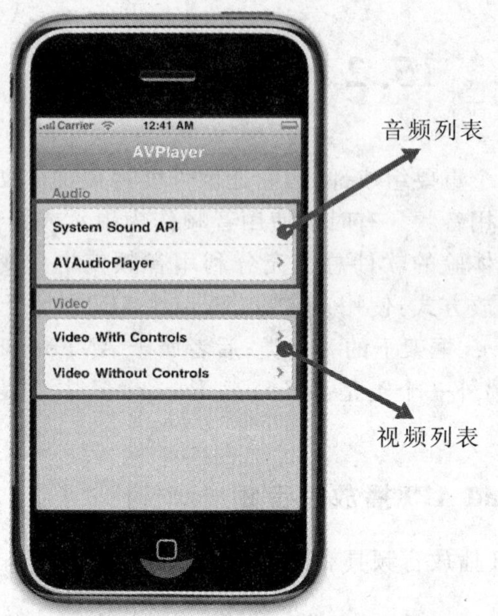


图 15-5 AVPlayerDemo 程序的最终效果

### 15.2.3 AVPlayerDemo 的音频部分

本程序的主要步骤有：

- 创建 AVPlayerDemo 项目；
- 构建音频播放列表；
- 实现音频播放功能，编译并运行程序。

**1. 创建 AVPlayerDemo 项目**

打开 Xcode，使用 Window-based Application 模板创建一个项目，将其命名为 AVPlayerDemo。

展开 Classes 文件夹，添加一个显示列表的根控制器 RootViewController 和一个用来播放音频的 AudioViewController 控制器；在开始代码编写之前，我们需要在项目中添加几个音频和视频文件（下一节需要用到视频文件，音频和视频文件可以在随书光盘中找到）以及两个播放音频的框架：AudioToolbox. framework 和 AVFoundation. framework；添加完毕之后，选中 AVPlayerDemoAppDelegate. h，添加如下黑体字所示的代码：

**代码 15.3 AVPlayerDemoAppDelegate. h 文件**

```
#import <UIKit/UIKit.h>
#import "RootViewController.h"
@interface AVPlayerDemoAppDelegate :
NSObject <UIApplicationDelegate> {
```

```
    UIWindow *window;
    RootViewController *root;
    IBOutlet UINavigationController *nav;
}
@property (nonatomic, retain) IBOutlet UIWindow *window;
@property (nonatomic, retain) RootViewController *root;
@property (nonatomic, retain) IBOutlet UINavigationController *nav;
@end
```

选中 AVPlayerDemoAppDelegate.m 文件，添加如下黑体字所示的代码：

**代码 15.4 AVPlayerDemoAppDelegate.m 文件**

```
#import "AVPlayerDemoAppDelegate.h"
@implementation AVPlayerDemoAppDelegate
@synthesize window;
@synthesize root;
@synthesize nav;
- (void)applicationDidFinishLaunching:(UIApplication *)application {
    self.root = [[RootViewController alloc]
                initWithNibName:@"RootViewController"
                bundle:nil];                                    ❶
    self.nav = [[UINavigationController alloc]
                initWithRootViewController:self.root];
    [window addSubview:nav.view];
    [window makeKeyAndVisible];
}
- (void)dealloc {
    [window release];
    [root release];
    [nav release];
    [super dealloc];
}
@end
```

对于以上代码你应该很熟悉，我们创建了一个导航控制器实例对象，并将 RootViewControl 的一个实例作为它的根视图控制器，如❶处代码所示。

**2. 构建音频播放列表**

选中 RootViewController. h 文件，添加如下黑体字所示的代码：

**代码 15.5　RootViewController. h 文件**

```
#import <UIKit/UIKit.h>
#import "AudioViewController.h"
@interface RootViewController : UIViewController {
    AudioViewController * audio;
    IBOutlet UITableView * myTable;
}
@property (nonatomic, retain) AudioViewController * audio;
@property (nonatomic, retain) UITableView * myTable;
@end
```

在 RootViewController. h 文件中，我们声明了一个表视图输出口 myTable，用它来构建音频和视频播放列表。现在我们就来构建本程序的播放列表界面，双击 MainWindow. xib 文件，往 Xib 窗口中拖入一个 Navigation Controller，同时为导航控制器连接到输出口 nav；然后，双击 RootViewController. xib 文件，往视图上添加一个 Table View，将它与 myTable 输出口连接，并为 Table View 指定委托和数据源。

接下来的工作是在界面中实现音频播放列表，选中 RootViewController. m 文件，添加如下黑体字所示的代码：

**代码 15.6　RootViewController. m 文件**

```
#import "RootViewController.h"
@implementation RootViewController
@synthesize myTable;
@synthesize audio;
- (void)viewWillAppear:(BOOL)animated {      ┐
        if (audio == nil) {                  │
            [audio release];                 │
            audio = nil;                     │ ❶
        }                                    │
        [audio.player pause];                │
}                                            ┘
- (void)dealloc {
    [myTable release];
    [audio release];
    [super dealloc];
```

```
}
#pragma mark -
#pragma mark dataSource Methods
-(NSInteger)numberOfSectionsInTableView:(UITableView
*)tableView {
        return 1;
}
- (NSInteger)tableView:(UITableView *)tableView
numberOfRowsInSection:(NSInteger)section {
   return 2;
}
- (NSString *)tableView:(UITableView *)tableView
titleForHeaderInSection:(NSInteger)section {
     //Audio
     NSString *sectionTitle;
     if (section == 0) {
         sectionTitle = @"Audio";
     }
    return sectionTitle;
}
- (UITableViewCell *)tableView:(UITableView *)tableView
        cellForRowAtIndexPath:(NSIndexPath *)indexPath {
     static NSString *myIdentifier = @"myIdentifier";
     UITableViewCell *cell =[tableView
     dequeueReusableCellWithIdentifier:myIdentifier];
     if(cell == nil) {
    cell = [[[UITableViewCell alloc]
                initWithFrame:CGRectZero
                reuseIdentifier:myIdentifier]
                autorelease];
     }
    //Audio
    if (indexPath.section == 0) {
        if(indexPath.row == 0) {
              cell.textLabel.text = @"System Sound API";
     }
    if(indexPath.row == 1) {
```

```
            cell.textLabel.text = @"AVAudioPlayer";
        }
    }
    [cell setAccessoryType:
        UITableViewCellAccessoryDisclosureIndicator];
    return cell;
}
#pragma mark Tabel View Delegate Methods
- (void)tableView:(UITableView *)tableView
        didSelectRowAtIndexPath:(NSIndexPath *)indexPath {
        //Audio
        if (indexPath.section == 0) {
            if (indexPath.row == 0) {
                if(audio == nil) {
                    audio = [[AudioViewController alloc]
        initWithNibName:@"AudioViewController"
                                    bundle:nil];
        }
        audio.audioName = @"Damn";
        audio.flag = YES;
        [self.navigationController
        pushViewController:audio animated:YES];
    }
    if (indexPath.row == 1) {
        if(audio == nil) {
            audio = [[AudioViewController alloc]
        initWithNibName:@"AudioViewController"
                        bundle:nil];
                }
                audio.audioName = @"XinJing";
                audio.flag = NO;
                [self.navigationController
                pushViewController:audio animated:YES];
        }
    }
}
@end
```

需要特别指出的是❶处代码的作用：为了防止内存泄露，每次返回根视图时，程序都需要释放 AudioViewController 实例对象，并且停止正在播放的音频。其他的代码用来构建音频播放列表，这些都是对表的一些基本操作，在前面章节已经介绍过，这里就不再赘述。

**3. 实现播放音频功能，编译并运行程序**

下面再来编写音频播放代码，选中 AudioViewController.h 文件，添加如下黑体字所示的代码：

**代码 15.7 AudioViewController.h 文件**

```
#import <UIKit/UIKit.h>
#import <AudioToolbox/AudioToolbox.h>       ❶
#import <AVFoundation/AVFoundation.h>
@interface AudioViewController : UIViewController
    <AVAudioPlayerDelegate>{
    NSString *audioName;
    Bool flag;
    SystemSoundID soundID;
    AVAudioPlayer *player;                   ❷
    IBOutlet UIButton *playBtn;
    IBOutlet UIButton *pauseBtn;
    IBOutlet UISlider *mySlider;
    IBOutlet UIButton *shortBtn;
}
@property (nonatomic, retain) NSString *audioName;
@property (nonatomic, retain) AVAudioPlayer *player;
@property (nonatomic, retain) IBOutlet UIButton *playBtn;
@property (nonatomic, retain) IBOutlet UIButton *pauseBtn;
@property (nonatomic, retain) IBOutlet UISlider *mySlider;
@property (nonatomic, retain) IBOutlet UIButton *shortBtn;
- (IBAction)playShortSound;
- (IBAction)sliderChange:(id)sender;         ❸
- (IBAction)play;
- (IBAction)pause;
@end
```

在此文件中，首先，在❶处添加了两个音频播放头文件；其次，在❷处添加了 AVAudioPlayerDelegate 协议，audioName 字符串型变量用来存储音频名，flag 对象用来控制界面上 playBtn、pauseBtn、mySlider 和 shortBtn 控件的显示，SystemSoundID 类型的变量用来注册声音文件；还声明了一个 AVAudioPlayer 对象，用来播放长音频；声明的 UIButton类型的输出口用来控制音频播放。最后，在❸处声明了 4 个操作方法：playShortSound、sliderChange:、play 和 pause，分别用来播放短音频、控制长音频播放进度、播放长音频以及暂停。

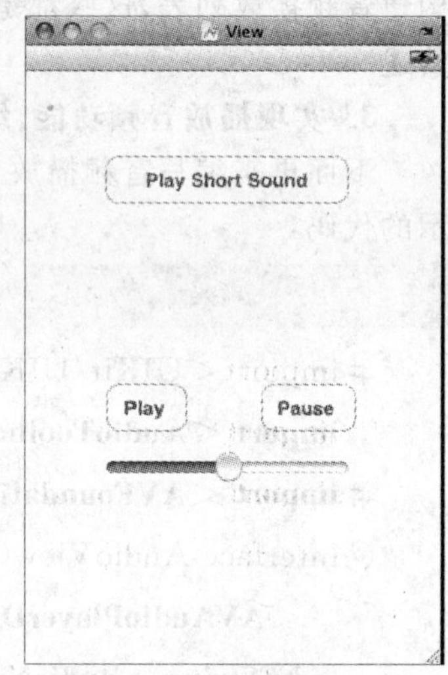

图 15-6　Audio 界面

接下来，需要构建音频播放界面，双击打开 AudioViewController. xib 文件，在 View 中添加 3 个 Button 和 1 个 Slider，调整它们的大小、位置和标题，最终界面如图 15-6所示。然后，为各个控件指定输出口和操作方法，输出口 shortBtn 连接到 Player Short Sound 按钮上，其他 3 个输出口分别与 play 按钮、pause 按钮和 Slider 控件连接，它们还需要分别与 play、pause 和 sliderChange:操作方法连接。最后，保存并关闭 Interface Builder，切换到 Xcode。

选中 AudioViewController. m 文件，添加如下黑体字所示的代码：

**代码 15.8　AudioViewController. m 文件**

```
#import "AudioViewController.h"
@implementation AudioViewController
@synthesize audioName;
@synthesize flag;
@synthesize player;
@synthesize playBtn;
@synthesize pauseBtn;
@synthesize mySlider;
@synthesize shortBtn;

// System Sound API 播放短音频
- (void)playShortSound {
```

```
    NSString * path = [[NSBundle mainBundle]
                             pathForResource:self.audioName
                             ofType:@"caff"];                    ❶
    NSURL * url = [NSURL fileURLWithPath:path];
   AudioServicesCreateSystemSoundID ((CFURLRef)url, &soundID); ❷
    AudioServicesPlaySystemSound(soundID); ❸
}
- (void)disposeSound {
        AudioServicesDisposeSystemSoundID(soundID);  ❹
}
//使用 AVAudioPlayer 播放长音频
- (IBAction)play {
    if (player == nil) {
        NSError * error = nil;
        NSString * path = [[NSBundle mainBundle]
                               pathForResource:audioName
                               ofType:@"mp3"];                 ❺
        NSURL * url = [NSURL fileURLWithPath:path];
        player = [[AVAudioPlayer alloc]
                               initWithContentsOfURL:url
                               error:&error];                  ❻
        player.delegate = self;
        }
    [player play];
    UInt32 category = kAudioSessionCategory_MediaPlayback;
    AudioSessionSetProperty
(kAudioSessionProperty_AudioCategory,
                             sizeof (category), &category);
    AudioSessionSetActive (true);
    }
}
- (IBAction)pause {
    [player pause];
```

```
        AudioSessionSetActive(false);
    }
    //slider 控制播放进度
    - (IBAction)sliderChange:(id)sender {
        UISlider * slider = (UISlider * )sender;
        player.currentTime = slider.value * player.duration;
    }
    - (vo id)viewWillAppear:(BOOL)animated {
        if (fl ag == YES) {
            playBtn.hidden = YES;
            pauseBtn.hidden = YES;
            mySlider.hidden = YES;
            shortBtn.hidden = NO;
        }                                          ❼
        if (flag == NO) {
            playBtn.hidden = NO;
            pauseBtn.hidden = NO;
            mySlider.hidden = NO;
            shortBtn.hidden = YES;
        }
    }
    - (void)dealloc {
        [self disposeSound];
        [audioName release];
        [player release];
        [playBtn release];
        [pauseBtn release];
        [mySlider release];
        [shortBtn release];
        [super dealloc];
    }
    @end
```

在 AudioViewController.m 文件的 playShortSound 方法中，❶处代码是用来获取音频文件的存储路径，并转化成 NSURL 对象；在❷处我们使用了 SystemSoundServices 框架中的 AudioServicesCreateSystemSoundID 方法注册音频，在❸处调用框架中的 AudioServicesPlaySystemSound 方法播放音频，播放完成之后，还需要调用❹处的 disposeSound 方法释放音频资源。

在使用 AVFoundation.framework 框架播放长音频时，首先，在❺处同样需要获取音频存储路径，并转换成 NSURL 类型；接下来在❻处使用此音频文件的路径初始化一个 AVAudioPlayer 类型的实例对象 player。play、pause 和 sliderChange 操作方法，分别用来控制长音频的播放、暂停和进度。

❼处的 viewWillAppear：方法用来控制界面上显示的元素，它根据用户在根控制器的列表中选择的内容设置 flag 值，从而呈现不同的控件。

上述工作完成之后，点击 Build and Go 按钮，编译并运行程序，效果如图 15-7 所示。

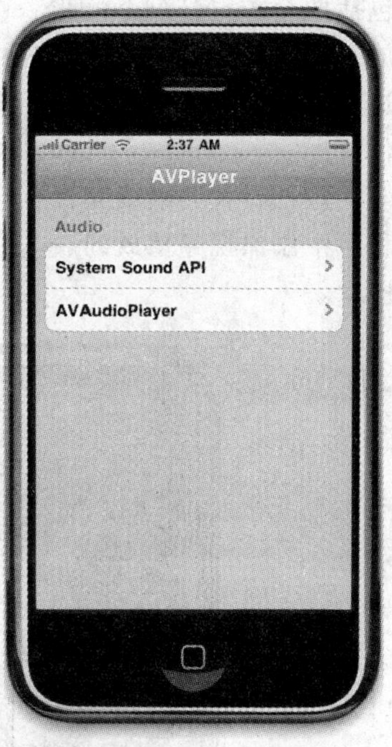


图 15-7 Audio 运行效果

### 15.2.4 其他音频播放框架

对于 iPhone/iPod 强大的音频处理功能来说，上面讲述的内容只是非常普通的一部分，如果想让我们的软件处理音频时表现得更加完美，还需要深入地学习和了解一些其他的音频播放框架。为了方便大家学习，在此列出几个重要的框架。

Audio Tool Box 框架提供了针对音频文件和音频流的回放和录音的服务。该框架也支持管理音频文件以及播放系统警告的声音。

Audio Unit 是 iPhone OS 提供的一个音频插件程序的核心集合，叫做音频单元，可以在任何应用程序中使用它。Audio Unit 框架为我们在应用程序中打开、连接和使用那些 iPhone OS 支持的音频单元提供了 C 语言的接口。

OpenAL(Open Audio Library)是自由软件界的跨平台音效 API(应用程序接口)，它可以表现多通道三维位置音效的特效，其 API 风格模仿自 OpenGL。

## 15.3 iPhone 视频

前面我们讲述了如何使用 iPhone/iPod 的照片库，以及如何播放音频资源，本节将介绍另一种重要的多媒体——视频。播放视频时主要使用的是 Media Player 框架，该框架包括两个类：MPMoviePlayerController 和 MPVolumeView。它们管理着整个视频的播放，我们只需要提供视频文件的 URL，并调用 play 方法播放视频就可以了。如果要使用这两个类，应在项目中加入 Media Player 框架，并引入 MediaPlayer/MediaPlayer.h 头文件。

这一节我们继续完善 15.2 节中的播放器程序，增加 AVPlayerDemo 的视频部分，在程序中添加视频播放列表，实现两种不同模式下的视频播放，程序效果如图 15-8 和 15-9 所示。

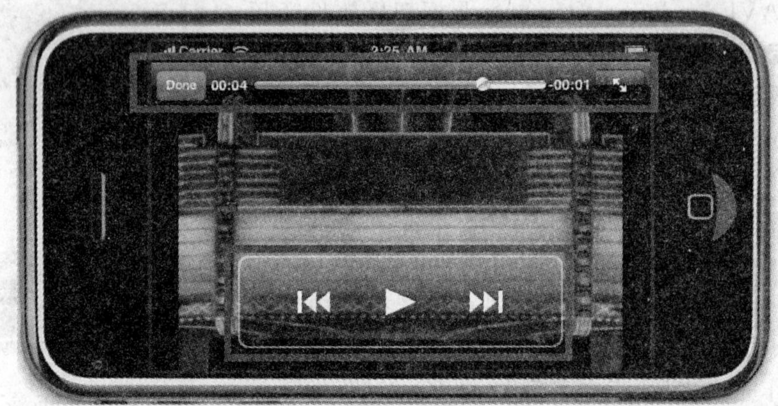


图 15-8　Video Play With Controls 模式

图 15-9　Video Play Without Controls 模式

本节需要完成以下工作：

- 修改根控制器，添加视频播放列表；
- 修改 VideoViewController.h 文件；
- 构建视频播放界面，实现视频播放功能。

**1. 修改根控制器，添加视频播放列表**

展开 Classes 文件夹，添加一个控制器 VideoViewController 用来播放视频文件。展开 Frameworks 文件夹，添加 Media Player 框架。

我们首先需要对根控制器做一点修改，在 RootViewController.h 中需要声明一个 VideoViewController 控制器对象 video。选中 RootViewController.h 文件，添加如下黑体字所示的代码：

**代码 15.9 RootViewController.h 文件**

```
#import <UIKit/UIKit.h>
#import "AudioViewController.h"
#import "VideoViewController.h"
@interface RootViewController : UIViewController {
    IBOutlet UITableView *myTable;
    AudioViewController *audio;
    VideoViewController *video;
}
@property (nonatomic, retain) IBOutlet UITableView *myTable;
@property (nonatomic, retain) AudioViewController *audio;
@property (nonatomic, retain) VideoViewController *video;
@end
```

在 RootViewController.m 文件中，添加如下黑体字所示的代码：

**代码 15.10 RootViewController.m 文件**

```
#import "RootViewController.h"
@implementation RootViewController
@synthesize myTable;
@synthesize audio;
@synthesize video;
- (void)viewWillAppear:(BOOL)animated {
    if (audio == nil) {
            [audio release];
            audio = nil;
    }
    if (video == nil) {
            [video release];
            video = nil;
    }
    [audio.player pause];
}
```

```
- (void)dealloc {
    [myTable release];
    [audio release];
    [video release];
    [super dealloc];
}
#pragma mark -
#pragma mark dataSource Methods
-(NSInteger)numberOfSectionsInTableView:(UITableView *)tableView {
    return 1;
    return 2;
}
- (NSInteger)tableView:(UITableView *)tableView
numberOfRowsInSection:(NSInteger)section {
    return 2;
}
- (NSString *)tableView:(UITableView *)tableView
titleForHeaderInSection:(NSInteger)section {
    //Audio
    NSString *sectionTitle;
    if (section == 0) {
        sectionTitle = @"Audio";
    }
    //Video
    if (section == 1) {
        sectionTitle = @"Video";
    }
    return sectionTitle;
}
- (UITableViewCell *)tableView:(UITableView *)tableView
        cellForRowAtIndexPath:(NSIndexPath *)indexPath {
    static NSString *myIdentifier = @"myIdentifier";
    UITableViewCell *cell =[tableView
    dequeueReusableCellWithIdentifier:myIdentifier];
    if (cell == nil) {
        cell = [[[UITableViewCell alloc]
                initWithFrame:CGRectZero
```

```
                reuseIdentifier:myIdentifier]
                autorelease];
    }
    //Audio
    if (indexPath.section == 0) {
        if (indexPath.row == 0) {
            cell.textLabel.text = @"System Sound API";
    }
    If (indexPath.row == 1) {
        cell.textLabel.text = @"AVAudioPlayer";
    }
    }
    //Video
    if (indexPath.section == 1) {
        if (indexPath.row == 0) {
            cell.textLabel.text = @"Video With Controls";
        }
        if (indexPath.row == 1) {
            cell.textLabel.text = @"Video Without Controls";
        }
    }
    [cell setAccessoryType:
        UITableViewCellAccessoryDisclosureIndicator];
    return cell;
}
#pragma mark Tabel View Delegate Methods
- (void)tableView:(UITableView *)tableView
    didSelectRowAtIndexPath:(NSIndexPath *)indexPath {
    //Audio
    if (indexPath.section == 0) {
        if (indexPath.row == 0) {
            if (audio == nil) {
                audio = [[AudioViewController alloc]
                        initWithNibName:@"AudioViewController"
                        bundle:nil];
            }
            audio.audioName = @"Damn";
```

```
                audio.flag = YES;
                [self.navigationController
                pushViewController:audio animated:YES];
            }
            if (indexPath.row == 1) {
                if (audio == nil) {
                    audio = [[AudioViewController alloc]
                            initWithNibName:@"AudioViewController"
                            bundle:nil];
                }
                audio.audioName = @"XinJing";
                audio.flag = NO;
                [self.navigationController
                pushViewController:audio animated:YES];
            }
        }
        //Video
        if (indexPath.section == 1) {
            if (indexPath.row == 0) {
                if (video == nil) {
                    video = [[VideoViewController alloc]
                            initWithNibName:@"VideoViewController"
                            bundle:nil];
                }
                video.videoName = @" iPhone ";
                video.flag = NO;
                [self.navigationController
                pushViewController:video animated:YES];
            }
            if (indexPath.row == 1) {
                if (video == nil) {
                    video = [[VideoViewController alloc]
                            initWithNibName:@"VideoViewController"
                            bundle:nil];
                }
                video.videoName = @" Cutscene ";
                video.flag = YES;
```

```
            [self.navigationController
            pushViewController:video animated:YES];
        }
    }
}
@end
```

**2. 修改 VideoViewController.h 文件**

选中 VideoViewController.h 文件,并添加如下黑体字所示的代码:

**代码 15.11 VideoViewController.h 文件**

```
#import <UIKit/UIKit.h>
#import <MediaPlayer/MediaPlayer.h> ❶
@interface VideoViewController : UIViewController {
    MPMoviePlayerController * player;
    NSString * videoName;
    BOOL flag;
}
@property (nonatomic, retain) MPMoviePlayerController * player;   ❷
@property (nonatomic, retain) NSString * videoName;
@property BOOL flag;
- (IBAction)play;
@end
```

在 VideoViewController.h 文件中,我们首先需要在❶处引入视频播放头文件;接下来在❷处声明了一个 MPMoviePlayerController 对象 player,同时,还声明了一个 BOOL 类型的 flag,用来标识视频播放时的两种播放模式,videoName 用来存储视频文件名,play 操作方法用来播放视频文件。

**3. 构建视频播放界面,实现视频播放功能**

界面非常简单,只有一个播放按钮。双击打开 VideoViewController.xib 文件,向 View 中添加一个 Button,同时为它连接操作方法 play。

保存修改,切换到 Xcode,选中 VideoViewController.m 文件,添加如下黑体字所示的代码:

**代码 15.12 VideoViewController.m 文件**

```
#import "VideoViewController.h"
#import <MediaPlayer/MediaPlayer.h>
@implementation VideoViewController
```

```
@synthesize videoName;
@synthesize player;
@synthesize flag;
- (IBAction)play {
    NSString * path = [[NSBundle mainBundle]
                          pathForResource:videoName
                          ofType:@"mov"];                          ❶
    NSURL * url = [NSURL fileURLWithPath:path];
    player = [[MPMoviePlayerController alloc]
              initWithContentURL:url];

    if (flag == YES){
            player.scalingMode = MPMovieScalingModeAspectFill;
            player.movieControlMode = MPMovieControlModeHidden;    ❷
        }
    [player play];

    [[NSNotificationCenter defaultCenter] addObserver:self
            selector:@selector(didFinishPlaying:)
            name:MPMoviePlayerPlaybackDidFinishNotification        ❸
            object:player];
}
-(void)didFinishPlaying:(NSNotification * )notification {
    if (player == [notification object]) {
        [[NSNotificationCenter defaultCenter]
        removeObserver:self
            name:MPMoviePlayerPlaybackDidFinishNotification
            object:player];                                        ❹
        [player release];
        player = nil;
    }
}
- (void)dealloc {
    [[NSNotificationCenter defaultCenter] removeObserver:self];
    [super dealloc];
}
@end
```

和播放音频一样，在 play 方法中，我们首先需要在❶处获取视频文件存储路径，将它转换成 NSURL 类型，然后，使用此视频文件的 URL 初始化一个 MPMovieController 对象 player；在❷处通过 movieControlMode 属性设置视频播放模式，一种是能控制视频播放的模式（如图 15-8 所示），一种是不能控制视频播放的模式（如图 15-9 所示）；在播放过程中，我们在❸处还需要通过消息中心 NSNotificationCenter 时刻监视视频是否播放完成，一旦接收到 MPMoviePlayerPlaybackDidFinishNotification 通知，将调用❹处的方法，释放相关资源。

编译并运行程序，在视频列表中选择 Play With Controls 时，效果如图 15-8 所示；选择 Play Without Controls 时，效果如图 15-9 所示。

# 15.4 小结

本章我们主要介绍了如何使用图像选取器选取 iPhone/iPod 照片库中的图像，通过 UIImagePickerController 类程序可以很轻松地访问到 iPhone/iPod 的照片库；此外，还介绍了 System Sound Services、AVFoundation 和 Media Player 等处理音频和视频的框架。通过学习本章知识，我们可以在以后的 iPhone 应用程序开发中充分利用多媒体资源，打造更加完美的软件。

在下一章中，将通过一个小游戏讲解 iPhone/iPod 中蓝牙的使用方法。

# 第16章 Bluetooth

## 本章内容

- 如何实现蓝牙连接
- 使用蓝牙传输数据

在上一章，我们了解了 iPhone/iPod 中多媒体的相关技术，它不仅具有拍摄、处理和查看照片的功能，还可以用来播放各种格式的音频和视频文件。iPhone/iPod 中的多媒体资源非常丰富，那么如何才能使设备之间共享这些资源呢？使用蓝牙技术将很容易解决这个问题。蓝牙(Bluetooth)是一种支持设备短距离(一般是 10m 以内)传输信息的无线电技术。本章将介绍如何使用蓝牙连接到其他 iPhone/iPod 设备，以及如何使用蓝牙传输数据。利用蓝牙技术，移动设备之间交换数据变得更加方便快捷，例如移动电话、个人数字助理(PDA)和笔记本电脑之间的信息交换。同样，蓝牙在 iPhone/iPod 设备上使用得也非常广泛，目前 App Store 中很多应用程序和游戏软件都用到蓝牙技术。图 16-1 是 App store 上一款使用蓝牙技术实现照片共享的免费软件 Bluetooth Photo Share。

图 16-1 Bluetooth Photo Share

# 16.1 GameKit 框架

在 iPhone SDK 3.0 及以后的版本中都增加了 GameKit.framework 框架，它是 iPhone SDK 中唯一提供了蓝牙接口的框架。GameKit.framework 包含以下三个类：GKPeerPickerController、GKSession 和 GKVioceChatService；它们分别对应三个重要的协议：GKPeerPickerControllerDelegate、GKSessionDelegate 以及 GKVoiceChatServiceDelegate。以上除了 GKVoiceChatService 类和它所对应的 GKVoiceChatServiceDelegate协议之外，其他两对是我们用来实现本章游戏的核心知识点。

在传输数据之前，需要在设备间建立蓝牙连接。为此我们需要创建一个 GKPeerPickerController 类的实例对象，用它来构建蓝牙连接时的选取界面。该类的属性主要有 connectionTypesMask、delegate 和 visible，分别用来设置网络连接类型、委托和选取界面是否可见；它的方法主要有 show 和 dismiss，分别用来显示和隐藏选取界面；GKPeerPickerConnectionType 常量用来设置网络连接类型。在委托中，我还需要实现 GKPeerPickerControllerDelegate 协议中声明的三个方法：peerPickerController：sessionForConnectionType：方法，用来创建一个连接会话；peerPickerController：didConnectPeer：toSession：方法，用来建立连接后隐藏选取界面；peerPickerControllerDidCancel：picker：方法，用来处理用户取消连接时的操作。它们都需要一个 GKSession 类型的对象作为参数。

同时，我们还需要实现 GKSessionDelegate 协议中 sendDataToAllPeers：withDataMode：error：方法和 receiveData：fromPeer：inSession：context：方法，分别用于发送和接收数据。图 16-2 是设备之间通过蓝牙发送数据的大致过程。

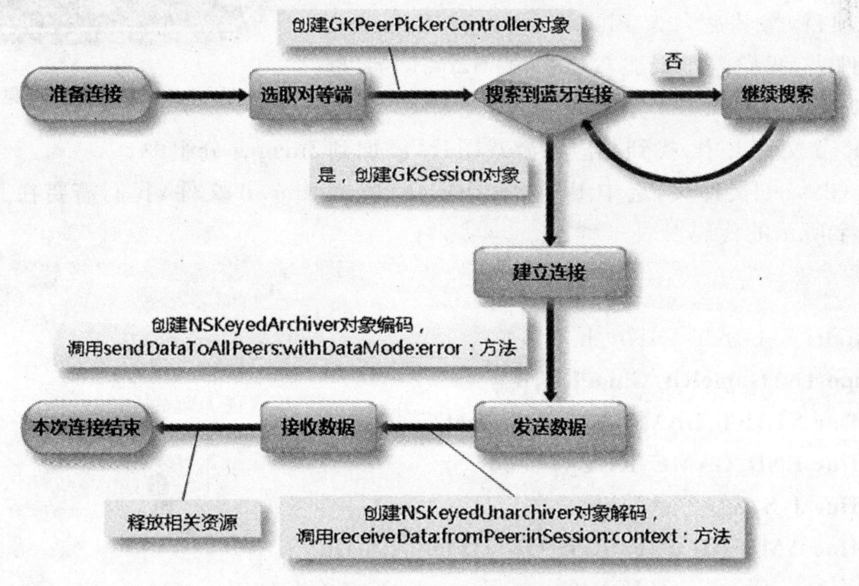


图 16-2 设备之间通过蓝牙发送数据

# 16.2 实现游戏

本章我们将实现一个简单的蓝牙对战游戏，并以此介绍使用蓝牙如何建立连接、传输数据、接收数据以及如何处理连接、断开等技术。它就是我们非常熟悉的"剪刀石头布"游戏，完成之后的效果如图16-3所示，相信你很清楚它的游戏规则，这里不再赘述。

图16-3 游戏运行效果图

本程序大体需要完成以下工作：

- 创建项目BlueTooth，在BlueToothViewController.h文件中添加相关协议及声明变量；
- 构建程序界面；
- 修改BlueToothViewController.m文件，创建GKSession实例对象并建立连接；
- 修改BlueToothViewController.m文件，实现游戏逻辑及传输数据；
- 修改BlueToothViewController.m文件，接收数据。

**1. 创建项目BlueTooth，在BlueToothViewController.h文件中添加相关协议及声明变量**

打开Xcode，使用View-based Application模板创建一个项目，将其命名为BlueTooth。展开Frameworks文件夹，我们需要添加GameKit.framework框架；同时新建一个Images分组，从随书光盘Chapter 16 Bluetooth文件夹中，找到相应的游戏图片，添加到Images分组中。

展开Classes文件夹，选中BlueToothViewController.h文件，我们需要在其中添加如下黑体字所示的代码：

**代码16.1 BlueToothViewController.h文件**

```
#import <UIKit/UIKit.h>
#import <GameKit/GameKit.h> ❶
#define START_GAME_KEY @"start"
#define END_GAME_KEY @"end"
#define J_S_B @"jsb"                                   ❷
#define AMIPHD_P2P_SESSION_ID @"AISIDE"
@interface BlueToothViewController : UIViewController
```

```
<GKPeerPickerControllerDelegate, GKSessionDelegate,
    UIAlertViewDelegate> {
    BOOL isHost;
    NSString *opponentID;
    GKSession *gkSession;
    int playerChoice;
    int opponentChoice;                                   ❸
    IBOutlet UILabel *playerTypeLabel;
    IBOutlet UILabel *opponentTypeLabel;
    IBOutlet UIButton *startButton;
    IBOutlet UIImageView *playerImage;
    IBOutlet UIImageView *opponentImage;
}
@property BOOL isHost;
@property int playerChoice;
@property int opponentChoice;
@property (nonatomic, retain) NSString *opponentID;
@property (nonatomic, retain) GKSession *gkSession;
@property (nonatomic, retain) UILabel *playerTypeLabel;
@property (nonatomic, retain) UILabel *opponentTypeLabel;
@property (nonatomic, retain) UIButton *startButton;
@property (nonatomic, retain) IBOutlet UIImageView *playerImage;
@property (nonatomic, retain)
                    IBOutlet UIImageView *opponentImage;
- (IBAction)buildConnection;
- (IBAction)choiceButtonClick:(id)sender;
- (void)showResult;
- (void)initGame;
- (void)start;                                            ❹
- (void)joinGame;
- (void)endGame;
- (void)invalidateSession:(GKSession *)session;
@end
```

在头文件中，需要在❶处引入 GameKit/GameKit. h 头文件，接下来我们在❷处定义了几个常量用来标识游戏开始、结束和两个 session 标识符(唯一标识本次会话)。

在❸处代码段中，添加了三个协议，并声明了一个布尔类型的变量 isHost，用来标记建立连接后是服务器端还是客户端；两个 int 型的变量 opponentChoice 和 playerChoice 分别用来存储用户选择的手型类型；oppoentID 用来存储对方的 session 标识符，有关 session 的知识稍后将介绍；两个 UILabel 类型的输出口 opponentTypeLabel 和 playerTypeLabel 用来显示游戏胜负信息；同时我们还声明了一个开始按钮 startButton；最后是两个UIImageView类型的输出口 playerImage 和 opponentImage，用来显示游戏双方所选择的手型。

在❹处代码段中，我们声明了 buildConnection 方法，用来建立蓝牙连接；choiceButtonClick:操作方法用来处理用户的选择操作；showResult 方法是程序的一个主要逻辑，用来判断双方的胜负；initGame、start、joinGame 和 endGame 分别用来初始化游戏、开始游戏、加入游戏和结束游戏；最后我们声明了 invalidateSession 方法，它用来结束本次会话。

**2. 构建程序界面**

接下来，我们需要构建游戏操作界面，双击 BlueToothViewController. xib 文件，根据图 16-4 所示，在视图中添加 4 个 Button、2 个 Label 以及 3 个 Image View，调整各个控件的大小和位置，并设置游戏相关图片，最后效果如后面的图 16-5 所示。

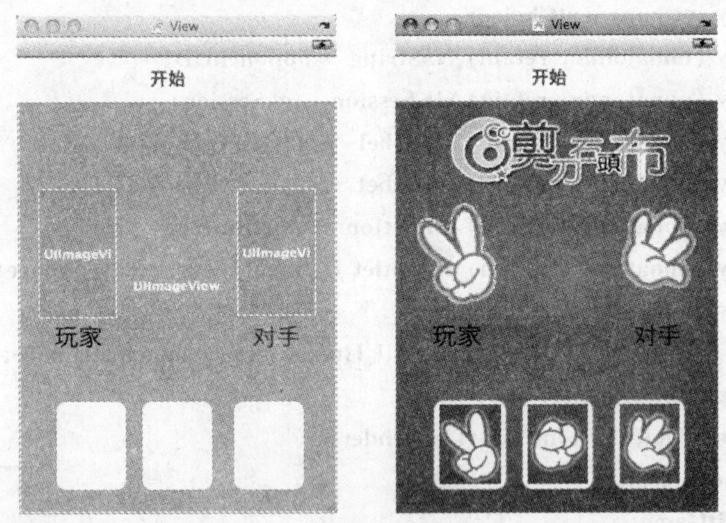


图 16-4 构建游戏界面

此外，还需要为相应的控件指定输出口和操作方法，startButton 连接 buildConnection 操作方法，其他三个按钮的 Touch Up Inside 事件均连接到 choiceButtonClick:操作方法。

**3. 修改 BlueToothViewController. m 文件，创建 GKSession 实例对象并建立连接**

每次建立蓝牙连接时，我们都需要使用 Game Kit 框架中的 GKSession 类为本次连接创建一个 session(会话)。session 主要用于建立蓝牙连接以及在游戏过程中传输数据。

下面通过程序说明如何创建 session 以及如何使用它建立连接。选中 BlueToothViewController. m 文件，并添加如下黑体字所示的代码：

**代码 16.2 BlueToothViewController.m 文件**

```
#import "BlueToothViewController.h"
@implementation BlueToothViewController
@synthesize isHost;
@synthesize playerChoice;
@synthesize opponentChoice;
@synthesize opponentID;
@synthesize playerTypeLabel;
@synthesize opponentTypeLabel;
@synthesize playerImage;
@synthesize opponentImage;
@synthesize startButton;
@synthesize gkSession;
- (IBAction)buildConnection {
        if (!opponentID) {
            isHost = YES;
            GKPeerPickerController *peerPickerController =
                [[GKPeerPickerController alloc] init];
            peerPickerController.delegate = self;
            peerPickerController.connectionTypesMask =
                GKPeerPickerConnectionTypeNearby;                    ❶
            [peerPickerController show];
        }
        else{
            [self endGame];
        }
}
#pragma mark -
#pragma mark GKPeerPickerControllerDelegate Methods
- (GKSession *)peerPickerController:(GKPeerPickerController *)controller
          sessionForConnectionType:(GKPeerPickerConnectionType)type {
          if (!gkSession) {
              self.gkSession = [[GKSession alloc]
                          initWithSessionID:AMIPHD_P2P_SESSION_ID
                              displayName:@"J_S_B"                    ❷
                              sessionMode:GKSessionModePeer];
          gkSession.delegate = self;
        }
        return gkSession;
}
```

```
- (void)peerPickerController:(GKPeerPickerController *)picker
    didConnectPeer:(NSString *)peerID
    toSession:(GKSession *)session {
        [picker dismiss];
        picker.delegate = nil;
        [picker autorelease];
}                                                                  ❸
- (void)peerPickerControllerDidCancel:(GKPeerPickerController *)picker{
    picker.delegate = nil;
    [picker autorelease];
    if(self.gkSession != nil) {
        [self invalidateSession:self.gkSession];                   ❹
        self.gkSession = nil;
    }
}
#pragma mark GKSessionDelegate Methods
- (void)session:(GKSession *)session
    didReceiveConnectionRequestFromPeer:(NSString *)peerID {       ❺
        isHost = NO;
}
- (void)session:(GKSession *)session peer:(NSString *)peerID
    didChangeState:(GKPeerConnectionState)state {
    switch (state)
    {
      case GKPeerStateConnected:
        [session setDataReceiveHandler: self withContext: nil];    ❻
        opponentID = peerID;
        isHost ? [self start] : [self joinGame];
        break;
    }
}
- (void)invalidateSession:(GKSession *)session {
    if(session != nil) {
        [session disconnectFromAllPeers];
        session.available = NO;
        [session setDataReceiveHandler: nil withContext: NULL];    ❼
        session.delegate = nil;
    }
}
- (void)dealloc {
    [playerImage release];
```

```
    [opponentImage release];
    [opponentID release];
    [startButton release];
    [gkSession release];
    [playerTypeLabel release];
    [opponentTypeLabel release];
    [super dealloc];
}
@end
```

在 buildConnection 方法❶处，我们首先需要判断是否已经连接到其他设备，如果 oppoentID 为空，将 isHost 值设为 YES，自身当作服务器，并使用 GKPeerPickerController 类创建一个对等选取器(GKPeerPickerController)实例对象，代码如下：

```
GKPeerPickerController * peerPickerController =
                    [[GKPeerPickerController alloc] init];
```

用代码 peerPickerController. delegate = self;设置对等选取器的委托，最后需要设置网络连接方式，并将对等选取器显示出来，代码如下：

```
peerPickerController. connectionTypesMask =
                    GKPeerPickerConnectionTypeNearby;
[peerPickerController show];
```

connectionTypesMask 是一个常量，表示网络连接类型，它有两种取值：GKPeerPickerConnectionTypeNearby 和 GKPeerPickerConnectionTypeOnline，前者表示使用蓝牙连接到设备，后者表示一些其他的网络连接方式。

在 GKPeerPickerControllerDelegate 协议的委托方法❷处，使用下面几行代码来创建一个 session 并且设置它的委托为 self：

```
gkSession = [[GKSession alloc]
            initWithSessionID:AMIPHD_P2P_SESSION_ID
                displayName:@"J_S_B"
                sessionMode:GKSessionModePeer];

gkSession. delegate = self;
```

initWithSessionID 参数要求是一个能唯一标识 session 的字符串。如果将它设置为 nil，sessionID 的取值将从应用程序束的标识符中随机产生一个。displayName 就是显示在选取界面上的名称，如果设置为 nil，将默认为设备名称。sessionModel 的取值有以下三种：

- GKSessionModelService，作为服务器端，它在网络中只向其他设备广播自身，而不主动搜索其他设备。
- GKSessionModelClient，作为客户端，它能主动搜索服务，而不会在网络中广播自身。

· GKSessionModelPeer，作为对等端使用，此时，它不仅是服务器端，还是客户端，能够在网络中广播自身，同时还能搜索其他设备，这是最常使用的一种模式。

建立连接后，将自动执行❸处 GKPeerPickerControllerDelegate 协议的 peerPickerController:didConnectPeer:方法，通过调用 GKPeerPickerController 类的 dismiss 方法，隐藏选择器界面。

如果对方接受了连接请求，就需要调用❺处 GKSessionDelegate 协议的 session:didReceiveConnectionRequestFromPeer:方法来处理连接请求，并设置 isHost 为 NO，表示它将作为客户端加入游戏。

然后，系统会自动调用❻处的 session: peer:didChangeState 方法，更改连接状态，根据 isHost 的值判断是以服务器角色开始游戏还是以客户端角色加入游戏中。

GKPeerPickerControllerDelegate 协议的 peerPickerControllerDidCancel:方法是用户取消连时调用的操作方法，如❹所示。它设置选取器对象的委托为 nil，并释放内存资源；同时还需调用❼处的 invalidateSession:方法，与所有的对等端断开连接，并停止传输数据，释放 session 对象所占的内存资源。

**4. 修改 BlueToothViewController. m 文件，实现游戏逻辑及发送数据**

在上一步通过创建 GKSession 实例对象实现了蓝牙连接，接下来需要完成的是游戏的逻辑，同时将数据封装、打包发送给接收方。

选中 BlueToothViewController. m 文件，添加以下代码：

**代码 16.3　BlueToothViewController. m 文件**

```
……
- (void)initGame {
        self.playerChoice = -1;
        self.opponentChoice = -1;                         ❶
}
- (void)start {
        [self initGame];
        NSMutableData *message = [[NSMutableData alloc] init];
        NSKeyedArchiver *archiver = [[NSKeyedArchiver alloc]
                        initForWritingWithMutableData:message];
        [archiver encodeBool:YES forKey:START_GAME_KEY];        ❷
        [archiver finishEncoding];
        NSError *sendErr = nil;
        [gkSession sendDataToAllPeers: message
                  withDataMode:GKSendDataReliable
                        error: &sendErr];
```

```
        if (sendErr) {
            UIAlertView *alert = [[UIAlertView alloc]
                          initWithTitle:@"发送数据失败"
                          message:[NSString stringWithFormat:
                                              @"%@",sendErr]
                          delegate:self
                          cancelButtonTitle:@"OK"
                          otherButtonTitles:nil];
            [alert show];
            [alert release];
        }
        [message release];
        [archiver release];
        [self showResult];
}
- (IBAction) choiceButtonClick:(id)sender {
    [self.view setUserInteractionEnabled:NO];
    self.playerChoice = [sender tag];
    [self showResult];
    NSMutableData *message = [[NSMutableData alloc] init];
    NSKeyedArchiver *archiver =[[NSKeyedArchiver alloc]
                       initForWritingWithMutableData:message];            ❸
    [archiver encodeInt:self.playerChoice forKey: J_S_B];
    [archiver finishEncoding];
    [gkSession sendDataToAllPeers: message
              withDataMode:GKSendDataUnreliable error:NULL];
    [archiver release];
    [message release];
}
- (void) showResult {     ❹
    if(self.playerChoice!=-1) {
        switch (self.playerChoice) {
            case 0:
            {
                [self.playerImage setImage:
                        [UIImage imageNamed:@"j0.jpg"]];
            }
```

```
                break;
            case 1:
            {
                [self.playerImage setImage:
                    [UIImage imageNamed:@"s0.jpg"]];
            }
                break;
            case 2:
            {
                [self.playerImage setImage:
                    [UIImage imageNamed:@"b0.jpg"]];
            }
                break;
            default:
                break;
        }
    }
    if(self.opponentChoice != -1) {
        switch (self.opponentChoice) {
            case 0:
                {
                    [self.opponentImage setImage:
                        [UIImage imageNamed:@"j0.jpg"]];
                }
                    break;
                case 1:
                {
                    [self.opponentImage setImage:
                        [UIImage imageNamed:@"s0.jpg"]];
                }
                    break;
                case 2:
                {
                    [self.opponentImage setImage:
                        [UIImage imageNamed:@"b0.jpg"]];
                }
                    break;
```

```
                default:
                    break;
            }
        }
        if(self.playerChoice != -1 && self.opponentChoice != -1) {
            [self.view setUserInteractionEnabled:YES];
            BOOL playerWins;
            if((self.playerChoice==0 && self.opponentChoice==1)||
            (self.playerChoice==1 && self.opponentChoice==2)||
            (self.playerChoice==2&& self.opponentChoice==0)) {
                playerWins = YES;
                playerTypeLabel.text=@"胜利";
                opponentTypeLabel.text = @"失败";
            }
            else if(self.playerChoice == self.opponentChoice) {
                playerTypeLabel.text = @"平局";
                opponentTypeLabel.text = @"平局";
            }
            else {
                playerWins = NO;
                playerTypeLabel.text=@"失败";
                opponentTypeLabel.text=@"胜利";
        }
            [self initGame];                                                ❺

            if (playerWins) {
            NSMutableData *message = [[NSMutableData alloc] init];
            NSKeyedArchiver *archiver =[[NSKeyedArchiver alloc]
                        initForWritingWithMutableData:message];
            [archiver encodeBool:YES forKey:END_GAME_KEY];                  ❻
            [archiver finishEncoding];
            [gkSession sendDataToAllPeers: message
                withDataMode:GKSendDataReliable error:NULL];
            [archiver release];

            [message release];
            }
```

```
    }
}
- (void)joinGame {
        [self initGame];
        [self showResult];           ❼
}
- (void)endGame {
        [self showResult];
        self.playerTypeLabel.text=@"";
        [self.playerImage setImage:nil];          ❽
        self.opponentTypeLabel.text=@"";
        [self.opponentImage setImage:nil];
}
……
@end
```

当用户接受请求并建立连接后，需要调用❶处方法初始化游戏：将双方的选择均赋值为-1，表示还没有选择。

在 start 方法中，服务器端首先需要给对方发送开始游戏信息，如代码段❷所示，该信息将被 NSKeyedArchiver 类型的实例打包成 NSData 对象，并调用 GKSession 类的 sendDataToAllPeers:withDataMode:error:方法，发送数据给所有的对等端，发送失败会弹出相应的错误信息。

❸处的 choiceButtonClick:方法用来处理用户的选择。首先，我们需要获取用户选择的手型，将其存储在一个 NSMutableData 类型的变量 message 中，并将其打包；然后调用 sendDataToAllPeers: withDataMode: error 方法发送给所有的对等端。withDataMode 属性表示网络中数据的传输模式，它的取值为 GKSendDataUnreliable 或者 GKSendDataReliable，前者是不可靠的，只以 UDP 形式发送一次数据，而不管是否发送成功；后者是可靠的，以 TCP 形式不断地发送数据，直到接收方成功接收到全部信息。

❹处的 showResult 方法是本游戏的逻辑部分，它首先会显示玩家选择的手型图片；然后，在❺处根据玩家的选择进行游戏胜负判断，如果你对"剪刀石头布"的游戏规则比较熟悉，相信会很轻松地理解这个逻辑，它是通过获取各个按钮的 tag 值(界面设计时我们已经设定不同的手型图片具有不同的 tag 值)来判断双方的胜负，并在❻处将胜负信息打包发送给对等方。

我们还需要一个 joinGame 方法❼，使得玩家能加入到其他人的游戏中；游戏结束时，需要调用❽处的 endGame 方法显示游戏结果。

**5. 修改 BlueToothViewController.m 文件，接收数据**

GKSessionDelegate 协议的 receiveData:fromPeer:inSession:context:方法是用来接收数据的。发送方发送数据之后，接收方需要使用此方法接收数据。

选择 BlueToothViewController.m 文件，并添加以下代码：

**代码 16.4 BlueToothViewController.m 文件**

```
- (void)receiveData:(NSData *)data fromPeer:(NSString *)peerID
        inSession:(GKSession *)session
          context:(void *)context {
    NSKeyedUnarchiver *unarchiver = [[NSKeyedUnarchiver alloc]
                          initForReadingWithData:data]; ❶
    if ([unarchiver containsValueForKey:J_S_B]) {
            opponentChoice = [unarchiver decodeIntForKey:J_S_B];
            [self showResult];
        }
        if ([unarchiver containsValueForKey:END_GAME_KEY]) {
            [self endGame];
        }
        if ([unarchiver containsValueForKey:START_GAME_KEY]) {
            [self joinGame];
        }
        [unarchiver release];
    }                                                              ❷
......
@end
```

接收到数据之后，首先在❶处创建 NSKeyedUnarchiver 类型的实例对象，用来解压接收到的数据包，根据接收到的不同结果，客户端将调用不同的处理方法，如❷处所示。

至此，本章的程序就完成了，编译并运行程序，将出现图 16-3 所示的效果。当点击"开始"按钮后将出现图 16-5 所示的搜索设备界面；搜索到其他设备之后，出现图 16-6 所示的选取对等端界面。

图 16-5 搜索设备界面

图 16-6 选取对等端界面

# 16.3 小结

在本章中，我们完成了一个蓝牙实例，并通过它学习了如何使用蓝牙建立无线连接、如何在设备间传输数据。相信通过本章的学习，你已经对蓝牙技术有了初步的了解。蓝牙是一种比较新的技术，本章介绍的只是冰山一角，如果你对这方面的技术感兴趣，想要了解更多，建议阅读其他资料，进行更深入的学习。

# 参考文献


1 Ali M. iPhone SDK 3 Programming. West Sussex: John Wiley and Sons, Inc., 2009

2 Pilone D , Pilone T. Head First iPhone Development. Sebastopol: O'Reilly Media, Inc., 2009

3 Zdziarski J. iPhone SDK Application Development. Sebastopol: O' Reilly Media, Inc., 2009

4 Sadun E. The iPhone™ Developer's Cookbook. Boston: Addison-Wesley Professional, Inc., 2009

5 Zirkle P , Hogue J. iPhone Game Development. Sebastopol: O'Reilly Media, Inc., 2009

6 Hillegass A. Cocoa(R) Programming for Mac(R) OS X (3rd Edition). Boston: Addison-Wesley Professional, Inc., 2009

7 Davidson J D. Learning Cocoa with Objective-C. Sebastopol: O'Reilly Media, Inc., 2002

8 Dudney B, Adamson C. iPhone SDK Development: Building iPhone Applications. North Carolina Dallas: Pragmatic, Inc., 2009

9 Dalrymple M, Knaster S. Learn Objective-C on the Mac. New York: Apress, Inc., 2009

10 Mark D, LaMarche J. Beginning iPhone Development: Exploring the iPhone SDK. New York: Apress, Inc ., 2009

11 Owens M. The Definitive Guide to SQLite. New York: Apress, Inc., 2006

12 Piper I. Learn Xcode Tools for Mac OS X and iPhone Development. New York: Apress, Inc., 2009

**图书在版编目（CIP）数据**

iPhone 应用程序开发指南(基础篇)/张英锋，刘超主编．—济南：山东科学技术出版社，2010(2011．重印)
(艾诗德移动技术丛书)
ISBN 978-7-5331-5603-9

Ⅰ.i... Ⅱ.①张... ②刘... Ⅲ.移动通信—携带电话机—应用程序—程序设计—指南 Ⅳ.TN929.53-62

中国版本图书馆 CIP 数据核字(2010)第 038768 号

艾诗德移动技术丛书
**iPhone 应用程序开发指南**
**(基础篇)**

丛书主编　王绪兵　彭楚夫
本书主编　张英锋　刘　超

**出版者：山东科学技术出版社**
地址：济南市玉函路 16 号
邮编：250002　电话：(0531)82098088
网址：www.lkj.com.cn
电子邮件：sdkj@sdpress.com.cn
**发行者：山东科学技术出版社**
地址：济南市玉函路 16 号
邮编：250002　电话：(0531)82098071
**印刷者：济南华东彩印有限公司**
地址：济南市商河彩虹路 2 号
邮编：251600　电话：(0531)84872167

**开本：**787mm×1092mm　1/16
**印张：**27.5
**版次：**2011 年 9 月第 1 版第 2 次印刷

**ISBN 978-7-5331-5603-9**
**定价：49.00 元(赠光盘一张)**